BIOLOGY AND QUATERNARY ENVIRONMENTS

based on the Symposium on Biological Problems in the Reconstruction of Quaternary Terrestrial Environments, held in Canberra, February 1978.

Edited by
D. Walker and J.C. Guppy

Australian Academy of Science
1978

Published by the Australian Academy of Science
P.O. Box 783, Canberra City, ACT 2601, Australia

ISBN 0 85847 050 0

Printed by Union Offset Co. Pty. Ltd. Canberra

CONTENTS

AUSTRALIAN ACADEMY OF SCIENCE

NATIONAL COMMITTEE FOR QUATERNARY RESEARCH

Dr J.M. Bowler	Mr E.D. Gill
Dr G.G. Beckmann	Professor E.S. Hills (Chairman)
Dr J.H. Calaby	Dr R. Jones
Dr R.W. Galloway	Mr E.B. Joyce
	Professor D. Walker

ORGANIZING COMMITTEE FOR THE SYMPOSIUM

Dr J.H. Calaby	Professor M.V. Macfarlane
Dr L.T. Evans	Dr A.G. Thorne
Dr J.N. Jennings	Professor D. Walker (Chairman)
Technical Secretary:	Joan C. Guppy

AUTHORS

J.H. Calaby, Division of Wildlife Research, CSIRO, Box 84 P.O., Lyneham, ACT 2602, Australia.

J. Chappell, Department of Geography, School of General Studies, Australian National University, Box 4 P.O., Canberra, ACT 2600, Australia.

Dr T.L. Crisman, Department of Environmental Engineering Sciences, University of Florida, A.P. Black Hall, Gainesville, Florida 32611, USA.

Professor M.B. Davis, Department of Ecology and Behavioral Biology, University of Minnesota, 108 Zoology Building, Minneapolis, Minnesota 55455, USA.

Dr I. Franklin, Division of Animal Production, Genetics Research Laboratories, CSIRO, Box 90 P.O., Epping, NSW 2121, Australia.

Professor M.V. Macfarlane, Department of Animal Physiology, The Waite Institute, Waite Road, Urrbrae, SA 5064, Australia.

Professor A.R. Main, Department of Zoology, University of Western Australia, Nedlands, WA 6009, Australia.

Mr H.A. Nix, Division of Land Use Research, CSIRO, Box 1666 P.O., Canberra City, ACT 2601, Australia.

Dr L. Rymer, Department of Biogeography and Geomorphology, Research School of Pacific Studies, Australian National University, Box 4 P.O., Canberra, ACT 2600, Australia.

Dr J.M.B. Smith, Department of Geography, University of New England, Armidale, NSW 2351, Australia.

Dr A.G. Thorne, Department of Prehistory, Research School of Pacific Studies, Australian National University, Box 4 P.O., Canberra, ACT 2600, Australia.

Dr N.M. Wace, Department of Biogeography and Geomorphology, Research School of Pacific Studies, Australian National University, Box 4 P.O., Canberra, ACT 2600, Australia.

Professor W.A. Watts, Botany Department, Trinity College, Dublin, Republic of Ireland.

Professor D. Walker, Department of Biogeography and Geomorphology, Research School of Pacific Studies, Australian National University, Box 4 P.O., Canberra, ACT 2600, Australia.

Dr R. Wells, School of Biological Sciences, Flinders University of South Australia, Bedford Park, SA 5042, Australia.

FOREWORD

The articles in this volume arose from a symposium on Biological Problems in the Reconstruction of Quaternary Terrestrial Environments organized under the auspices of the Australian Academy of Science and held in Canberra on 21 and 22 February 1978. All but two are the edited texts of the papers which were presented in abbreviated forms at the meeting. The first paper (Chappell) is intended to provide a frame of reference about Quaternary environmental changes which is not derived from biological data. The fourteenth paper (Rymer) was written after the symposium in order to try to illuminate the application of uniformitarianism, a subject hotly discussed during several sessions. The final paper (Walker) is one person's attempt to isolate the three or four major issues which the meeting exposed or stressed. Papers 2 to 6 inclusive in this volume present some of the problems as Quaternary research workers see them. Papers 7 to 13 inclusive are partly experimentalists' responses to such problems and partly their views on the significance of Quaternary history for the distributions, physiologies, anatomies and behaviours of plants, animals and man today.

The symposium was designed to bring together Quaternary research workers and those in related biological disciplines. It was hoped that, by doing so, the insights of modern experimental biology might be brought to bear on some of the difficulties experienced in reconstructing terrestrial environments from fossils. At the same time it was anticipated that exposure to the kind of data accumulating about the Quaternary Period might prove useful to experimental biologists seeking to set their results in a longer time context. In the event 127 persons participated in the symposium and discussion was lively.

ACKNOWLEDGEMENTS

The Organizing Committee is grateful to the staff of the Australian Academy of Science for assistance in arranging the meeting. It also wishes to thank the secretarial and technical staffs of the Department of Biogeography and Geomorphology of the Australian National University and Mr T. Baumann of the Department of Human Geography, Australian National University, who revised or drew many of the figures. Dr S.L. Duigan helped design the cover.

Professor D.J. Anderson, Dr W.D.L. Ride and Dr H.C. Tyndale-Biscoe kindly acted as session chairmen while Dr R.L. Kirk and Dr G.B. Sharman read contributions from two authors who were unable to be present.

The symposium was financed by the Australian Academy of Science and by contributions from the following industrial organizations whose support is greatly appreciated:

Australian Paper Manufacturers Limited
Broken Hill Proprietary Company Limited
Comalco Limited
Conzinc Riotinto of Australia Limited
M.I.M. Holdings Limited
North Broken Hill Limited
Utah Foundation
Western Mining Corporation Limited

Professor E.S. Hills and the late Sir Maurice Mawby gave much-needed assistance in the early stages of the symposium's organization.

Permission to use previously published figures and tables is acknowledged from the publishers, editors and authors of the following publications:

Geol. Soc. Am. Memoir 145, 1976. Kellog, T.B. (figure in this volume 1.5).

Arctic and Alpine Research, 5, 1973. Andrews, J.T. (figure in this volume 1.8).

Invest. Indiana Lakes and Streams, VI, No.1, 1964. Mueller, W.P. (figure in this volume 4.1).

Archiv für Hydrobiologie, 10, 1966. Goulden, C.E. (figure in this volume 4.3).

Trans. Conn. Acad. Arts and Science, 44, 1972. Tsukada, M. (figure in this volume 4.4).

Invest. Indiana Lakes and Streams, VI, No.2, 1964. De Costa, J.J. (figure in this volume 4.6).

Trans. Conn. Acad. Arts and Science, 44, 1972. Tsukada, M. (figure in this volume 4.7).

Commentationes Biologicae, 35, 1970. Alhonen, P. (figure in this volume 4.8).

Ecology, 55, 1974. Kerfoot, W.C. (figure in this volume 4.10).

Ph.D. Thesis, Indiana University, 1965. Iovino, A.J. (figure in this volume 4.13).

Verh. Internat. Verein Limnol., 19, 1975. Warwick, W. (figure in this volume 4.14).

Hydrobiologia, 48, 1976. Clair, T. and Paterson, C.G. (figure in this volume 4.15).

Cold Spring Harbor Symp. Quant. Biol., 24, 1959. Kurten, B. (figure in this volume 7.1).

Handbook of Vegetation Science. Part VII. Vegetation and Environment, 1974. Scott, D. (figure in this volume 11.1).

1. CHRONOLOGICAL METHODS AND THE RANGES AND RATES OF QUATERNARY PHYSICAL CHANGES

by J. Chappell

SUMMARY

The limits of resolution for interpreting past physical changes are set, firstly by imperfections in our knowledge of physical processes acting in the landscape, secondly by problems of interpreting past changes of conditions and processes from stratigraphic and geomorphologic evidence, and, thirdly, by the limits of our ability to date past events by radiometric and other means. These matters are reviewed in order to show why the moving picture of Quaternary physical changes, in its detail and in many regions, is still comparatively remote from us.

Physical processes themselves are variable and warrant description in probabilistic terms. Probable recurrence times of events of large magnitude and relatively profound physical effects are not well known for many processes affecting the landscape. When this fact is coupled with the intrinsic uncertainty of dating methods, which increases with distance in past time, the result is that the courses and mechanisms of Quaternary physical changes remain obscure unless careful use is made of uniformitarian arguments plus modern and historical data. The question of variability, against change of mean conditions, is discussed, especially for climate and is extended to particular cases of landscape change.

Chronologic methods, critical for analysis of Quaternary history, are reviewed in some detail, particularly radiocarbon dating which provides the basis not only for interpretation of the last 40,000 years but also for construction of analogies for changes in more remote Quaternary times. From this base, changes of sealevels, icecaps, climates, and other major physical elements are discussed.

INTRODUCTION TO CHRONOLOGIC METHODS

Through Quaternary time the earth has experienced repeated changes of the processes acting at its surface, reflecting changes of atmosphere and ocean climates. These changes, which have affected the biosphere, have been extensively investigated geologically since the time of Darwin, Lyell and their contemporaries. The generally accepted philosophy is uniformitarian, holding that meteorologic and geologic processes in the past have continuously been similar to those today, and that at no time has the earth's surface been altered by cataclysms far surpassing any historic event. However, certain deposits, mainly sedimentary, frequently have been thought difficult to explain in uniformitarian terms, and have been staple fare for a few scientists who have appealed to successive vast catastrophes to explain Quaternary geology. While current opinion places such catastrophists nearer to a lunatic fringe than to the center of science, the various deposits to which they refer illustrate the difficulty of adopting a very rigid uniformitarian stance (Rymer, this volume). Basically, this is because the weather must be characterised by its variability as well as by its mean condition, and geomorphic processes likewise. Most of these processes are understood from probability relationships between events of different magnitude; figure 1.1a illustrates how rainfall should be characterised, for example. The problem of magnitude-frequency concepts is more complex in many geomorphic processes, which often change discontinuously with an increase of hydro-meteorologic 'input' (figure 1.1b). Magnitude-frequency concepts are fundamental, however, Only from the probabilities can the likelihood be estimated of concatenations of events so extreme that their cumulative effect constitutes a 'catastrophic' change of landscape. For many processes the historic period of observation is insufficient to estimate statistics of largest-magnitude events, and in many instances our experimental models only now are beginning to be able to predict these.

These facts are important for interpreting rates and ranges of Quaternary physical changes, as here we focus more sharply than when viewing the more remote geologic past. In particular, for periods from 10^5 to 10^6 years duration we are concerned with detail over 10^1 to 10^3 years, which is the time range of recurrence intervals of 'extreme' events in climatic and landscape phenomena. Clearly the uniformitarian concept must be held in perspective with magnitude-frequency ideas when considering Quaternary environmental changes. It should also be pointed out that magnitude-frequency descriptions of geomorphic phenomena based on modern instrumental data may have limited applicability to many episodes of Quaternary-age changes when, at times of more major or rapid climatic changes, landscapes are likely to have been extensively in disequilibrium.

These points should be kept in mind when framing questions about ranges and rates of physical changes, especially when such changes are being investigated with modern dating methods as adjunct tools. While physical changes, rapid and slow, are mainly investigated stratigraphically, the chronology of change rests principally on radiometric dating. The latter techniques have accuracies from 1.5% to 20%, varying with method and with sample material. Therefore, we cannot directly measure the time-interval of change within a few tens of years for stratigraphies much older than a few thousand years, or change-intervals of a few thousand years much beyond 100,000 years, at the best. The only stratagem for identifying rates of rapid changes beyond about 15,000 years ago is to argue from very recent stratigraphic analogies guided, ideally, by known probability relationships between events of different magnitudes. The frequent question of whether a particular change (climatic, say) was globally synchronous or whether it marched across successive regions cannot be answered through analogies from young stratigraphies. If the march of change occurred within the time range of the dating uncertainty, the question cannot be answered, except in cases where singular events can be widely used as time horizons. Examples, each of which has limitations, include widespread volcanic ash falls, rapid changes of ocean water volume and rapid excursions of the magnetic field.

Leaving aside special cases employing singular events, strategies for establishing chronologies fall into three groups: direct methods, radiometric dating and physico-chemical relative age methods. Many are applicable only to restricted ranges of materials and sites, sometimes to particular stratigraphic situations which by themselves throw little light on regional physical changes. The three groups are discussed in turn. I am concerned to show what the methods can do, rather than how they are done.

DIRECT CHRONOLOGY

Historical records.

Instrumental records are limited mainly to meteorologic and hydrologic parameters and, except for a few northern hemisphere stations, rarely extend beyond a few tens of years to a hundred years or so. However, considerable information about the physical environment which is distributed through official logs, journals, personal diaries and more diffuse historical material potentially can yield more valuable information about changes over the last several hundred years (Manley 1953). Systematic reduction of this information by computer 'content analysis' to trustworthy records of physical parameters is current state of the art (Moodie and Catchpole 1976). It is principally from such historical material that H.H. Lamb (1968) has compiled his excellent study of historical Little Ice Age climatic variations. Analysis of geomorphic phenomena from historical records so far is limited principally to explosive vulcanism (e.g. Lamb 1970) and to earthquake activity. Selected cases of physical changes studied in this way will be discussed later, together with examples which have their chronologies established by other methods.

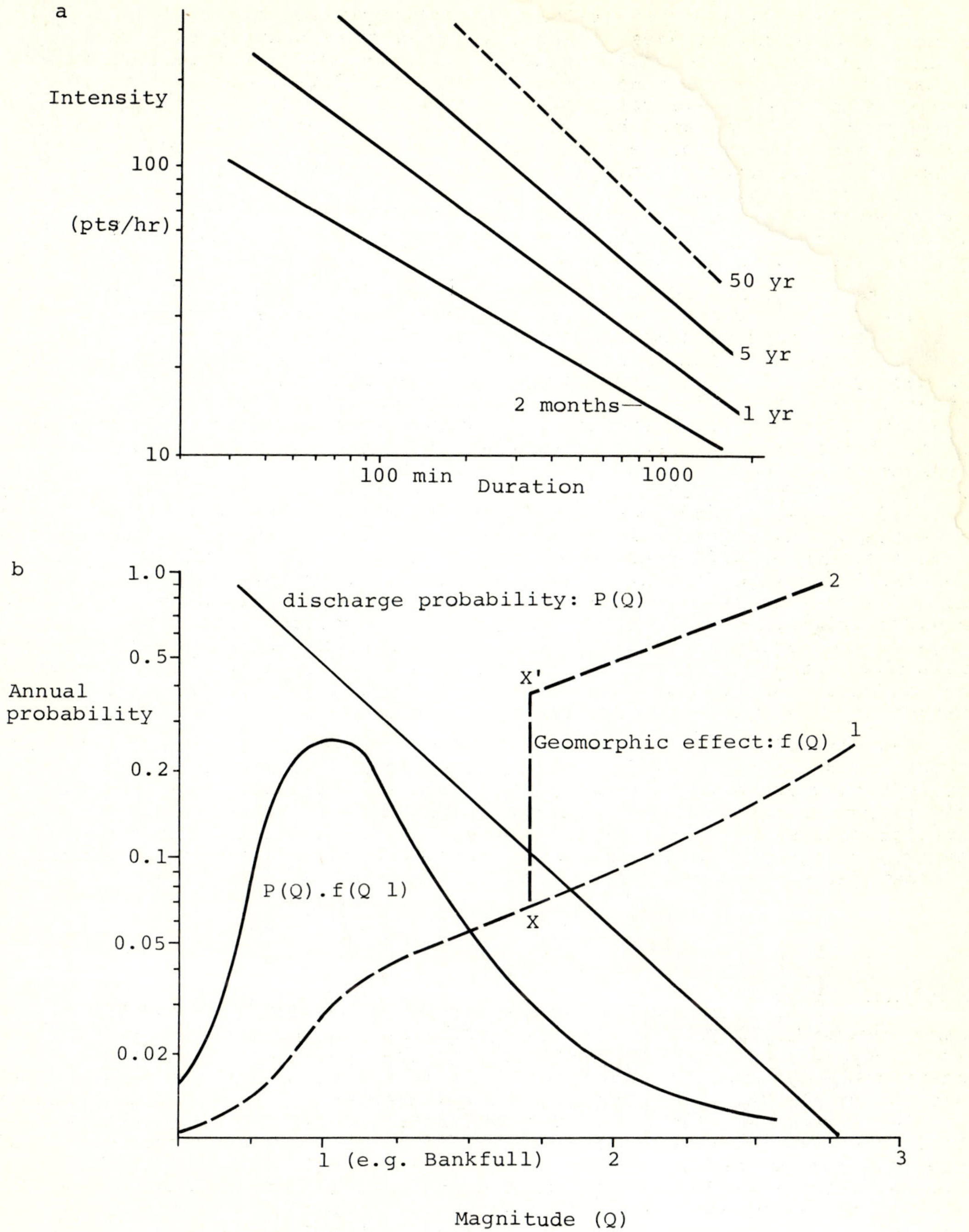

Figure 1.1 Probabilistic representation of physical processes:

(a) Intensity-duration-recurrence interval graph of rainfall.

(b) Net geomorphic activity is the product of process-response f(Q) to hydro-meteorologic input x probability curve P(Q) of the input. f(Q) may be continuous function (curve 1) or discontinuous (curve 2). P(Q).f(Q) is illustrated for case 1.

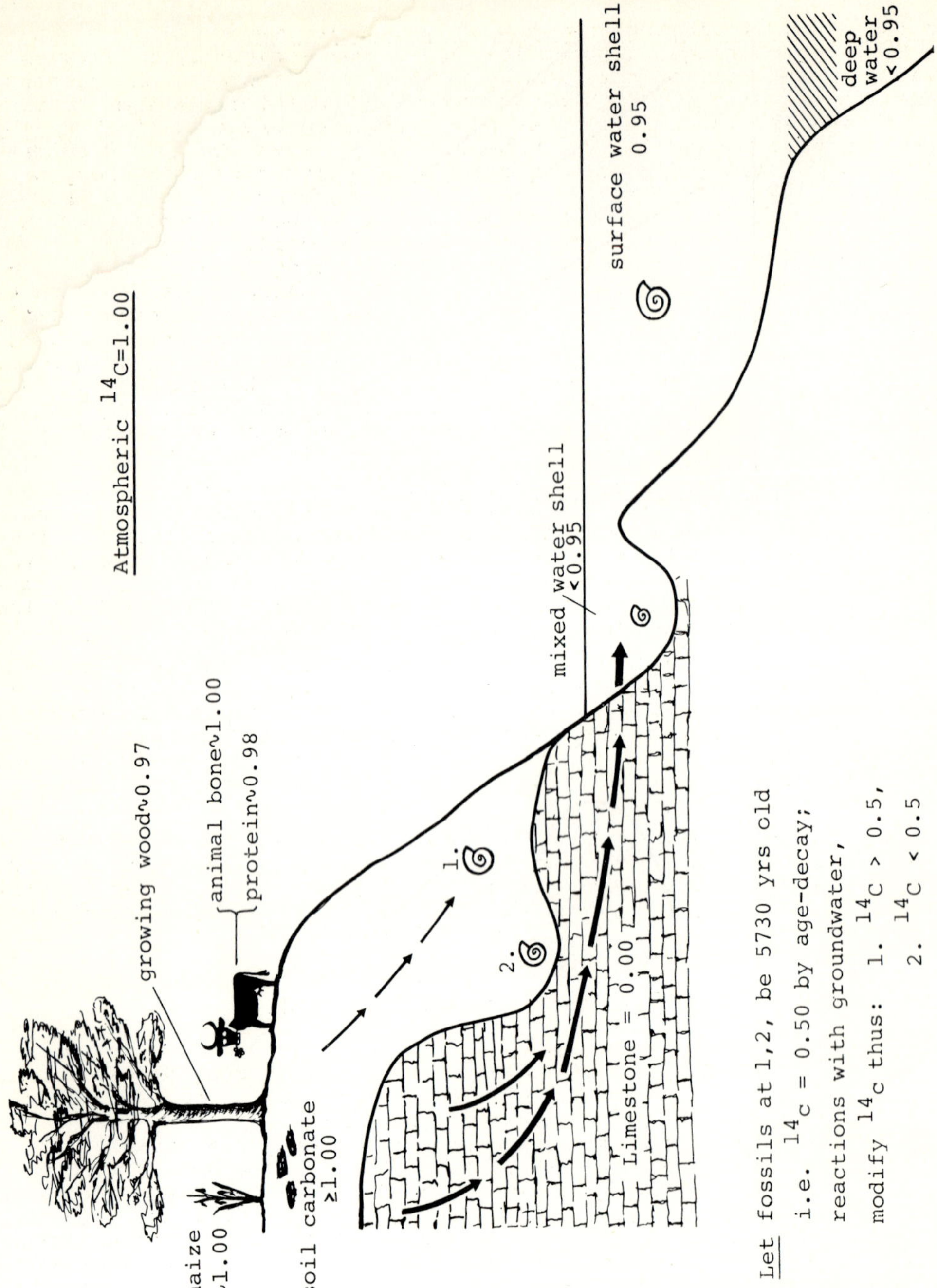

Figure 1.2

Factors affecting radiocarbon dating: initial content of ^{14}C varies between materials according to isotope fractionation as well as the reservoir in which they develop. Additionally, fossil material may become contaminated due to post-formation chemical reaction with carbon bearing solutions.

Physical and biologic records showing annual layers.

Tree growth rings, glacial ice, and varve sediments in lakes in annually frozen terrain not only can record variations of environmental conditions (usually meteorologic and hydrologic factors in some combination) but also can provide chronologies potentially accurate to the year, by careful counting of annual layers. The classical varve studies in Sweden and Finland by de Geer and Sauramo provided the first firm chronology of the pattern of geographic changes of the last deglaciation in Scandinavia. These varve sequences have been successively refined, correlated with other records of environment changes such as are found palynologically in bog sediments, and cross-checked by radiocarbon dating (Tauber 1970). Perhaps the most successful single study of a varve sequence comes from the Lake of the Clouds in Minnesota (Stuiver 1970a) which has provided comparison of radiocarbon and varve chronologies to 10,000 years B.P. (discussed again below). The range of the method for absolute chronology is restricted to the period since the last glacial climax (i.e. about the last 15,000 years at most), although in isolated instances varve counts have provided estimates of the duration of more ancient Pleistocene episodes (e.g. Turner and Shackleton 1967).

Dendrochronology, based on tree-ring counting, has progressed considerably since classical studies of older historical building materials, especially in the case of the Arizonan bristlecone pine which has provided the most comprehensive basis for cross-checking the radiocarbon dating method over the last 8000 years (La Marche and Harlan 1973) as discussed below. Paleoclimatic interpretation of variations within the bristlecone pine growth pattern (La Marche 1974) are discussed later.

Layer-counting in long cores from glacial ice, usually icecap glaciers in high latitudes, provides a chronology for changing meteorologic patterns which are geochemically recorded within the ice. Chronologic errors occur, however, due to diastems (non-depositional intervals) and ice-fabric ambiguities, so that layer counts in Greenland ice cores taken down to historic volcanic ashes can differ by several years between cores (Dansgaard et al. 1977).

Summarizing, these annually-layered records provide excellent chronologies for themselves, and contain evidence of hydro-meteorologic variations of a detailed kind although interpretation is often complex. However, they have not the wide stratigraphic applicability of the best radiometric dating techniques, to which we now turn.

RADIOMETRIC DATING METHODS

Radiocarbon dating.

This has proved to be the most useful of all dating methods applied in Quaternary studies. Although its range is limited to the last 40,000 years, i.e. latest Quaternary time, its wide applicability stems from the fact that all living organisms contain carbon and when preserved as fossil material potentially are dateable. In addition, inorganic carbonate deposits likewise in principle can be dated. Because of its extensive usage in archaeology and Quaternary studies, with materials ranging from the organic constituents of soils to deep soil sediments, considerable attention has been paid to errors which may attend radiocarbon ages, arising not from measurement of the radiocarbon content of the sample material but from contamination of the sample by carbon-carrying solutions surrounding the sample material in its field site. Additionally, differences occur between broad categories of carbonaceous or carbonate materials, arising from (i) initial variations of ^{14}C in the environmental reservoirs within which the materials lived or were precipitated, and (ii) differential uptake or precipitation of ^{14}C relative to common carbon (^{14}C) reflecting bio- or geo-chemical differences between materials. These factors are illustrated in figure 1.2; this is not the place to review them in technical detail but certain guidelines for the reader of any radiocarbon - based chronologies should be stated, so that he or she may evaluate age-relationship statements based on such results.

^{14}C half-life. Ages reported by most laboratories are based on a half-life of 5568±30 years, but some use a value of 5730±30 years which is considered to be more accurate. An age calculated on the larger value may be multiplied by 1.030 for comparison with ages based on the shorter half-life (Polach 1975; Stuiver and Polach 1977).

Isotopic fractionation differences between different materials during their formation cause differences in apparent ages of up to several hundred years. When this is corrected for, laboratories declare that their age estimates are based on $D^{14}C$ measurement. If the age is indicated as based on $D^{14}C$ correction must be applied (Polach 1975; Stuiver and Polach 1977).

Reservoir differences. Particularly important when comparing marine samples one with another, or with terrestrial materials, these can cause differences in apparent age commonly of several hundred years and ranging to over 1000 years in extreme cases. Corrections can be applied when the characteristic ^{14}C activity of the modern reservoir, relative to the atmosphere, is known. In cases where this factor is significant, the user must carefully take account of the way in which the laboratory preparing the date has accounted for it (see reviews by Mangerud (1972) and Olsson (1974)). Radiocarbon ages conventionally should be reported without incorporating the reservoir effect (Stuiver and Polach 1977).

Contamination of the sample after its preservation by carbon rich in, or depleted in, ^{14}C. When sample material is chemically complex, contains several different constituents, or is chemically simple but of an intrinsically reactable nature (e.g. mollusc carbonate), this can be a sizeable problem, the elucidation of which is the joint responsibility of the laboratory and the scientist using the dates. Literature on the matter is quite large (e.g. Polach 1975), but a simple guideline is that the reader must place most credence on studies where contamination (preferably the lack of it) is mentioned explicitly. Sometimes a literature search is necessary to find the first or most complete description of a given set of dates.

Statistical measurement error. If the foregoing potential errors are properly reckoned with, it is statistical measurement error which sets precision limits for radiocarbon dating, or for any other type of radiometric dating method. This error arises from the random nature of the radioactive decay process and reflects sampling error of decay event-counting for both the specimen and modern reference standard as well as spectrometer background noise (Polach 1975). This error, usually 2-3% of sample age but sometimes more, is calculated for each sample from its measurement statistics and is stated with each published date as age error. This is not the total error range but the 1 standard deviation (1σ) value of an approximately Gaussian error distribution. Age-comparisons must be made in this statistical light, and I remind the reader of my introductory remarks about statements about physical changes at time scales of 10^1 to 10^3 years.

I have dwelled on these qualifications of radiocarbon dating, not only to set perspectives on this important technique, but also because analogous remarks pertain to all radiometric dating. Statistical and contamination matters are similar for all methods discussed below and will not be repeated *in extenso*. There is one further important qualification on radiocarbon dating, namely that radiocarbon years are not astronomical years. In radiocarbon dating, as in any other radiometric method, the initial amount of the radio-isotope in a sample at the time of its formation must be known before an age can be calculated from its measured ^{14}C activity. The ^{14}C level in the atmosphere has varied in the past, partly because magnetic field variations have affected the rate of ^{14}C production in the upper atmosphere. The discrepancies between radiocarbon and astronomical time have been measured back to about 8000 years BP by comparing bristlecone pine tree ring chronology with radiocarbon dates taken at intervals in the same trees, showing that in general radiocarbon ages between 5000 and about 8000 sidereal years are about 600 to 800 years too young. A similar exercise with the varve sequence from Lake of the Clouds, Minnesota, indicates a similar discrepancy to 10,000 years (Stuiver 1970a, 1970b; La Marche and Harlan 1973). This in no way affects comparison between radiocarbon dates from different sites, but must be remembered when radiocarbon results are compared with other dating methods.

Uranium-series: (U-disequilibrium, $^{230}Th/^{234}U$, ionium, U-He)

Based on measurement of certain members of the uranium decay series, principally isotopes of U, Th, and sometimes Pa, these methods have the advantage over radiocarbon of reaching further back into Pleistocene time, typically 250,000 years for $^{230}Th/^{234}U$, the most widely used method (also known as the ionium method). They have the disadvantage of being applicable only to a limited suite of materials and, for many of these, establishment of an age requires measurement, not only of several isotope activity values, but also determinations from many samples from a given site. Marine materials are better suited than terrestrial; because Th is insoluble in sea water its initial level is zero in marine organisms. Consequently, any organism that accumulates U in its skeleton potentially can be dated from the relative amounts of U and its decay-daughter, Th. Corals in particular accumulate U and the most satisfactory, most reproductible, dates have been obtained from ancient coral reefs (e.g. Ku et al. (1974) on a 120,000 year-old Hawaiian reef).

Statistical errors usually are 6-9% for $^{230}Th/^{234}U$ but contamination problems outweight statistical imprecision. Although the potential for good results with corals is high, their aragonite skeletons are prone to react with, and be contaminated by, any U-series ions in solution, such as in groundwater. Well-established criteria which allow contamination to be identified must be stated with all creditable results (Chappell et al. 1974).

Of other marine materials, deep sea sediments in particular have been dated by U-series methods, although geochemical problems are greater than for corals (Turekian 1965), and results have been disputed to such an extent (e.g. Broecker and van Donk 1970) that deep sea core chronology no longer relies much on this method. Marine molluscs in important stratigraphic contexts likewise have been measured frequently. However, as molluscs virtually exclude uranium in life, the relatively high U concentration in fossil molluscs reflects post-death uptake of the element, usually with some addition of Th. Hence the ages based on this method intrinsically are unsatisfactory although by determining multiple samples from a given site some satisfactory results have been obtained, albeit with confidence levels significantly lower than for corals.

Terrestrial materials for which U-series methods have been mainly used are limestone cave speleothems (stalagmites and stalactites). These, like tree rings and ice cores, contain in their growth structure a record of hydro-meteorologic variations identifiable geochemically. Although more complex from the U-series dating point of view, results so far are encouraging, and speleothems are becoming a significant source in Quaternary paleo-hydrology (e.g. Duplessy et al. 1972; Ford et al. 1976). Other materials for which the method is still in its early stages include marine phosphate deposits, fossil bones and inorganic carbonates in lagoonal environments.

The U-He variant of U-series dating has the potential of reaching far beyond the Quaternary Era. Lack of contamination is an imperative, and so far this method has been applied only to fossil corals, principally from Barbados (Bender 1973). While coral is the best material, the problem of finding uncontaminated samples in ancient reefs older than 200,000 years is so great that applications are limited.

Summarising overall, U-series dating has yielded particularly useful chronologies of fossil coral reefs, despite unresolved questions of detail (H.H. Veeh pers comm.). Geochemical problems in its other applications are being tackled and reliable estimates have been achieved in several studies.

Potassium-Argon dating

Based on decay of ^{40}K to ^{40}Ar with a half-life of 1.2×10^{10} years, the method can range across the whole of geologic time. It can be applied to any potassium - bearing rocks which have reached temperatures sufficient to shed all previous ^{40}Ar. In the Quaternary context it is usually volcanic rocks interbedded with stratigraphically significant sediments which are dated by this method. There are difficulties in applying it to volcanic materials much younger than 5×10^5 years old, due firstly to difficulty of detection of small amounts of radiogenic argon against a much larger background of atmospheric argon absorbed in the rock and, secondly, to the problem of whether all prior argon was lost during eruption (McDougall et al. 1969). In favourable cases reliable ages from rocks as young as 50,000 years have been obtained (McDougall 1971), though most Quaternary rocks determined have been older than about 5×10^5 years. The number of Quaternary stratigraphic sites dated is small, especially compared with radiocarbon dating, and Flint (1971) in his review of Quaternary geology mentions only about a dozen cases. The statistical error of the method is small, about 1.5% to 2% for good samples.

A greater contribution has been made by joint K-Ar and paleomagnetic studies of long sequences of lavas. These have yielded a record of magnetic reversals which has become a tool of great power in geology generally, and in Quaternary studies has enabled correlations to be made between diverse stratigraphies which otherwise have not been dated, such as loess sequences on one hand and marine sedimentary sequences on the other (Cox 1969. Most of the instants of magnetic field change have been dated to within about 3% of their age, some yet more accurately.

Fission-track dating (particle track analysis)

The method rests on the fact that when certain isotopes disintegrate by natural fission, the fission fragments make damage-tracks within crystalline and other materials. The age of crystallisation can be determined by counting such tracks within a crystal and measuring the concentration of isotopes involved. Since its development, the technique has been applied to very many problems ranging far outside geologic dating (Fisher 1975). Within geology it is extensively used for volcanic materials with ^{238}U as the main fissioning isotope (Wagner 1973). It is rarely useful for ages less than about 10^5 years because ^{238}U fission-rate is low and the statistical error is rather large (about $\pm 10\%$). The method has the advantage of being applicable to volcanic materials inaccessible to K-Ar dating due to low potassium-levels or large air-argon background, because it can be used even in rather weathered materials in which sufficient single crystals (usually of zircon) or volcanic glass fragments survive. Hence, it is particularly useful for providing age estimates, albeit with somewhat large statistical error, in volcanic tephras, which are often useful horizons in stratigraphic correlations (as, for example, in New Zealand; Seward 1975).

Thermoluminescence dating (TL)

Thermoluminescence occurs in rock-forming minerals when heated as a result of displacement of electrons, trapped at an energy level some distance above the valence band, to a lower level. Electrons are displaced to the trapping level in the first place by radiation from natural radio-isotopes plus cosmic rays; total radiation level is stated as dose rate. For sites which have been heated to temperatures sufficient to displace all trapped electrons (e.g. above 500^0C for quartz), the time elapsed since the heating is proportional to measurable thermoluminescence of a sample divided by dose rate. The method clearly is useful for archaeologic materials such as pottery and hearths (e.g. Zimmerman 1971). As long as dose rate can be measured accurately, resolution is good (± 2-3%) for materials ranging to many thousands of years old. In complex geologic contexts, dose rate measurements requires assessment of many factors (Bell 1978) and total error can be as high as about 12% (e.g. Huxtable and Aitken 1977). So far the method has only rarely been applied to non-archaeologic stratigraphy, although it clearly has potential for sites where geologic baking has occurred, such as by lava flows.

PHYSICO-CHEMICAL RELATIVE AGE METHODS

The notion that slow chemical processes might be used to estimate relative ages of fossil materials was first seriously explored by measuring fluorine content of bones, on the grounds that hydroxy-apatite in bones traps fluorine ions from ground water. Exposure of the Piltdown hoax was supported in this way, demonstrating that mandible and skull bones had very different origins (Oakley 1950). The great problem with this, and all other suggested chemical methods of relative dating, is that chemical reactions are affected not only by temperature but also by pH and dissolved ion concentrations of solutions at the reaction site.

The method in this group with potentially the widest application to Quaternary problems is that using amino acid racemisation in similar materials from geochemically similar sites. Biologic amino acids, the basic components of proteinaceous matter, are of the L-form and change with time to an equilibrium mixture of D- and L- forms in the racemisation process. Conversion rates differ between amino acid species and, although rate varies with conditions at the reaction site, the process occurs through first order kinetics in simple systems. Clearly, a stratigraphic sequence containing biologic materials and dateable to a certain depth by radiocarbon, might have its chronology extended by (i) measuring D/L ratios of a slow-converting animo acid at several points within the dated sequence, and (ii) using this calibrated time-dependency of D/L ratio to estimate ages beyond that. The method has been closely tested in several contexts, ranging from deep sea cores to mummified tissues. Wehrmiller and Hare (1971), for example, demonstrated that the assumption of uniform reaction rate-constant holds approximately to about 400,000 years B.P. in deep-sea cores, but that rate-constant apparently diminishes beyond that. Similar time-dependency of rate-constant has been demonstrated for other materials and the explanation lies partly at least with chemical complexities within the total amino acid system in a sample, as well as with past changes of the sample environment (Dungworth 1976; Schroeder and Bada 1976). As these authors explain, however, sufficiently comprehensive analysis of the amino acid suite within samples for a set of similar sites can provide the basis for reckoning with such problems. Although the method so far has been in the proving stages rather than in wide use for Quaternary dating, the potential seems good for many stratigraphic situations where some calibration by a radiometric method is possible.

PERSPECTIVE ON QUATERNARY PHYSICAL HISTORY

Absolute dating of Quaternary deposits and landforms has advanced remarkably in the last two decades and time-correlations between physical changes in different regions will continually improve as the various dating methods are applied and cross-referenced to each other. However, it must be remembered that the fundamental method of identifying past physical changes is stratigraphic, aided by sedimentologic, petrographic, paleontologic, geomorphic and geochemical studies. Correlations of relative history between different regions or sequences usually are proposed on similarities of patterns of change, aided by singular events. With no reference to dating methods, deep sea cores, for example, are correlated around the globe through (i) rapid and major changes of the $^{16}O/^{18}O$ ratio, such as the major zigzag about 3-5m down most cores, (ii) the first point of change from normal to reversed magnetisation and (iii) the point of appearance of the foraminifer *Globorotalia truncatulinoides*. Yet more detailed correlations are made from similarities between the complete isotopic, magnetic, and paleontologic signatures down the length of the cores. It is from absolute dating, of course, that we know that the 3 singular events referred to occurred about 0.12, 0.7, and

1.8 million years ago respectively.

The impression may have been gained, from my review of methods, that a salient difficulty lies in finding materials suitable for dating at each important site. While true, a greater exercise, if not a greater difficulty, is the establishment of the stratigraphic and geomorphic histories themselves to the level of detail needed for good inter-regional correlations. A large example which well illustrates the general problem is to do with relationships between the major glacial episodes in Europe and North America, namely the Wurm, Riss, Mindel, Gunz and Donau on the one hand and the Wisconsinan, Illinoian, Kansan and Nebraskan on the other. While later stages of the Wurm and Wisconsinan glaciations long have been fairly compared, earlier relationships are a much-mooted matter because the patterns of successive ice advances have been imperfectly known. In this case, absolute dating has yielded little clarification owing to scarcity of datable materials beyond the radiocarbon range, except to throw serious doubt on the hypothesis of simple pairing of major glacial episodes across the Atlantic (Cooke 1973). Relationship between these great glacial episodes, themselves identified through countless man-years of field work, and the ocean-climate variations known in detail from deep sea cores, is quite unclear, particularly beyond 120,000 years ago.

The problem in understanding Quaternary physical changes is this. The nature of some processes, their means and magnitude-frequency variabilities, are known from instrumental and historical observations. For many other processes our direct records are all too short to supply such statistics. For several, it is hard to estimate how their magnitude-frequency distributions would alter when strong climatic changes occur, even given data from contemporary man-induced landscape disequilibria. Thus it is unsurprising that our minds are speculative and our estimates qualitative when we are confronted with evidence of extraordinary activity in the landscape, such as massive piles of mudflows in an area where they do not occur today, or vast fossil fields of coastal dunes in humid regions. Most fruitful for understanding physical rates in situations analogous to these are studies set within the last 30,000 years or so, where radiocarbon dating, our best chronologic tool, can supply a time base. Although dramatic events appearing in this frame can sometimes be used to extend magnitude-frequency descriptions of process, changes will generally be interpretable only as mean rates. From this base, detailed interpretations may be extended, by using analogies, at least to the last interglacial climax 130,000 to 120,000 years ago and possibly to300,000 years ago. As things stand at present, however, studies beyond the radiocarbon range can only rarely be used to extend estimates of physical change-rates. Some cases where change-rates exceeding anything known from the last 15,000 years have been claimed from much older times are very questionable. Examples are discussed below.

Perspectives of the greater Quaternary on land are vague by comparison with the last 130,000 years and especially with the last 30,000 years. From the oceans the record is considerably more clear, as exemplified by the recent and excellent CLIMAP studies (Cline and Hays 1976). There are good indications that the very long and semi-continuous glacial loess records from Czechoslovakia (Kukla 1970) and from China (Liu 1966) show strong resemblances to the deep-sea Quaternary climatic record. However, the quandary which I mentioned about intercontinental glacial correlations illustrates that the complete moving picture is somewhat remote from us.

In discussing rates of physical changes, it is most important to dissect the biologic input from the purely physical input to stratigraphic and other evidence of change. Not only is much of the interpretation biologically based, through paleontology, but also landscape instability is usually interactive with biologic change and often a consequence, not a cause, thereof. The sciences must be interactive also. For example, evidence for past episodes of rapid slope denudation can only be interpreted in dynamic terms through collaboration of the geologist and paleobotanist. In saying this, I indicate why the following account of changes, their rates and ranges, proceeds from those systems which are essentially physical towards those in which physical and biological dynamics jointly are involved; as we proceed the degree of certainty about the purely physical input to processes of change diminishes.

ATMOSPHERIC CLIMATE: VARIABILITY AND CHANGE

Direct records.

Atmospheric conditions at any place vary on time scales from seconds to more than 10^5 years. Weather is atmospheric variability on time scales ranging up to a week or two, and originates hydro-dynamically irrespective of the atmosphere's boundary conditions. Climate is the kind of weather characteristic of a fraction of a year, usually a particular month or season, and is reckoned statistically for a given year-fraction over a run of many years. Just how many years of records are necessary to establish climatic variability and distinguish climatic change is a moot point. The notion of change implies continuous drift of the mean of a given climatic index, or of its running mean taken over an interval of some number of years. To distinguish the necessary interval for defining climate, its mean and variability, is something of a problem.

Theoretically, if land and ocean surface thermal conditions at any given time of year were the same every year, observed variability from year to year would be weather variability, arising hydro-dynamically in the atmosphere. However, the earth's surface is thermally conditioned by meteorologic processes and has a thermal memory ranging from months for the land through years to even hundreds of years for different parts of the ocean, according to its circulation. Interaction between this memory and the atmosphere greatly increases climatic variability, leaving aside any possible variations of solar flux through the upper atmosphere. Because of the range of thermal memory time in different parts of the oceans, there is no well-agreed time-base for establishing trends of climatic indices and thence variability around a stable mean. On the basis of experience with 20th-century records it is agreed that at least many tens of years statistics are needed to establish intrinsic climatic variability. Yet, judging the time base for reckoning climatic variability on the basis of stability of a filtered form of the index, such as a running mean or any low-pass filter, becomes weaker when the filter base is much longer than a century or so, because variability at such scales may camouflage a relatively short-term change which is consequential from the biologic point of view.

It is probably the example of the historical Little Ice Age which best illustrates practical discrimination of change from variability. Temperature and rainfall for the last 1000 years in central England have been painstakingly compiled by Manley and by Lamb; figure 1.3 shows graphs of their winter mean values (from Lamb 1972). The decline of both indices in the 14th century, and the rise in the 19th century certainly constitute climatic change, no matter whether judged from a social, or crop-growth, or general vegetational point of view (Lamb 1968, 1972). The excursion in both indices between about 1470 and 1570 A.D. likewise should be judged as a change.

Expanding from this example, we can discuss ranges and rates of change. The shift in temperature and rainfall means from about 1250 to 1400 A.D. are about 1^0C and 14% respectively, which are about the same as standard deviations of temperature and rainfall over central England over a 30-year period (data in Lamb 1972); i.e. that total climatic change of the Little Ice Age was no more than the likely difference between mean values for successive winters. The rate of change is much less. Whereas variability is likely to be 1^0C between successive years, the change-rate over 1250-1400 A.D. is somewhat less than 10^{-2} deg./year. When the time-base over which a climatic change is reckoned is increased to 10,000 years (the interval between the last glacial maximum and complete deglaciation) the mean rate of change known from geologic records is about an order of

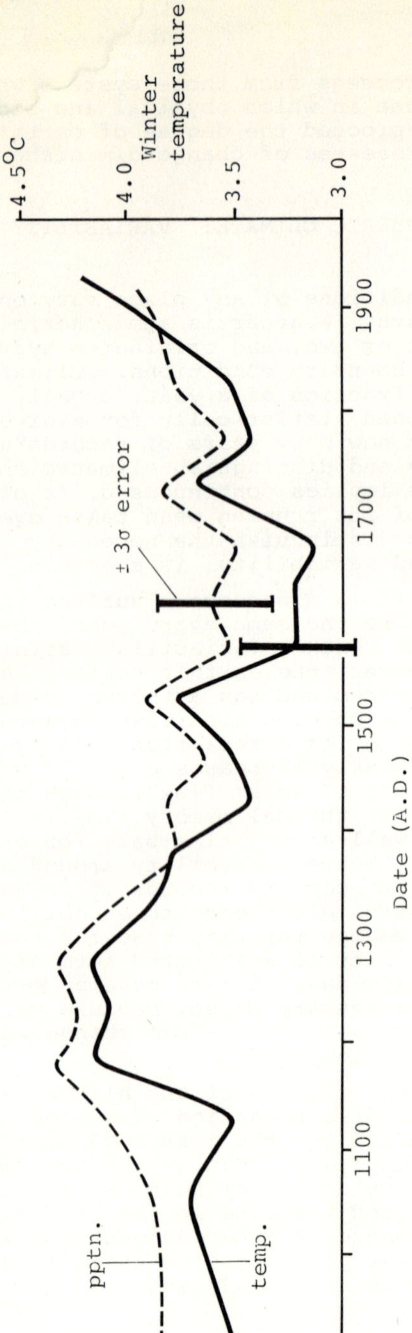

Figure 1.3

Climatic variability or climatic change? Temperature and precipitation in winter, Central England, from historical data of several kinds (after Lamb 1972).

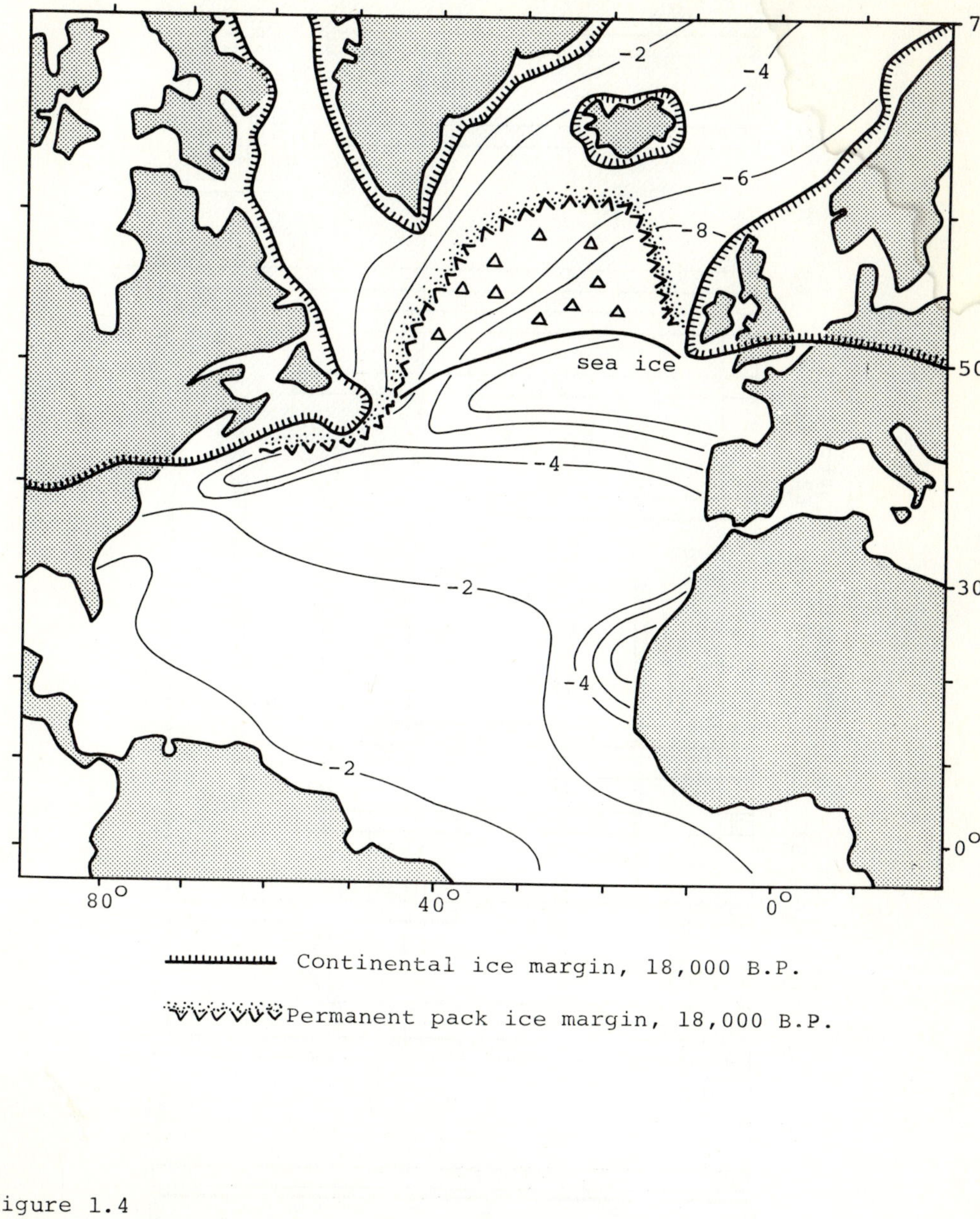

Figure 1.4

Regional variation of Late Quaternary temperature changes: temperature differences between 18,000 yrs B.P. and present in northern Atlantic surface waters (after McIntyre et al. 1976).

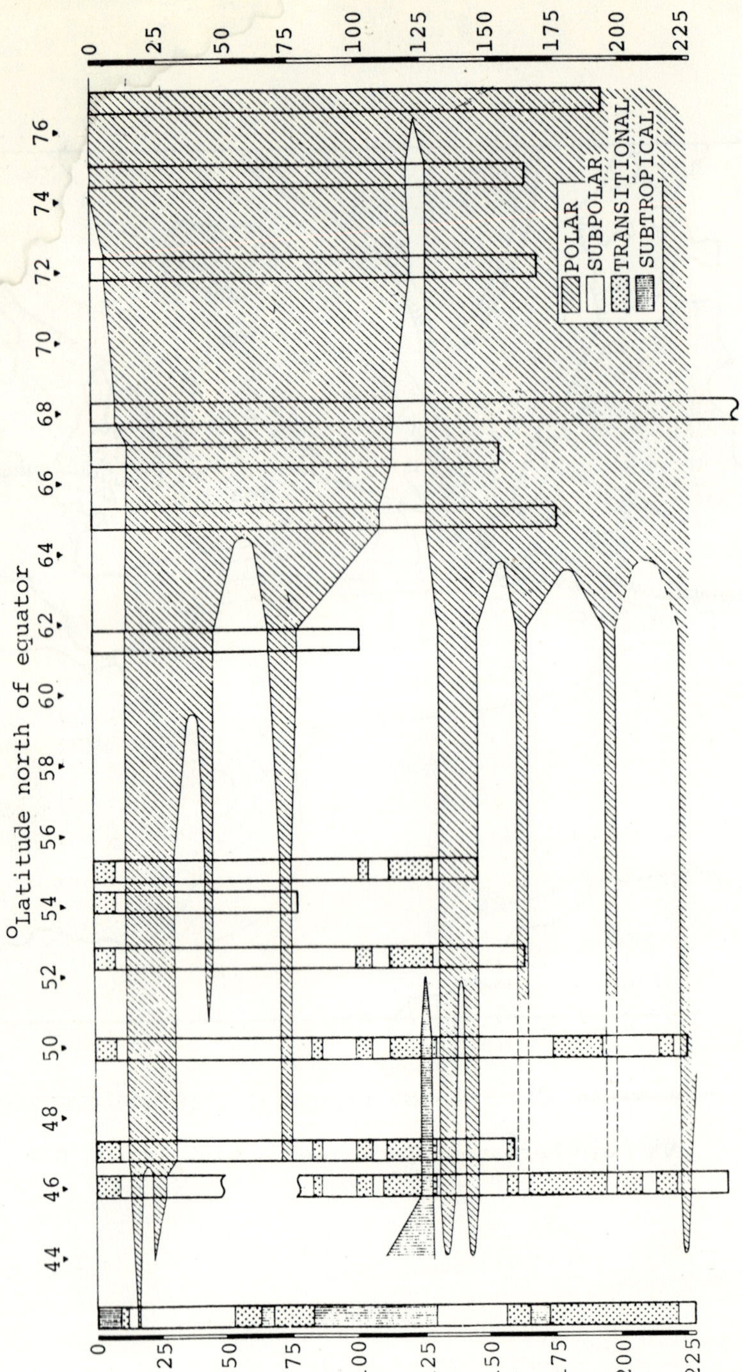

Figure 1.5

Meridional migrations of fauna and surface water masses in north-east Altantic-Norwegian sea over last 225,000 years (Kellog 1976).

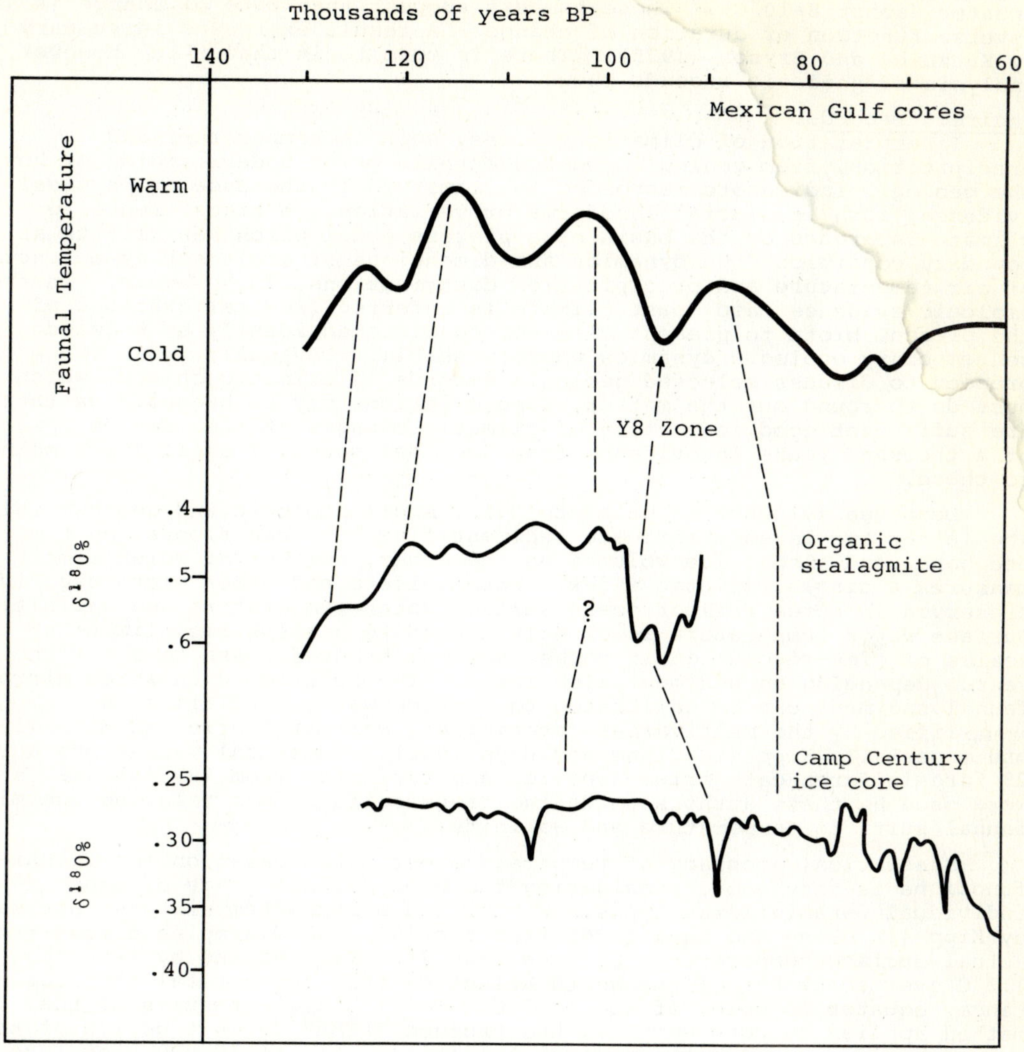

Figure 1.6

Illustrating difficulties of comparing different indicators of paleoclimates: paleo-faunal temperature estimates in Gulf of Mexico (Kennett and Huddlestun 1972), paleo-hydrologic changes indicated by $\delta^{18}O$ French stalagmite (after Duplessy et al. 1971), and paleotemperatures (?) indicated by $\delta^{18}O$ in Greenland ice core (after Dansgaard et al. 1971).

magnitude smaller again, although the magnitude of the change itself is greater (about 8-10°C). These numbers suggest that rate of change is an inverse function of duration of change - a result expressed in summary by Kutzbach and Bryson (1975). There is more to it than this, however, as analysis of geologic records shows.

Indirect records.

Interpretation of climatic indices, such as temperature and precipitation, from geologic records entails prior understanding of how the geologic indicators responded to climate. In the case of physical evidence, such as glacial landforms or variations, a transference to climate is argued on the basis of a process model which has climate as a boundary condition; the dynamics and dimensions of a glacier as a function of air temperature and precipitation distributions, for example. In the biologic evidence case, past climate is inferred from calibrations of the present biota to present climate, conditioned ideally by knowledge of contemporary ecologic dynamics where possible. Obviously, when I now proceed to discuss selected geologic records of climatic change, which I must do to round out the matter, some questions may be begged. As there are sufficient good indicators of climatic changes at time scales upwards of a thousand years in evidence from deep sea cores, I shall stick mainly to these.

Deep sea evidence is paleontologic, sedimentologic and geochemical. The latter two largely indicate sediment flux to ocean floors, on the one hand, and global ice volumes on the other, neither of which simply measures a single climatic index. Foraminifera and other micro-organisms preserved in cores reflect ocean surface water temperature and salinities. Surface water temperature is closely linked to atmospheric climate at the scales of time-resolution in cores (several hundred years to a few thousand years, depending on sedimentation rate). The accuracy with which micro-faunal sediment can be calibrated to surface water conditions is exemplified by the multivariate statistical work of Imbrie and Kipp (1971) and especially Kipp (in Cline and Hays 1976). Numerical variations across 29 foraminiferan categories (species and variants) from 197 Atlantic sites were used by these authors to define transfer functions relating the micro-faunal suite to temperature and salinity.

Statistical accuracy of temperature estimates based on the transfer functions is very good, considering the wide oceanic range of many individual foraminiferan types. Winter sea surface temperatures estimated by Kipp (in Cline and Hays 1976) from the 191 seabed samples differ from actual surface temperatures by less than 2°C over 99% and by less than 1.2°C over about 85% of the North Atlantic; this for a total temperature range, equator to pole, of about 27°C. To illustrate results of the method applied to core samples, the Project CLIMAP reconstruction of temperature differences between North Atlantic winter of 18,000 years ago and present winter is shown in figure 1.4 (from McIntyre et al. 1976, fig. 17). This indicates the level of detail about ranges of temperature differences accessible between two time planes through deep sea core analysis; rates of change recognisable through each of several cores are illustrated in figure 1.5 (from Kellog 1976). Major changes at some sites within about 7000 years are apparent, such as the transition from polar to subtropical faunas between 46°N and 50°N about 130,000 years ago. Ruddiman and McIntyre (1976 pp 127-128) discuss in more detail the interpretation of similar strong variations in deep-sea cores.

Although geologic records of several kinds suggest changes possibly even more rapid, caution must prevail when ranges as well as rates are being inferred. Consider the records in figure 1.6, lying in the time range 80,000 to 130,000 years BP, and showing, respectively, temperature changes in Mexican Gulf cores (Kennett and Huddleston 1972), $^{16}O/^{18}O$ variations in a stalagmite from southern France (Duplessy et al. 1972). The respective time scales differ in reliability and that for the ice core in particular is dubious (Chappell 1977). The rapid excursion in each record around 90,000 - 100,000 years BP (connected by dashed line in figure

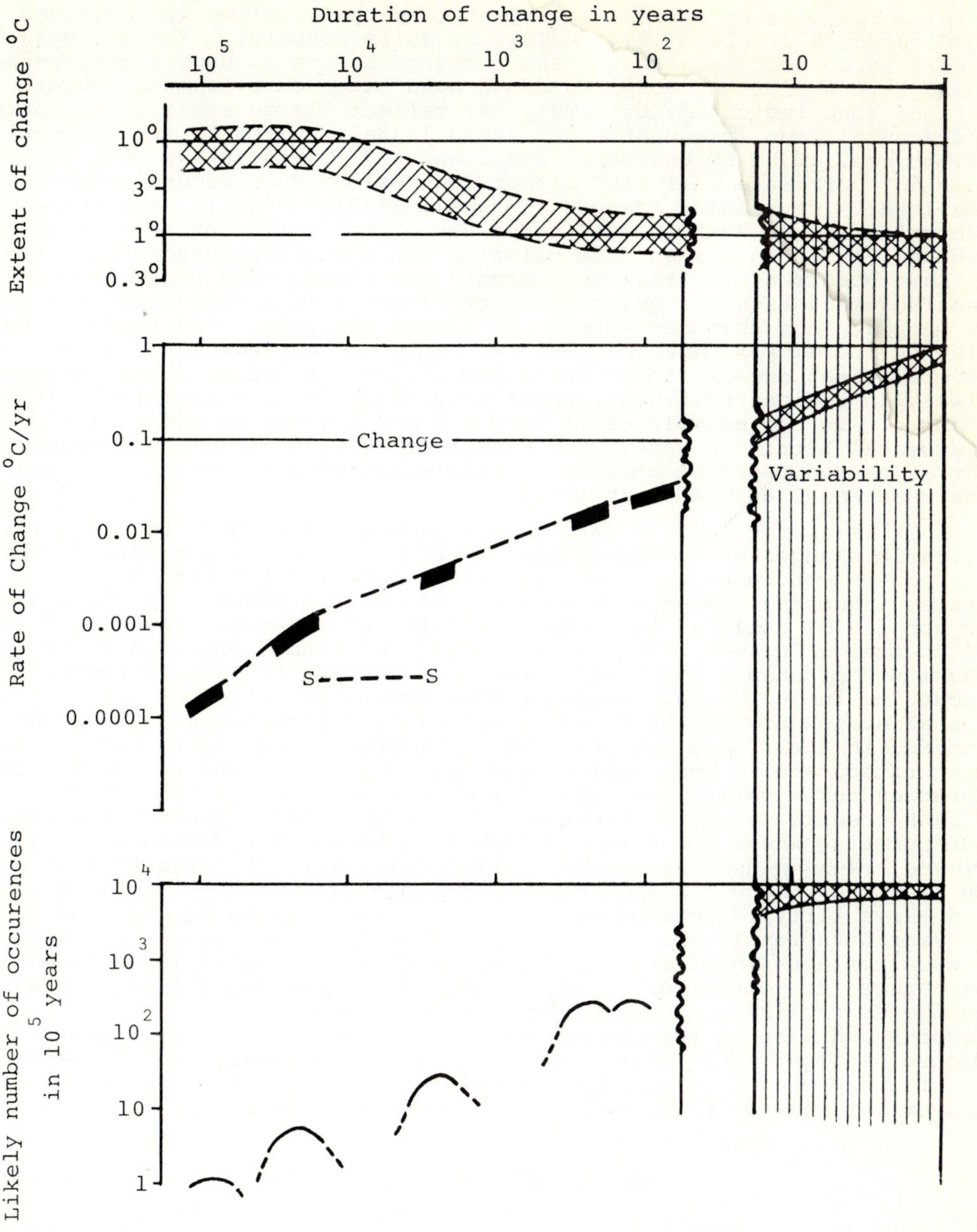

Figure 1.7

Magnitude-rate-likelihood estimates of climatic change based on a wide range of chronologically reliable records similar to those in figures 3,4,5, plus similar rate-duration estimates by Kutzbach and Bryson (1975). Peaks in likelihood curve reflect quasi-periodic changes.

1.4) has been seen by some as a very rapid global cooling (Kennett and Huddleston 1972; Flohn 1974). This is quite doubtful. The ice core record is suspect because the sharp change occurs in highly compressed basal layers of the icecap and could have happened over much longer periods than indicated, or simply may reflect "large scale topographic and accumulation irregularities" (Budd 1975a). Climatic interpretation of the stalagmite is equivocal (Emiliani 1971), and in any case its pattern does not correlate well with either the Mexican Gulf record or with known sea level (ice volume) changes (Chappell 1977). Finally, the sharp excursion in the Mexican Gulf cores cannot be considered as more than a regional event in sea surface climates until similar changes are proven to have occurred simultaneously around the world. The point is that magnitude and rate changes of a given climatic indicator have been greater in many regions than hemispheric or global averages. For example, both figures 1.4 and 1.5 indicate temperature in the northeastern Atlantic up to several times greater than many other places. Rapid or large changes are likely in regions critically positioned relative to circulation pattern changes. To the example of figures 1.4 and 1.5 may be added that of the historical Little Ice Age, when migration of troughs in the circum-polar vortex caused regional changes of precipitation substantially exceeding the mean zonal change (Lamb 1968).

It is impractical to extend this summary of methods for identifying climatic changes by reviewing chapter and verse of case studies. The section is concluded by synthesising the better estimates of rates and ranges, choosing temperature as the best-defined index. Kutzbach and Bryson (1975) outlined the inverse relationship between rate and duration of change; in figure 1.7 I have used a similar frame, extended to include magnitude of change and likely number of changes of given duration occurring in 10^5 years. Rate-duration curves for regional and hemispheric/ global scales are sketched, based on selected paleotemperature studies indexed in the figure caption. These curves are nearer to maximum than to mean rates, and in some near-equatorial ocean areas the regional rate-duration curve falls lower than the global mean curve. The Project CLIMAP oceanic reconstructions (Cline and Hays 1976) provide a guide for interpreting where the curve is likely to be for many ocean areas (e.g. figure 1.4 suggests that a rate-duration curve for the Sargasso Sea should be near the line S-S in figure 1.7). The division between climatic change and variability, discussed earlier, is shown blurring between 30- and 60-year duration levels in figure 1.7. The upturn in rate in the variability region reflects inevitability of variability as against lower likelihood of change; the curve for likely number of occurrences reflects this. The estimates of likely number of occurrences are based on paleotemperature records and also on records of glacial and ice cap changes indexed in the figure caption. The latter warrant further discussion as they represent major changes of physical environments, not only near the glaciers themselves but also in world sea levels and temperatures.

GLACIAL AND SEA LEVEL CHANGES

Glaciers and ice sheets

Glaciers vary in size as climate changes. However, in common with other physical systems in the landscapes, glacial changes do not track exactly with climate changes owing to impedance of the ice body. Response varies, not only with glacier or ice sheet size, its altitude distribution with respect to air temperature and the duration of a climatic change, but also with internal temperature and basal conditions (a relevant text is Paterson 1970). Although glacier physics is highly developed, glacial variations corresponding to climatic changes are known largely from geomorphic and geologic evidence rather than by prediction from physical modes alone. Physical and biogeographic effects of glaciers extend well beyond the ice margin, for example by deposition of glacial drift (a wide suite of materials ranging from moraines through mudflows to fluviatile outwash sediments) and of glacial loess. Migrations of glacial margins

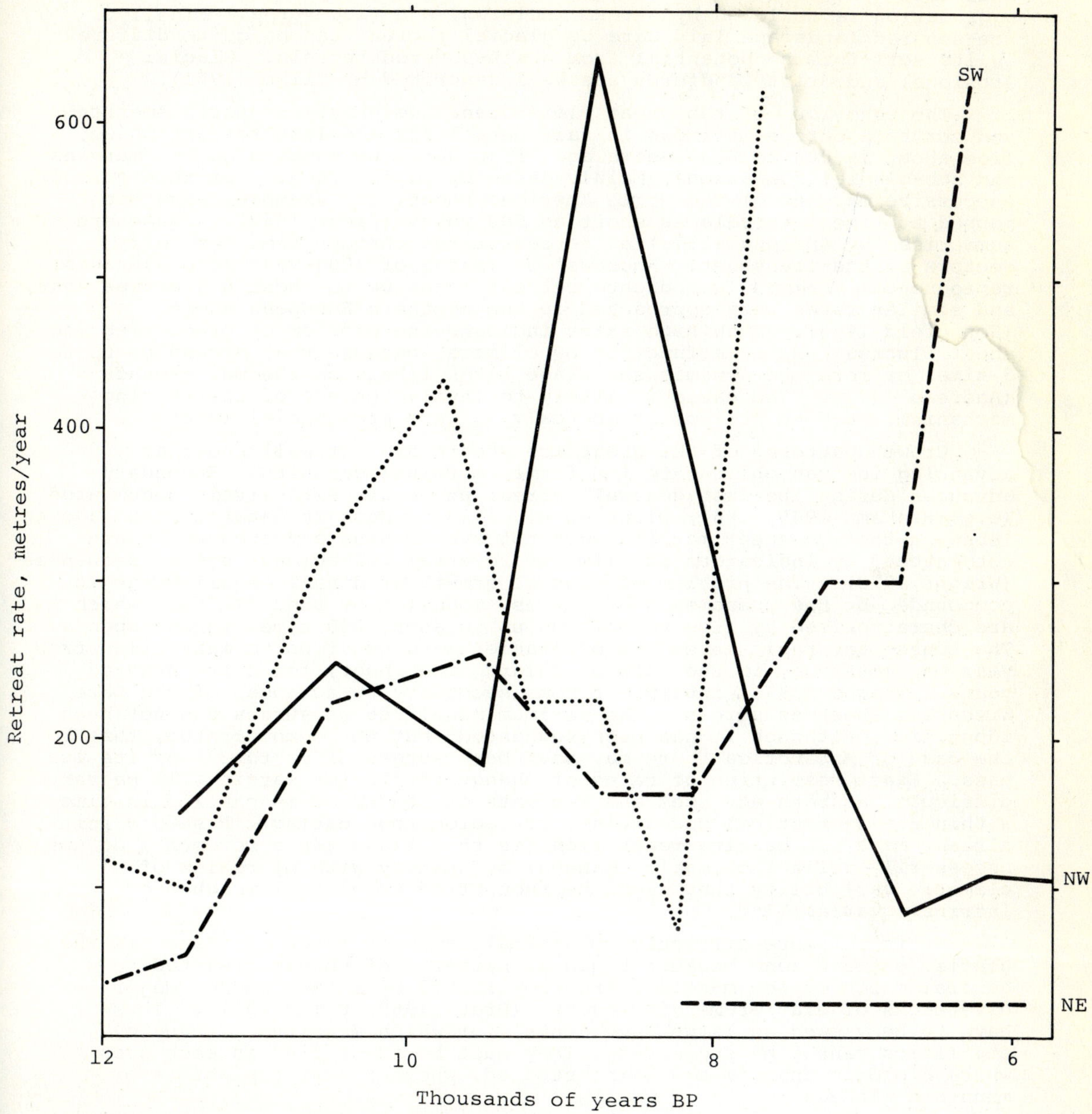

Figure 1.8
Retreat rates of different sectors of Late Wisconsinan ice sheet (Andrews 1973).

change the distributional relationships between these deposits and their associated rivers and streams; groundwater levels may also change when aggradation is followed by stream incision, and vice versa. Finally, ice-scoured landscape laid bare by glacial retreat can be quite different in its soil-forming potential from drift-covered terrain. Glacial erosional and drift landforms are well described by Flint (1971).

The behaviour of the great Pleistocene ice sheets of North America and northern Europe is known in fair detail for the last retreat phase, from about 15,000 to 6000 years ago, from detailed mapping of ice-margins and sub-glacial formations, mainly dated by radiocarbon. For this episode, successive margins of the North American sheet, for example, have been mapped at time intervals as short as 500 years (Prest 1969). These are summarised by Andrews (1973) as retreat rates through time for four sectors of the ice sheet. Andrews' estimates of 1000-year rate averages, reproduced in figure 1.8, indicate retreat rates up to about 650 metres year, and similar rates were approached in the northern European sheet (Zonnefeld 1973). Such high rates indicate the problem of predicting ice sheet fluctuations as a function of climate, because they exceed by up to 5 times or more the dynamic shrinkage rates likely on thermal grounds (Andrews 1973). The cause is likely to involve change of ice-wasting mechanism, such as periods of ice-calving into pro-glacial lakes.

Growth patterns of the great ice sheets are not well known as advancing ice can obliterate its former marginal deposits. Secondary advances during the last general retreat phase are extensively documented (e.g. Rankama 1965, 1967; Black et al. 1973). Seen in detail, re-advancing lobes in the ice sheet margins were not synchronous and frequently are not matched by indicators of climatic reversal in regional pollen sequences (Wright 1973). The problem of glacial growth or dynamic equilibrium is compounded by the phenomenon of surges, modelled by Budd (1975b), which are characterised by flow velocities up to about 200 times normal speeds. The consequent rapid movements of glacier margins of up to many kilometres/year may continue, in the case of a large ice sheet, for a few hundred years. Wright (1973) interprets some secondary re-advances of the Late Wisconsin sheet as surges. The maximum magnitude of surges has not been identified, although it has been suggested, but not demonstrated, that in the case of Antarctica there may have been surges of up to 30% of its ice mass. Exact resolution of rates of change of glacier margin will be very difficult, as both advances and retreats of 100 km or so can lie in time within the statistical uncertainty of radiocarbon dating. A second point, already made but bearing repetition, is that these rapid movements do not necessarily reflect climatic changes; by analogy with my review of climatic variability they might be thought of as glacial weather or intrinsic variability.

It is yet more difficult to generalise about rates of change in the glacial outwash zone because regional patterns of change, related to general march of ice margins, are complicated by rather sudden major diversions of glacier-margin rivers, (Bretz 1969; Flint 1971). These have to be viewed as large 'accidents' for which magnitude-frequency statistics cannot be suggested. They must be identified in each area where biologic changes are being studied, whether near ice sheets or mountain glaciers.

<u>Sea-level changes.</u>

Rather rapid and large sea-level variations have affected all the world's coastlines in Quaternary times. Especially in regions where continental shelves are broad, the repeated march of shorelines back and forth across tens of thousands of kilometers not only has affected connections between continents and islands, but also, by its very movement, has imposed an ecologic stress in coast lands different from that pertaining with stable sea-level. In some areas, wave climate and tidal ranges also varied.

As with climate, sea-level changes can be reckoned at global and local levels. Different from climate, global sea-level change is not the mean

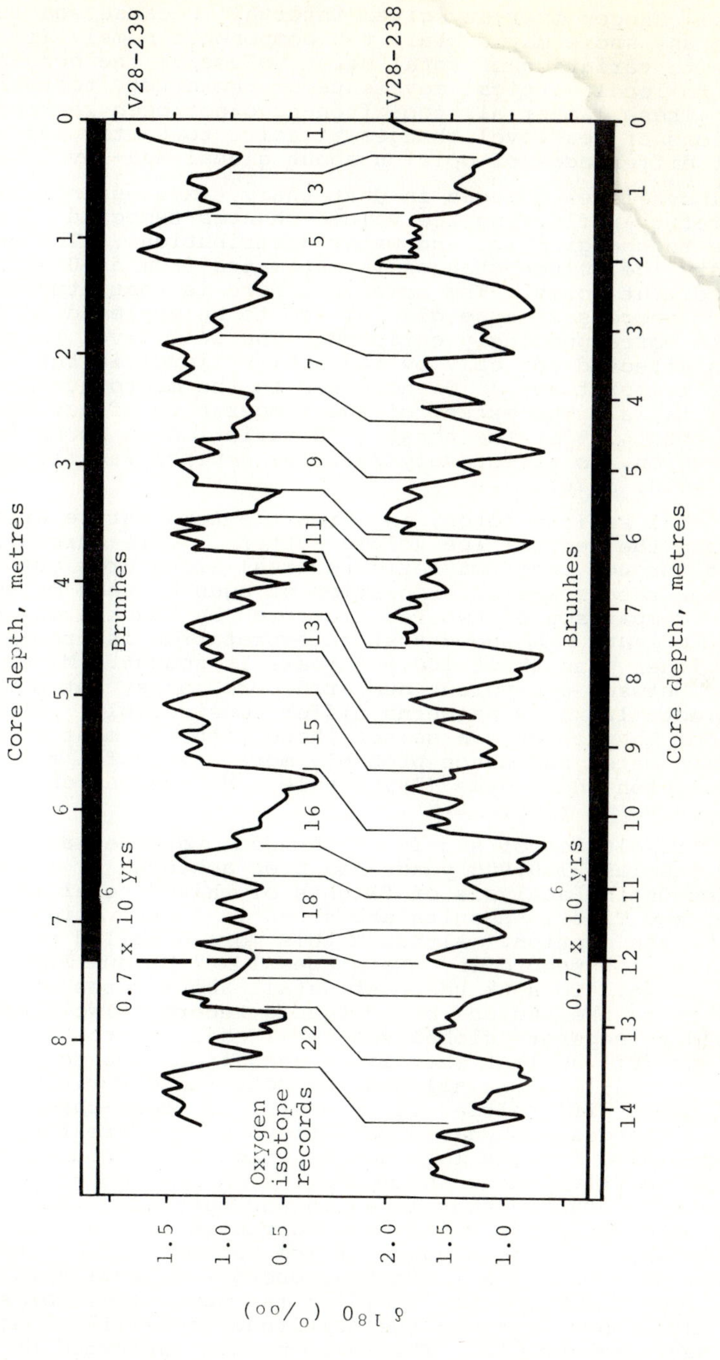

Figure 1.9

$\delta^{18}O$ variations in two equatorial west Pacific cores provide a generalised record of ice volume changes seemingly not masked by significant paleotemperature variations, although comparison of core zone thicknesses reveals substantial fluctuations of sedimentation rate (after Shackleton and Opdyke 1976).

of all local changes over any given interval, because sea-level change relative to any shore may contain two components namely (i) eustatic changes due to variation of total water volume in the oceans and (ii) shifts due to local vertical movements of the crust, tectonic and isostatic movements. These latter all too often have not been reckoned with during investigations of sea-level changes relative to particular shores, leading to stagnant differences of opinion about global sea-level changes.

Global sea-level changes in Quaternary times essentially are glacio-eustatic, reflecting ice volume changes tempered by global isostatic adjustments to changing ice and water distributions. The general pattern of Quaternary ice volume changes is estimated from $\delta^{18}O$ variations in deep sea sediment cores. The principle here is that atmospheric distillation processes cause glacial ice to be depleted in $\delta^{18}O$ and the oceans to be correspondingly enriched. The $\delta^{18}O$ level in ocean microfaunal sediment is affected not only by the ambient level in the surrounding ocean, but also by temperature; $\delta^{18}O$ increases in the microfaunal skeletons as temperature falls. The extent of the temperature effect varies geographically from about 30% of the total $\delta^{18}O$ variation in cores from the Caribbean region, to effectively zero for central Pacific cores (Shackleton and Opdyke 1973, 1976).

Equatorial Pacific cores, in which the temperature effect is minimal, best indicate the general ice volume pattern. Core chronologies are established through their magnetic reversal signatures but variations of sedimentation rate makes interpolation of ages between magnetic reversals uncertain. Comparison of two Pacific cores by Shackleton and Opdyke (1976), reproduced in figure 1.9, shows that interpretation is problematic when resolution finer than about 100,000 years is sought. Major $\delta^{18}O$ changes are similar between equivalent-numbered core zones, but both zone thicknesses and detailed $\delta^{18}O$ patterns differ considerably. Statistical comparison of many cores can select those with the most constant sedimentation rates and hence probably more meaningful records; on this basis, Shackleton and Opdyke (1976) select V28-238 as the better of the two (lower curve in figure 1.9).

Better resolution of sea-level changes over the last 250,000 years and especially the last 130,000 years has been achieved by close stratigraphic and U-series dating analysis of flights of raised coral reefs in Barbados, New Guinea, and Timor. Results are shown in curve a, figure 1.10; references are given in the caption. Although this more detailed pattern is crudely similar to the ^{18}O record in core stages 1 to 7 (figure 1.9), differences in detail are large. To seek the best detail about rates of sea-level changes, we again turn to the radiocarbon dateable record. Two problems are met: (i) the need to separate global sea-level changes from local vertical movements and (ii) ancient shorelines aged from 6000 to >40,000 years are below present sea-level in all but the most rapidly rising regions. Because of these difficulties it is useful to cross-check interpretations of sea-level changes with dated variations of the great northern ice sheets, especially for the last deglaciation from about 17,000 to 6500 BP. The exercise is illustrated in figure 1.10 curve b, showing age-depth positions of dated submarine shorelines in northeastern U.S.A. and eastern Australia. The shoreline depths are corrected for isostatic movements (Chappell 1974b). The mean curve of ice volume change since 17,000 BP agrees quite well with the isostasy-reduced sea level data, as shown. Curve b of figure 1.10 probably typifies the upper limit of sea level change rates and magnitudes, amongst the multitude of oscillations shown in curves 1 and 2 of figure 1.9. The maximum rate indicated by curve b of figure 10 is about 15 metres/1000 years; rate-estimates over much shorter intervals are not permitted by the data. The problem of reconciling sea-level, ^{18}O, and known icecap variations prior to 27,000 BP is still substantial, as a comparison of records back to 45,000 BP by Chappell and Veeh (in press) shows. In short, although the gross history of ice volume changes is known, at about the 20,000 year scale of resolution there are differences between various forms of evidence, and beyond about 40,000 BP the chronology of regional glacial changes is poorly known.

Shorter-term global changes of sea-level have been much debated from evidence of coastal deposits younger than 6000 years. There are numerous cases described where beach ridge complexes, cheniers and estuarine formations can best be explained in terms of relative sea-level changes of up to ±2 metres or so (e.g. figure 1.10c). So far, however, it has not been demonstrated by unequivocal radiocarbon age correlations that such relative movements are indeed global changes of sea-level (despite many claims to the contrary) and the question remains as to whether these represent anything more than local tectonic, sediment-compactional. or isostatic movements. It certainly has not been shown that local or regional patterns of Holocene sea-level movements (such as in figure 1.10c) have any connection with the climatic and minor glacial changes over the last 6000 years identified by Denton and Karlen (1975).

To state the position summarily, sea-level changes over time scales of 10^3 to 10^6 years can be inferred from curves in figures 1.9 and 1.10. tempered by (i) isostatic considerations (Chappell, 1974b; Clark, et al. in press), and (ii) the differences between forms of evidence illustrated in figure 1.10c. On tectonically mobile coasts the course of relative sea-level change includes a vertical component which must be reckoned with separately. Finally, detailed movements within the last 6000 years, at time-scales 10^2 to 10^3 years, cannot be inferred but must be separately investigated for each coastal region.

TECTONIC AND ASSOCIATED PHENOMENA

Tectonic movements, volcanic and seismic activity along the mountain and island chains associated with boundaries of great crustal plates (e.g. Le Pichony 1968; Isacks et al. 1968) have affected Quaternary physical environments in these regions. Horizontal movements have mean rates up to about 1 metre/10 years, and vertical movements up to about 1 metre in 250 years are known. Although as much as 3000 times slower than horizontal and vertical migrations of climatic zones during the ice age oscillations, the tectonic movements generally have continued without reversal, accumulating substantial displacements over Quaternary time in some places. Uplift of certain islands is likely to have altered the available range of habitats. For example, in northeast New Guinea uplift rates up to 3.5 metres/1000 years have been determined by U-series dating of raised coral reefs (Chappell 1974c), which amounts to over 6 km uplift for the total Quaternary time. Hence, even though uplift is partly counteracted by down-wearing, it is likely that northeast New Guinea is now significantly higher than at the end of the Pliocene, perhaps by a few kilometres. Similar remarks may be widely applicable to island and mountain chains in all tectonic (seismic) zones. However, uplift rates fo far are established for relatively few places, e.g. Timor (0.5m/1000 years), New Zealand (up to at least 1 m/1000 years), Ryukyu Islands (up to 1 m/1000 years), and Barbados (up to 0.5 m/1000 years).

These longer-term vertical and horizontal movements are accompanied by geologic 'weather', in the form of earthquake and volcanic activities, causing environmental stresses especially through landslides, debris flows and volcanic tephra fallout. Statistics of these phenomena, which are concentrated in tectonic zones, are known from historical records, although these data are generally insufficient to estimate recurrence intervals for earthquakes and eruptions of the largest magnitudes. Regional volcanic records in several countries are being extended by radiocarbon dating of tephra deposits, although magnitude-frequency analysis on this basis is still some distance away, even in the best-studied regions such as New Zealand and Japan. Finally, it is an open question whether global mean intensity of volcanic and seismic "weather" has remained constant or varied in Quaternary times.

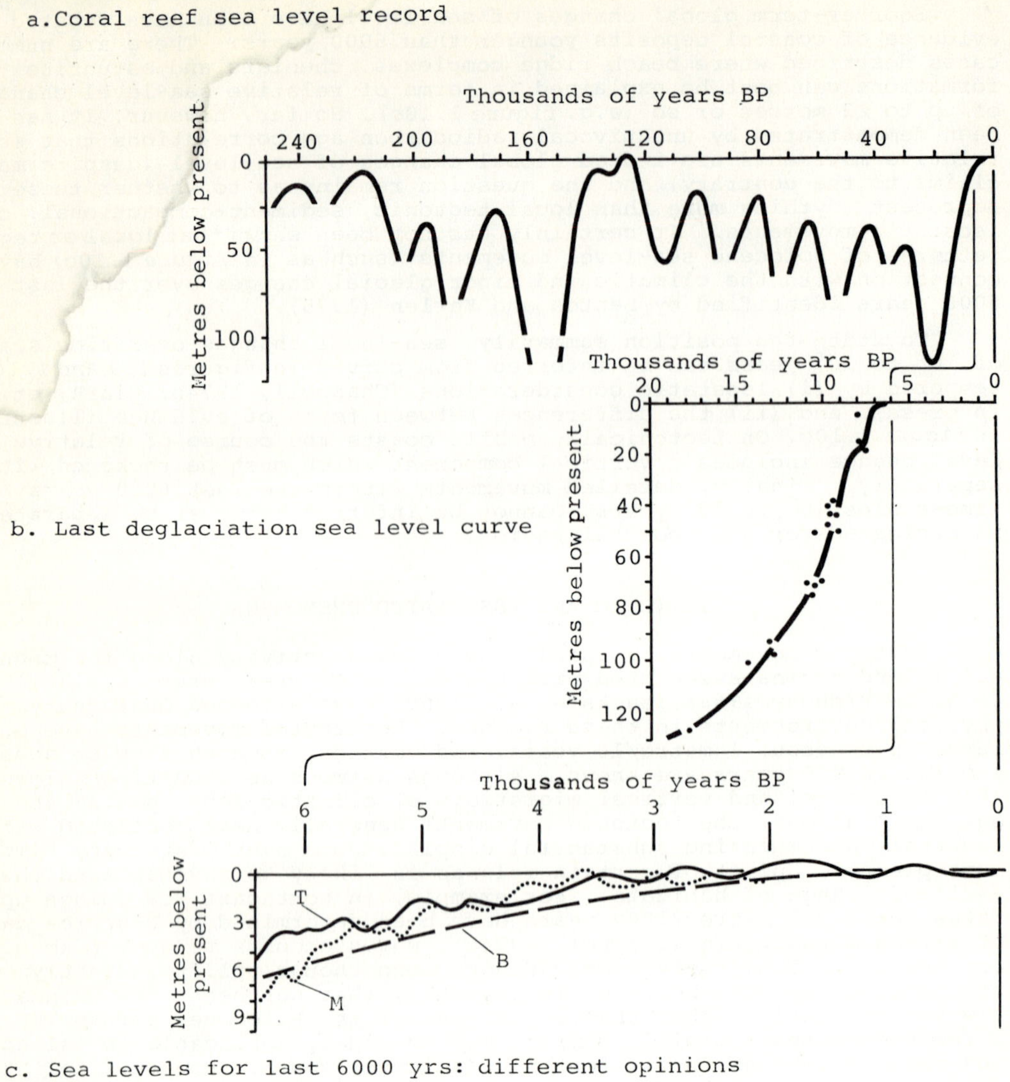

Figure 1.10

(a) Sealevel record for last 250,000 years from $^{230}Th/^{234}U$-dated raised coral reefs (Chappell 1974a) differs in detail from $\delta^{18}O$ records for same period (cf. figure 1.9).

(b) Sealevel change due to last deglaciation (dots: Chappell 1974b) compares well with Bloom's (1971) estimate of diminishing ice volume.

(c) Different radiocarbon-dated curves of sealevel changes over last 7000 years differ significantly: T - after Tooley (1974) from northwest England; M - after Mörner (1967) from southern Sweden; B - after Bloom (1970) from Caroline Is.

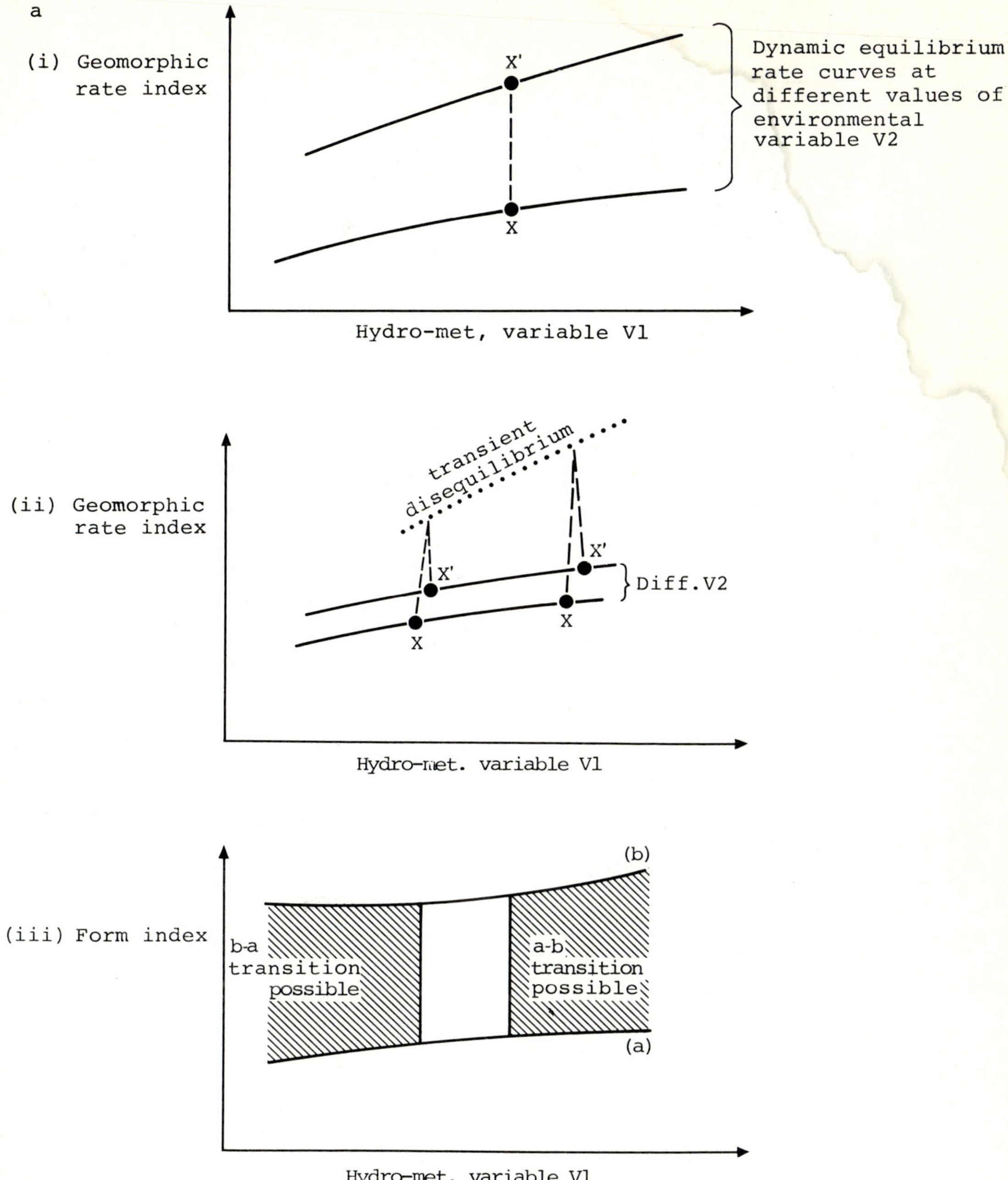

Figure 1.11

Illustrating patterns of geomorphic readjustment to disequilibrium induced by climatic changes.

(a) The rate of a geomorphic process varies with hydro-meteorologic input under given biological equilibrium (solid lines in (i) and (ii)). When the biological factor changes the process may jump to a new equilibrium (line $x - x^1$ in (i)) e.g. stream channel adjustment, or it may overshoot to a rate exceeding that associated with any equilibrium condition (ii) e.g. accelerated slope denudation. Additionally a landscape element may be naturally polymorphic irrespective of biologic variation, with transitions between the various rates occurring at particular ranges of the input hydro-meteorologic variables (iii).

b

(i) Surf beach behaviour

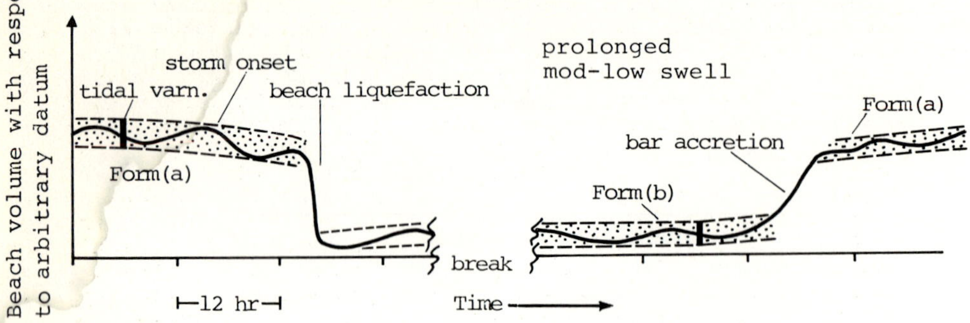

(ii) Glacier growth (Budd 1975)

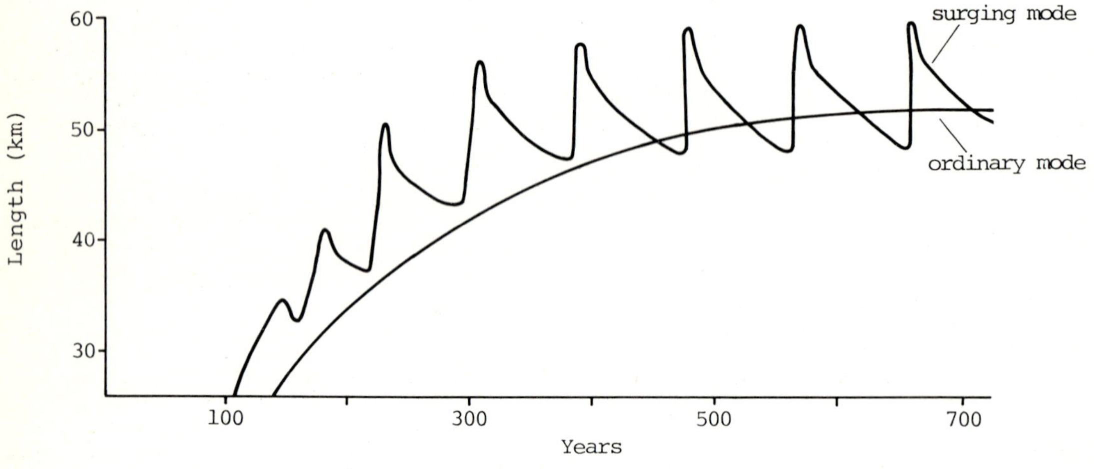

(b) (i) Natural polymorphic behaviour of a beach illustrating principle a(iii). (ii) Two modes of re-equilibration to a new hydro-meteorologic regime illustrated by a glacier, which may change smoothly (illustrating a(i)), or through a sequence of surges (after Budd 1975), illustrating a(i) with a(iii) superimposed.

LANDSCAPE CHANGES INVOLVING BIOLOGIC INTERACTION

Glaciers and sea-levels are physical systems the changes of which are independent of dynamics of the biosphere. Climate, the other physical system discussed above, is also independent except for a feedback effect from desert dust, which may subtly affect global radiation balance and hence climate. However, broad categories of landscape elements such as hillslopes, floodplains and drainage networks are affected by climatic changes not only through hydrologic effects, but also by responses of the biota. The importance of the latter can outweigh the hydrologic factor in engendering geomorphic change as is illustrated, for example, by extensive mobilisation of coastal dunes in Australia and New Zealand following the arrival of white men and their livestock. (In many of these cases it can be shown that mean water tables did not change when the dunes mobilised). Further, when geomorphic changes are mediated by biotic change, it is most hazardous to infer a rate of climatic change from the apparent rate of geomorphic change. Biogeomorphic changes will sometimes appear rapid because they represent a chance concatenation of 'extreme events' in the magnitude-frequency distribution of hydro-meteorologic phenomena, a matter of chance which can be quite independent of the rate of change of any parameter of the probability distribution involved. For example, in areas initially sub-humid and in which the climate drifts towards semi-aridity over hundreds or thousands of years, gully erosion is not initiated by slow lowering of the average water table but by relatively infrequent heavy storms occurring at the appropriate stage of declining vegetation density. Once triggered, denudation may accelerate through a positive feedback effect acting to disrupt the vegetation.

This last point is important. Once disequilibrium is initiated, the pathway to a new equilibrium can be independent of rate of climatic change and is regulated by the dynamics of the landscape system itself. Two examples are illustrative, both stemming from modern human interference. Coastal dunefields in humid regions have often advanced hundreds to thousands of metres inland from a narrow strip of initiation, killing previous vegetation by direct encroachment and by allowing free passage of saltspray in surface breezes on to salt-intolerant plants previously protected by the shore-zone vegetation. Secondly, rapid soil erosion on steep hillslopes is self-propagating and acts through soil-slip and slumping, resulting in scarified hills, and, in colder climates, runaway scree slope development.

Figure 1.11 illustrates the general problem of readjustment to disequilibrium and its caption elaborates the concept. The important question of the rate of change to a new equilibrium after the 'initiating catastrophe' (see figure 1.11, line x-x) needs to be understood for each type of geomorphic system. This is a tall order at present. For some purely physical systems the course and duration of recovery to the initial situation can be reckoned, as can the course towards equilibrium with new climatic conditions. Figure 1.11b(i) illustrates this for the cases of rapid erosion and recovery of a beach, as a short time-scale example, and figure 1.11b(ii) illustrates glacial readjustment as a longer time-scale example. However, for most landforms it is not yet possible to make quantitative statements about disequilibrium response. The problem is complicated, of course, by the fact that the magnitude-frequency distributions, which describe landform dynamics, are unlikely to be the same in disequilibrium as in equilibrium. The example of a prograding New Zealand coastline in figure 1.12 illustrates the point. The upper graph shows shoreline migration in probabilistic terms based on historical surveys; the lower shows the course of net progradation over the last 6000 years. Clearly, the upper graph will not apply to the phases of rapid progradation, such as occurred between 1500 and 2500 years BP.

In the general case, where geomorphic processes and the biosphere interact, it is hard to generalise about dynamics of geomorphic adjustment to hydro-meteorologic change. An active 'geomorphic locus' such as a beach, a river, or a hillside slump, may exhibit magnitude-frequency

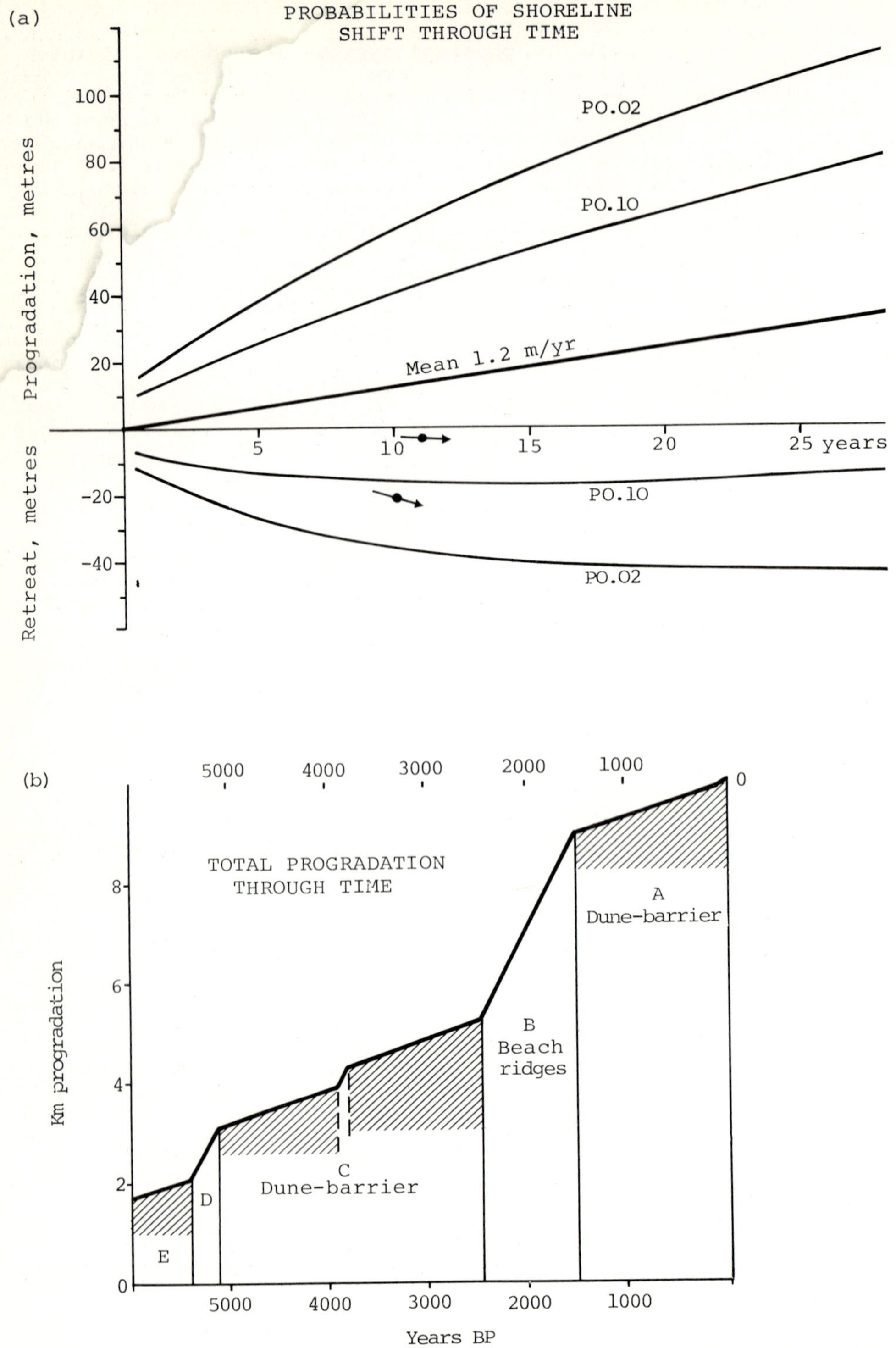

Figure 1.12

Probabilistic description of contemporary geomorphic behaviour often is unlikely to hold for the past - (a) probabilities of shoreline movement in Bay of Plenty, New Zealand, based on historical surveys; (b) dramatic changes of mean rate have occurred on same coast in Holocene times (after Chappell and Pullar, in press).

behaviour which is fairly stationary

(1) when rates of hydro-meteorologic change are slow (e.g. rates shown in figure 1.7 for changes persisting for about 10^4 years), and

(2) as long as vegetational cover remains fairly constant in size and density.

However, the dynamics are difficult to estimate for the case where hydro-meteorologic inputs are changing rapidly, as well as for the case where a threshold is crossed so that geomorphic behaviour changes (e.g. in the cases shown in figures 1.11 and 1.12).

In some cases, changes of form, or position, or dynamics of the "action locus" may be associated with a single extreme event, such as major river channel shifts (e.g. historical shifts of the Yellow River; or late Quaternary changes of southeastern Australia rivers). In other cases, response may be gradual owing to the influence of a memory factor, such as the thermal memory of frozen ground or the geo-mechanical memory expressed in isostatic rebound. Between these extremes lie a wide range of bio-geomorphic interactions where the memory effect of the biota is conditioned by hydro-meteorologic events of high magnitude, such as in slope wasting in the humid tropics. In order to speak at all accurately about rates of bio-geomorphic changes at local (or regional scales, therefore, pains must be taken, firstly to identify probabilistic dynamics of a comparable modern landform or landscape element, and secondly to identify interdependance of physical processes and the biota.

These remarks indicate why I am reluctant to extract geomorphic rate data from the many existing syntheses of regional physical changes in Late Quaternary times. In each study region, the analysis of patterns and rates of bio-geomorphic changes requires pooling of expert opinion in dating, statistical, geomorphic, and biologic fields.

REFERENCES

Andrews, J.T. 1973 The Wisconsin Laurentide ice sheet: dispersal centers, problems of rates of retreat, and climatic implications. Arctic and Alpine Research, 5, 185-199.

Bell, W.T. 1977 Thermoluminescence dating of Aboriginal fireplaces from Lake Mungo N.S.W. Paper delivered at International Symposium on Archaeometry and Archaeological Prospection, Philadelphia, March 1977.

Bender, M.L. 1973 Helium Uranium Dating of Corals. Geochimica et Cosmochimica Acta, 37, 1227-1247.

Black, R.F., Goldthwait, R.P. and Willman, H.B. (eds) 1973 The Wisconsinan Stage. Geological Society of America Memoir 136, 334 pp.

Bloom, A.L. 1970 Paludal stratigraphy of Truk, Ponape, and Kusare, Eastern Caroline Islands. Geological Society of America Bulletin, 81, 1895-1903.

Bloom, A.L. 1971 Glacio-eustatic and isostatic controls of sea level since the last glaciation. pp 355-379 in Late Cenozoic Glacial Ages (ed K. Turekian), Yale University Press, New Haven.

Bowler, J.M., Hope, G.S., Jennings, J.N., Singh, G. and Walker, D. 1976 Late Quaternary climates of Australia and New Guinea. Quaternary Research, 6, 359-394.

Bretz, 1969 The Lake Missoula floods and the channelled scabland. Jour. Geol., 77, 505-543.

Broecker, W.S. and van Donk, J. 1970 Insolation changes, ice volumes, and the ^{18}O record in deep sea cores. Review of Geophysics and Space Physics, 8, 169-198.

Budd, W.F. 1975a The evidence of long term climate change from $^{16}O/^{18}O$ isotope ratios in deep ice cores. Unpub. paper supplied to the Australian Academy of Science ad hoc Committee on Climatic Change.

Budd, W.F. 1975b A first simple model for periodically self-surging glaciers. Jour. Glaciol., 14, 3-21.

Chappell, J. 1974a Relationships between sealevel ^{18}O variations and orbital perturbations during the past 250,000 years. Nature, London, 252, 199-202.

Chappell, J. 1974b Late Quaternary glacio- and hydro-isostasy, on a layered earth. Quaternary Research, 4, 405-428.

Chappell, J. 1974c Geology of coral terraces on Huon Peninsula, New Guinea: a study of Quaternary tectonic movements and sealevel changes. Geological Society of America Bulletin, 85, 553-570.

Chappell, J. 1978 Status of various theories for Upper Quaternary ice ages. pp 211-225 in Climatic Change and Variability: a Southern Perspective (eds A.B. Pittock, L.A. Frakes, D. Jenssen, J.A. Peterson and J.W. Zillman), Cambridge University Press, Cambridge.

Chappell, J., Broecker, W.S., Polach, H.A. and Thom, B.G. 1974 Problem of dating Upper Pleistocene sealevels from coral reef areas. In Second International Coral Reef Symposium vol 2, 563-571, Queensland University, Brisbane.

Chappell, J. and Pullar, W.A. Variability of coastal progradation at Rangitaiki Plains, New Zealand. New Zealand Jour. Geol. Geophys. (in press).

Chappell, J. and Veeh, H.H. $^{230}Th/^{234}U$ age support of an interstadial sealevel of -40m at 30,000 yrs B.P., in disagreement with deep sea core ^{18}O results. Nature, London (in press).

Clark, J.A., Farrell, W.E. and Peltier, W.R. Global changes in postglacial sealevel - a numerical calculation. Quat. Research, 8 (in press).

Cline, R.M. and Hays, J.D. (eds) 1976 Investigations of Late Quaternary Paleo-oceanography and Paleoclimatology, Geol. Soc. America Memoir 145 462 pp.

Cooke, H.B.S. 1973 Pleistocene chronology: long or short? Quaternary Research, 3, 206-220.

Cox, A. 1969 Geomagnetic reversals. Science, 163, 237-244.

Dansgaard, W., Johnson, S.J., Clausen, H.B. and Langway, C.C. 1971 Climatic record revealed by the Camp Century ice core. pp 37-56 in Late Cenozoic Glacial Ages (ed K. Turekian), Yale University Press, New Haven.

Dansgaard, W., Gundestrup, N., Hammer, C., Johnson, S.J. and Reeh, N. 1977 The climatic significance of stable isotopes in polar glaciers (abstract). pp 585-588 in Polar Oceans (ed M.J. Dunbar), Arctic Institute of North America, Calgary.

Denton, G.H. and Karlen, W. 1977 Holocene glacial and tree-line variations in the White River Valley and Skolai Pass, Alaska and Yukon Territory. Quaternary Research, 7, 63-111.

Dreimannis, A. and Goldthwait, R.P. 1973 Wisconsinan glaciation in The Huron, Erie, and Ontario lobes. pp 71-106 in The Wisconsinan Stage (eds R.F. Black, R.P. Goldthwait and H.B. Willman), Geological Society of America Memoir 136.

Dungworth, G. 1976 Optical configuration and the racemisation of amino acids in sediments and fossils - a review. Chemical Geol., 17, 135-153.

Duplessy, J.C., Labeyrie, J., Lalou, C. and Nguyen, H.V. 1972 Le mesure des variations climatiques continentales. Quaternary Research, 1, 162-174.

Emery, K.O. and Garrison, L.E. 1967 Sea levels 7000 to 20,000 years ago. Science, 157, 684-687.

Emiliani, C. 1971 The last interglacial: Paleotemperature and chronology. Science, 171, 571-573.

Fisher, D.E. 1975 Geoanalytic applications of particle tracks. Earth Science Review, 11, 291-335.

Flint, R.F. 1971 Glacial and Quaternary Geology, Wiley, New York.

Flöhn, H. 1974 Background of a geophysical model of the initiation of the next glaciation. Quaternary Research, 4, 385-404.

Ford, D.C., Harmon, R.S., Schwarcz, H.P., Wigley, T.M.L. and Thompson, P. 1976 Geo-hydrologic and thermometric observation in the vicinity of the Columbia icefield, Alberta and British Columbia, Canada. Jour. Glaciol, 16, 219-230.

Huxtable, J. and Aitken, M.J. 1977 Thermoluminescent dating of Lake Mungo geomagnetic polarity reversal. Nature, London, 265, 40-41.

Imbrie, J. and Kipp, N.G. 1971 A new micropaleontological method for quantitative paleoclimatology: application to a Late Pleistocene Caribbean core. pp 71-182 in Late Cenozoic Glacial Ages (ed K. Turekian), Yale University Press, New Haven.

Isacks, B., Oliver, J. and Sykes, L.B. 1968 Seismology and the new global tectonic. Jour. Geophys. Res., 73, 5855-5900.

Kennett, J.P. and Huddlestun, P. 1972 Late Pleistocene paleoclimatology, foraminiferal biostratigraphy and tephrochronology, Western Gulf of Mexico. Quaternary Research, 2, 38-69.

Kellog, T.B. 1976 Late Quaternary climatic changes: evidence from deep-sea cores of Norwegian and Greenland Seas. pp 77-111 in "Investigations of Late Quaternary Paleo-oceanography and Paleoclimatology" (eds R.M. Cline and J.L. Hays), Geol. Soc. America Memoir 145.

Ku, T-L, Kimmel, M.A., Easton, W.H. and O'Neil, T. 1974. Eustatic sea level 120,000 years ago on Oahu, Hawaii. Science, 185, 959-962.

Kukla, G.J. 1970 Correlations between loesses and deep-sea sediments. Geol. Fören, Stockholm Förh., 92, 148-180.

Kutzbach, J.E. and Bryson, R.A. 1975 Previously unpublished material on p 13 in Understanding Climatic Change. U.S. Committee for GARP, National Academy of Sciences, Washington D.C.

LaMarche, V.C., Jr 1974 Paleoclimatic inferences from long tree ring records. Science, 183, 1043-1048.

La Marche, V.C., Jr. and Harlan, T.P. 1973 Accuracy of tree-ring dating of bristlecone pine for calibration of the radiocarbon time scale. Jour. Geophys. Res., 78, 8849-8856.

Lamb, H.H. 1968 The Changing Climate. Methuen, London.

Lamb, H.H. 1970 Volcanic dust in the atmosphere with chronology and assessment of its meteorological significance. Phil. Trans. Roy. Soc. London, 266, 425-533.

Lamb, H.H. 1972 Climate, Present, Past and Future, Vol. 1. Methuen, London.

Le Pichon, X. 1968 Seafloor spreading and continental drift. Jour. Geophys. Res., 73, 3661-3698.

Liu Tung-Sheng (ed) 1966 Composition and Texture of the Loess of China, the Science Press (Academia Sinica) Peking, China 132 pp. (in Chinese)

McDonald, B.C. 1971 Late Quaternary stratigraphy and deglaciation in eastern Canada. pp 331-353 in Late Cenozoic Glacial Ages (ed K. Turekian), Yale University Press, New Haven.

McDougall, I. 1971 The Geochronology and evolution of the young volcanic island of Reunion, Indian Ocean. Geochemica et Cosmochemica Acta, 35, 261-288.

McDougall, I., Polach, H.A. and Stipp, J.J. 1969 Excess radiogenic argon in young subaerial basalts from the Auckland volcanic field, New Zealand. Geochèmica et Cosmochemica Acta, 33, 1485-1512.

McIntyre, A., Kipp, N.G., Be, A.W.H., Crowley, T., Kellog, T.B., Gardner, J.V., Prell, W. and Ruddiman, W.F. 1976 Glacial North Atlantic 18,000 years ago: a CLIMAP reconstruction. pp 43-76 in Investigations of Late Quaternary Paleo-oceanography and Paleoclimatology (eds R.M. Cline and J.D. Hays), Geol. Soc. America Memoir 145.

Mangerud, J. 1972 Radiocarbon dating of marine shells, including a discussion of apparent age of recent shells from Norway. Boreas, 1, 143-172.

Manley, G. 1953 The mean temperature of central England 1698-1952. Quart.J. Roy. Met. Soc., 79, 242-261.

Moodie, D.W. and Catchpole, A.J.W. 1976 Valid climatological data from historic sources by content analysis. Science, 193, 51-53.

Mörner, N.A. 1969 The Late Quaternary history of the Kattegatt sea and the Swedish west coast. Sverig. geol. Unders. Afh. C., 640, 1-487.

Oakley, K.P. and Hoskins, C.R. 1950 New evidence on antiquity of Piltdown Man. Nature, Lond., 165, 379-382.

Olsson, I.U. 1974 Some problems in connection with evaluation of ^{14}C dates. Geol. Fören. Stockholm Forh., 96, 311-320.

Paterson, W.S.B. 1969 The Physics of Glaciers. Pergamon Press, Oxford.

Polach, H.A. 1975 Radiocarbon dating as a research tool in archeology - hopes and limitations. pp 255-298 in Proc. Symposium on Ancient Chinese Bronzes and Southeast Asian metal and other Archaeological Artifacts (ed N. Barnard), Nat. Gallery of Victoria, Melbourne.

Prest, V.K. 1969 Retreat of Wisconsin and Recent ice in North America. Map 1257A, Geological Survey of Canada, Ottawa.

Rankama, K. 1965 and 1967 The Quaternary, Vols. 1 & 2. Interscience, New York.

Ruddman, W.F. and McIntyre, A. 1976 Northeast Atlantic paleoclimatic changes over the past 600,000 years. pp 111-146 in Investigations of Late Quaternary Paleo-oceanography and Paleoclimatology (eds R.M. Cline and J.D. Hays) Geol. Soc. America Memoir 145.

Schroeder, R.A. and Bada, J.L. 1976 A review of the geochemical applications of the amino acid racemisation reaction. Earth Science Review, 12, 347-391.

Seward, D. 1975 Fission track ages of some tephras from Cape Kidnappers, Hawke's Bay, N.Z. N.Z. Journal Geology and Geophysics, 18, 507-510.

Shackleton, N.J. and Opdyke, N.D. 1973 Oxygen isotope and paleomagnetic stratigraphy of equatorial core V28-238 oxygen isotope temperatures and ice volume on a 10^5 and 10^6 year time scale. Quaternary Research, 3, 39-55.

Shackleton, N.J. and Opdyke, N.D. 1976 Oxygen isotope and paleomagnetic stratigraphy of Pacific core V28-239 late Pliocene to latest Pleistocene. pp 449-464 in Investigations of Late Quaternary Paleo-oceanography and Paleoclimatology (eds R.M. Cline and J.D. Hays), Geol. Soc. America Memoir 145.

Stuiver, M. 1970a Long-term ^{14}C variations. pp 197-213 in Radiocarbon variations and absolute chronology (ed I.U. Olsson), 12th Nobel Symposium, Almquist and Wiksell, Stockholm.

Stuiver, M. 1970b Tree ring, varve, and ^{14}C chronologies. Nature, London, 228, 454-455.

Stuiver, M. and Polach, H.A. 1977 Discussion of reporting ^{14}C data. Radiocarbon, 19, 355-363.

Tauber, H. 1970 The Scandinavian varve chronology and ^{14}C dating. pp 173-196 in Radiocarbon variations and absolute chronology (ed I.U. Olsson), 12th Nobel Symposium, Almquist and Wiksell, Stockholm.

Tooley, M.J. 1974 Sealevel changes during the last 9000 years. Geographical Jour., 140, 18-42.

Turekian, K.K. 1965 Some aspects of the geochemistry of marine sediments pp 81-126 in Chemical Oceanography (eds J.P. Riley and G. Skirrow), Academic Press, London.

Turner, C. and Shackleton, N.H. 1967 Correlation between marine and terrestrial Pleistocene successions. Nature, London, 216, 1079-1082.

Wagner, G.A. 1973 Anwendung anätzbarer Partikelspuren in Geochronologie und Geochemie. Max Planck Institut, Heidelberg.

Walcott, R.J. 1972 Past sea levels, eustasy and deformation of the earth. Quaternary Research, 2, 1-14.

Wehrmiller, J.F. and Hare, P.E. 1971 Racemisation of amino acids in marine sediments. Science, 173, 907-909.

Wright, H.E., Jr. 1973 Tunnel valleys, glacial surges, and subglacial surges of the Superior Lobe, Minnesota. pp 251-276 in The Wisconsinan State (eds R.F. Black, R.P. Goldthait and H.B. Willman), Geological Society of America Memoir 136.

Zimmerman, D.W. 1971 Thermoluminescent dating using fine grains from pottery. Archaeometry, 13, 29-52.

Zonneveld, J.I.S. 1973 Some notes on the Last Deglaciation in Northern Europe compared with Canadian conditions. Arctic and Alpine Research, 5, 223-228.

2. CLIMATIC INTERPRETATION OF POLLEN IN QUATERNARY SEDIMENTS

by Margaret B. Davis

SUMMARY

Paleoclimatic interpretation of the fossil pollen record uses both the distribution limits of species and their abundances. In each case the relationship of the plants to modern climate is used to interpret past occurrences. Problems arise in using pollen to determine accurately either the distribution of a species in space or its abundance, but the real limitation of the method seems to be lack of understanding of the role of climate in controlling plant species. The distribution data seem to give more precise paleoclimate estimates, but the approach leads to consideration of each species in isolation. The use of abundances, because it relies so heavily on interpretation of pollen assemblages, places more emphasis on the entire community and its relation to regional climate. Here the major difficulty comes in understanding unique communities that differ from modern analogues. They may differ because of biological factors such as immigration lag yet reflect a climate fully analogous to modern climate. Or they may indicate a unique combination of climatic conditions that has evoked a unique vegetation response. Complicating these interpretations are interactions among species, which can produce changes in abundance that mimic the effect of climatic change. Species interactions, although difficult to unravel from climatic effects in the fossil record, have potential for showing how changes in the physical environment influence the outcome of competition in plant communities.

INTRODUCTION

Reconstructions of ancient climate that are based on distributions of plants or animals rely on the generalization that each species tolerates a particular set of climatic conditions. The occurrence of fossils implies ancient climate within that particular species' range of tolerance. The assumed relationship of a species to climate is shown in figure 2.1a, where the abundance of an organism is plotted against a climatic parameter. Where conditions are optimal the organism will be abundant (region A), where conditions are marginal the organism will be rare (regions B and B'), and at the upper and lower limits of tolerance (points C and C') the organism will not survive.

Typically, interpretations of fossils use only a part of the information shown in figure 2.1a. The limits of the distribution are emphasized when interpretations are based on presence or absence. For example the presence of a fossil may be used to indicate a paleotemperature within the range C to C'. In practice, one is often dealing with one edge of the range of a fairly widespread species, thus arguing, for example, that the temperature exceeded C (fossil present) or was less than C (fossil absent). In order to gain more specific information, investigators sometimes use two or more species with overlapping tolerances (figure 2.1b). The presence of species 1 alone indicates temperature ranging between C and D, species 1 and 2 together temperatures ranging between D and C'. The classic example in palynology is Iversen's (1944) study of the changing occurrence of holly, ivy and mistletoe in the Danish Holocene. He concluded that mean temperatures in the coldest month four thousand years ago were above -2°C (the lower limit for ivy, which was present) but below -0.5^{o}C (the lower limit for holly, which was absent), while mean temperatures during the warmest month were above 17°C (the lower limit for mistletoe, which was present). Since that time the climate has become more maritime; temperatures in the coldest month now average above -0.5^{o}C and both holly and ivy are present, while the mean July temperature has fallen below 17°C, and mistletoe has become extinct. Iversen based his conclusions on summer and winter temperatures in regions where each species

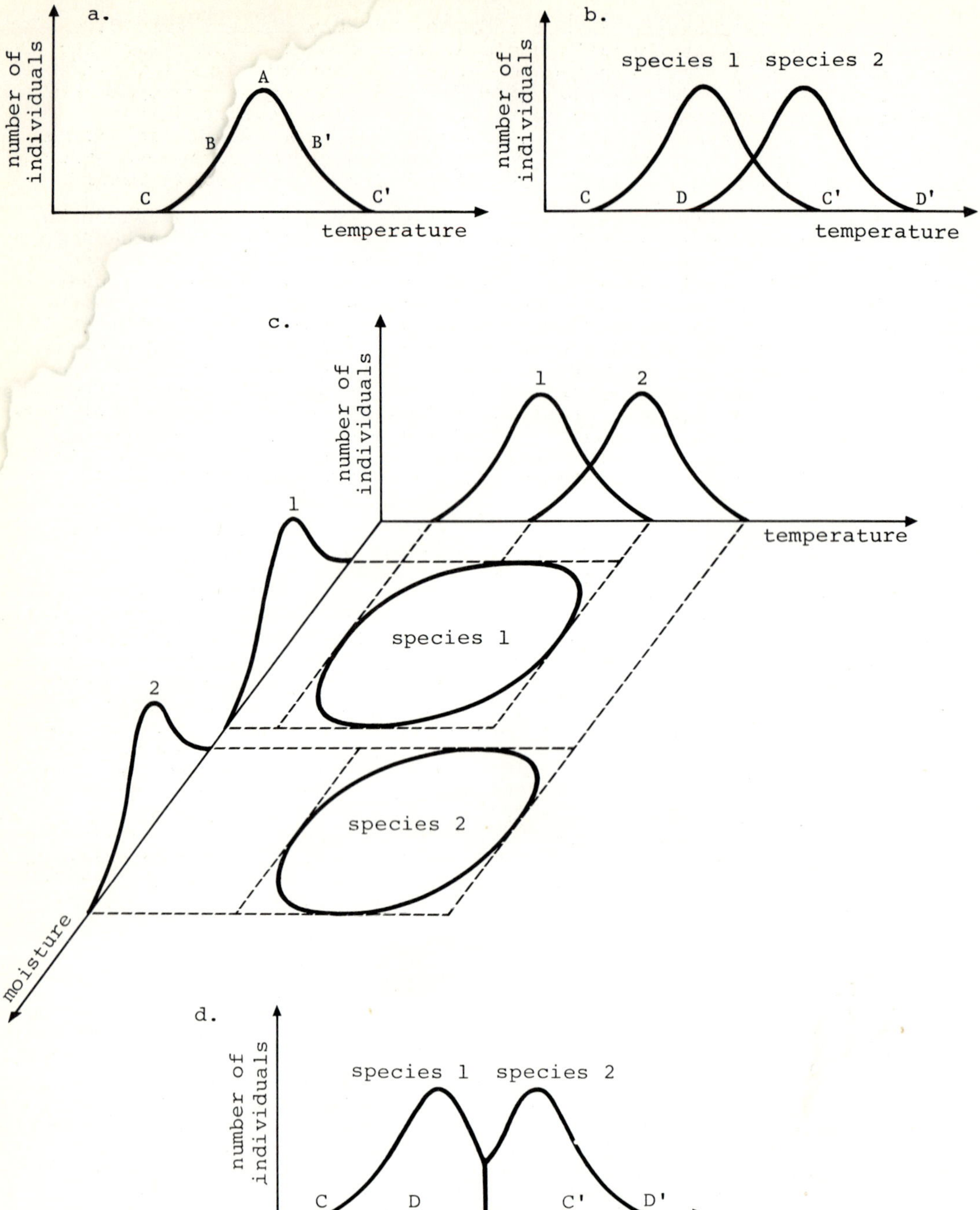

Figure 2.1 (a) Hypothetical relationship between the abundance of an organism and an environmental factor. (b) Hypothetical relationship between the abundances of two non-interacting species and an environmental factor. (c) Hypothetical relationships between the abundances of two, non-interacting, species and two environmental factors; the abundances are projected onto a surface that represents the two gradients; the elipses outline the occurrences of the two species. (d) Hypothetical relationships between abundances of two species and a single environmental factor; species 1 and species 2 interact; each excludes the other over a portion of the environmental gradient.

occurs today. He observed damage to holly and ivy along their respective range limits during an unusually cold winter, thus demonstrating that the two species have different ranges of tolerance for the same limiting factor. This last point is important; figure 2.1c shows how species with overlapping tolerances for one factor can still be separated in the environment by a second, perhaps unknown, niche dimension. Identification of the environmental variable limiting both species is therefore essential for interpretation of co-occurrences of species in the fossil record.

Competition between species poses problems in this methodology that seldom are discussed by paleoecologists. If species interact, they may never co-occur even though their potential limits overlap (figure 2.1d). Potential and realized tolerance limits are seldom distinguished when paleoecologists attempt to define the relationship of a species to the physical environment. Tree species clearly compete with one another in forests, although the degree to which a species can completely exclude competitors is not really known. Even without complete exclusion a species might become so rare that it appears "virtually" absent from the region. One might conclude erroneously that it cannot tolerate the climate there. Forest communities have changes through time, raising the possibility that the realized niche for a particular species has expanded as it was released from competition or has contracted as new competitors have entered the community.

The potential for climatic information is much greater when abundances of species are considered. Using abundances, one can utilize all the points along the curve between C and C' (figure 2.1a) instead of the limits of distribution only. Most palynologists rely heavily on interpretations of abundance. The quantitative implications of pollen abundance give pollen a major advantage over traditional paleobotanical materials such as leaves, seeds and wood. Although pollen is notoriously imprecise as an estimator of population size (Davis et al. 1978), one can deduce that a species is rare or that is it very abundant without precise information. Much greater precision is necessary to know whether a species is present or absent, especially as many plants are rare near the limits of their geographic range. Consequently pollen is more useful in giving information on abundance than in pinpointing the exact limits of distribution. I shall endeavour to show in this essay, however, that both presence/absence, and abundance can be determined from the fossil pollen record, and that each kind of information has potential for determining climates of the past. The limitations of the method seem to reside in the incompleteness of our understanding of the relationship of plants to their biological and physical environment.

DISTRIBUTION LIMITS AS CLIMATIC INDICATORS

Pollen is usually dispersed far beyond the range of the plant. Consequently, the mere presence of a few pollen grains cannot be construed as evidence that the plant grew nearby. The dispersal of pollen of *Ephedra* 1200 km across the central plains of the United States from the southern Rockies to Minnesota is a spectacular example of this phenomenon (Maher 1964). Nevertheless, there are circumstances where pollen can be used to trace species distributions. The consistent presence of pollen from small, low-growing, insect-pollinated plants (ivy, for instance) might be construed as evidence of presence provided that surface sediments show that modern dispersal occurs close to parent plants. Pollen frequencies supported by the presence of macrofossils might also be used, although in this case it is the macrofossil, not the pollen, that forms the basis for interpretation. Under special circumstances, it has been possible to demonstrate a quantitative relationship between pollen frequencies and a species range limit. An example of this is shown in figure 2.2, a map showing the percentages and influx rates for beech (*Fagus grandifolia*) pollen in Upper Michigan. The pollen measurements were made in modern lake sediment. A map of the distribution of beech in the modern forests, as measured in the U.S. Forest Service Inventory, is

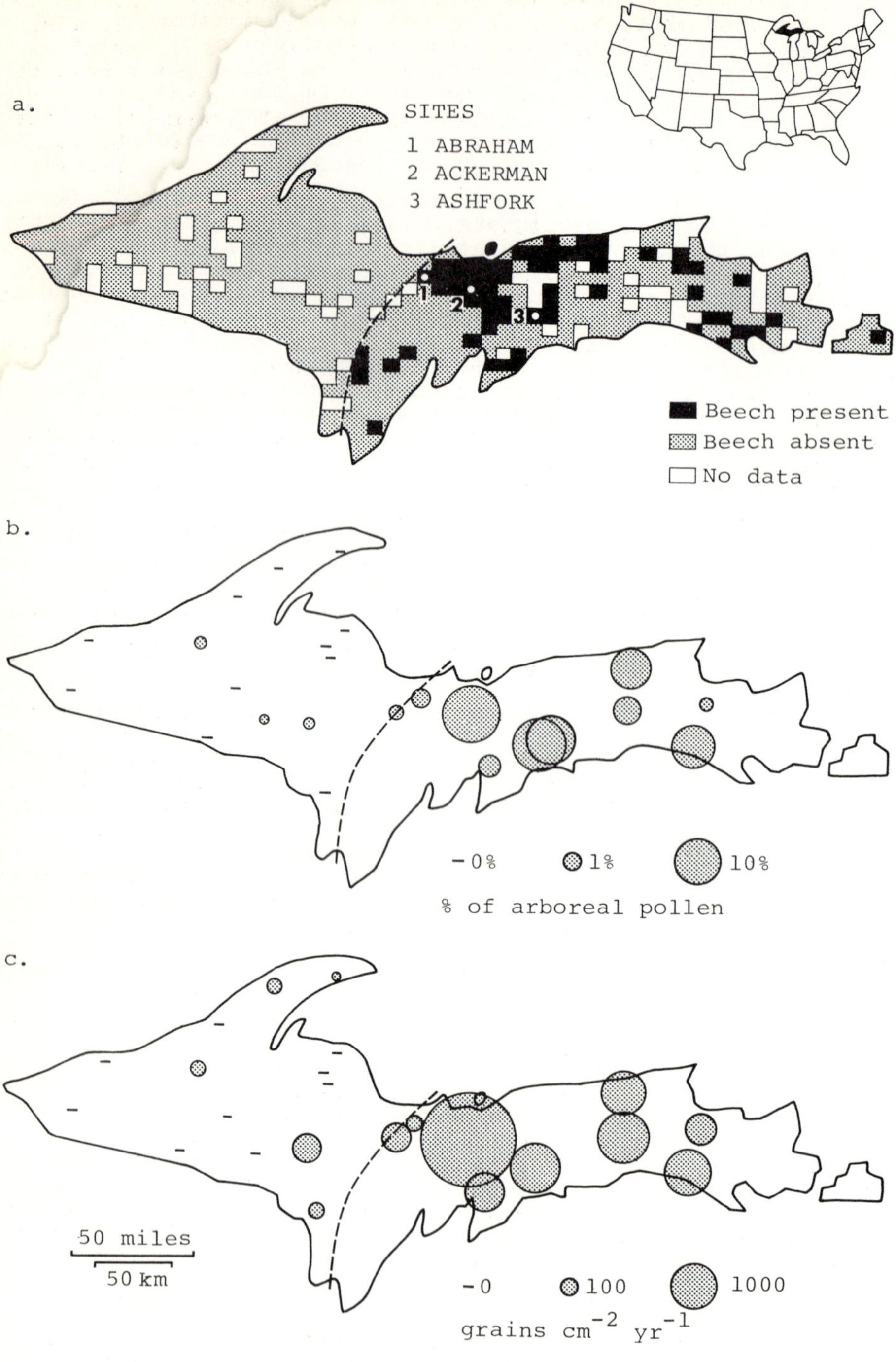

Figure 2.2 Maps of Upper Michigan, U.S.A. showing (a) Distribution of beech trees in modern forests; dashed line represents the western species limit. (b) Beech pollen percentages in surface sediment from a number of lakes. (c) Average beech pollen influx since the time of settlement, about 70 years ago, at the same sites.

shown for comparison. The dashed line indicates the western distribution limit for beech. The beech boundary is unusually sharp: beech is abundant (basal area greater than 14 m^2 ha^{-1}) only a few miles from its absolute limit. Except for beech, the forests are similar in composition on both sides of the boundary. Many lakes west of the boundary contain no beech pollen at all; those that do, contain less than 1%, or accumulation rates less than 200 grains cm^{-2} yr^{-1}. In contrast all lakes east of the boundary show beech pollen in percentages greater than 1% and accumulation rates greater than 200 grains cm^{-2} yr^{-1}.

Fossil pollen was studied in sediments from the three lakes indicated in figure 2.2a. Beech pollen was absent from all three sites until 3500 years ago, when it appeared at the easternmost site (figure 2.3). At this time the percentage rose to exceed 1%, indicating the presence of beech trees nearby. Beech distribution then remained stable for about 3000 years. About 500 years ago, when the Little Ice Age began in much of the northern hemisphere, beech extended its range 43 km to the west. A 35 km extension to the westernmost site followed within a few centuries. (The date of extension is imprecise because it was estimated by extrapolating ages of sediment between pollen culture horizons a few cm below the surface and radiocarbon dates at 1-1.5m depth.) The westernmost site, Abraham Lake, is at the very limit of beech today. The rapid extension of range to this site, averaging 175m yr^{-1}, seems to represent a response to climatic change associated with the "Little Ice Age" documented by resurgence of glaciers in many parts of the world within the last four or five centuries (Porter and Denton 1967).

These changes in the geographical limits for beech in Upper Michigan should provide climatic information, indicating that climatic conditions at the two westernmost sites, formerly unsuitable for the survival of beech, are now within the range of tolerance for the species. Although this conclusion seems justified, stating it in specific climatic terms is more difficult. The particular climatic factors that limit the distribution of beech in Michigan at the present time are unknown. The situation is all too typical of the dearth of autecological information on the very plants we should like to use as climate indicators. Geographical patterns of precipitation, snow depth and temperature, recorded by the U.S. Weather Bureau over the last 40 or more years in this region, are not correlated with the distribution pattern for beech. It seems likely that beech is affected by the magnitude of the differences in temperature between summer and winter. This difference increases westward and is therefore correlated with the species boundary. Moisture at critical seasons of the year is another possibility. Intuitively this seems important because beech is abundant right along the shore of Lake Superior, when frequent fogs may influence the moisture regime. On the other hand, the lake may be exerting a tempering effect on seasonal temperature differences. The western limit of beech at present coincides with the boundary between Paleozoic shales and limestones to the east of Pre-Cambrian basement rocks to the west. The coincidence suggests that soils are important, yet climate must also influence beech, otherwise we cannot explain the easterly location of the limit of beech 500 years ago nor the extension of range since that time. An interaction of climate and soils as the control on the present distribution of beech cannot be ruled out. The history of beech shows how it might be possible to use a changing species limit as a sensitive assay of climatic change, monitoring subtle changes that have occurred from one century to the next, provided one has understanding of the modern relationship of the organism to climate and soils sufficient for interpretation.

The geographical limit for beech distribution in Upper Michigan, despite ambiguities in interpretation, appears at least to be in equilibrium with the climate of this century. There is no evidence of movement at the present time, as trees at the very limit of distribution near Abraham Lake are producing few seedlings and sprouts relative to beech in forests in the central part of the species' range (Davis unpublished data; Ward 1961). However, there are many examples in the Holocene record of species boundaries that were not in equilibrium with climate for many thousands of years, due to lag in the extension of ranges

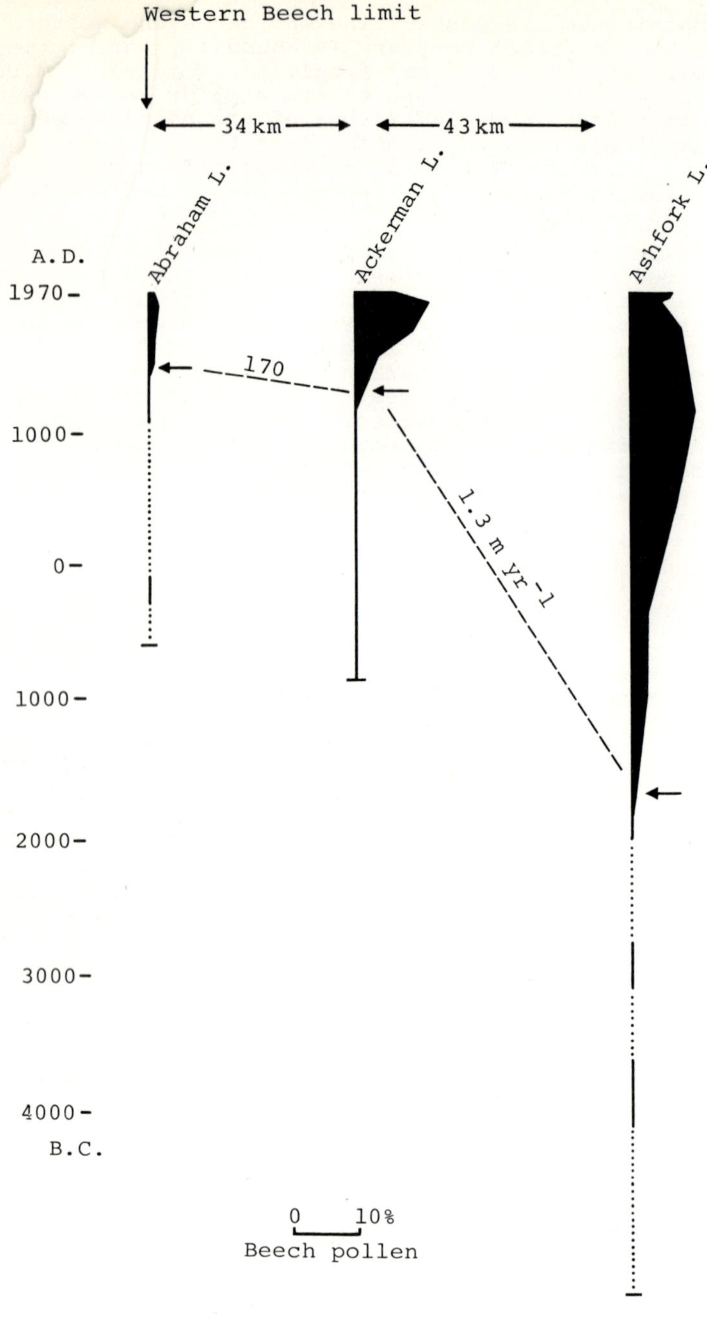

Figure 2.3 Beech pollen percentages plotted against time at the three sites indicated in figure 2.2a.

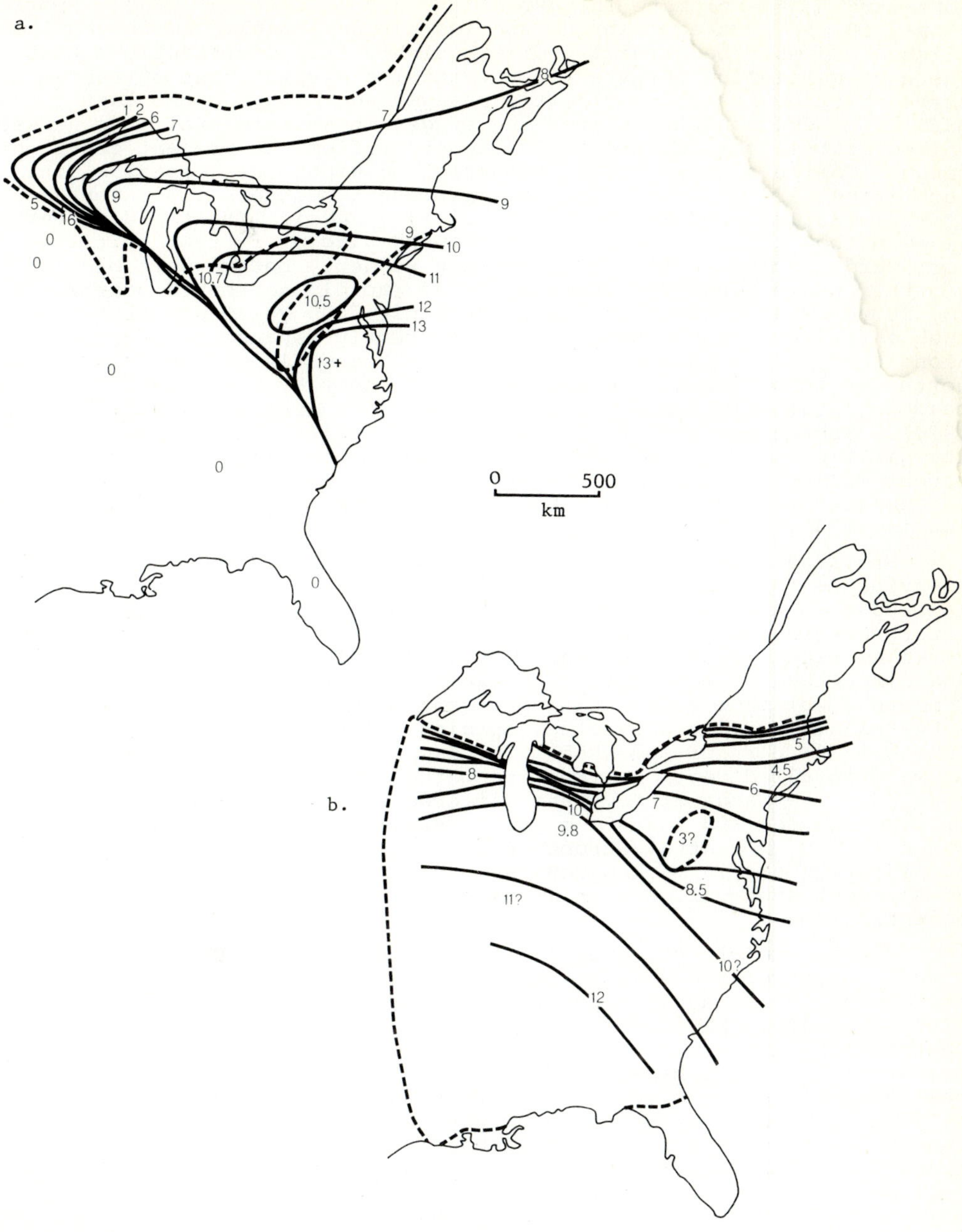

Figure 2.4

(a) Migration of white pine. (b) Migration of hickory. The dashed lines enclose the modern ranges of the species. Solid lines represent the frontier of each species at 1000-year intervals during the late-Pleistocene and the Holocene. Small numbers on the maps represent radiocarbon ages in thousands of years B.P. at sites where the rational limit was exceeded. (Redrawn from Davis 1976).

northward as the climate changed at the end of the last glacial period (Davis 1976). Non-equilibrium species boundaries give no information on climate and must be carefully distinguished from distributions that do. The non-equilibrium resulted from extensive displacement of ranges of trees in response to the climatic changes accompanying the last glacial maximum. Deciduous trees grew in refuges in southcentral or southwestern United States, several thousand kilometres from the present centre of their range. Conifers grew in the southeast and along the coastal plain. These localities were 1000 km south of the present southern limit for jack pine (*Pinus bansiana*) and white spruce (*Picea glauca*). The rapid temperature rise in the first five millenia following glacial retreat initiated rapid northward movement into territory that suddenly became suitable for tree growth. White pine, for example, moved rapidly northward starting 13,000 years ago, reaching a western limit 7000 years ago that remained stable for several thousand years, apparently because moisture was limiting along the prairie margin (figure 2.4a). Meanwhile it continued to expand northward into Canada, limited apparently only by its rate of migration, reaching stations beyond its present range 5000 years ago (Terasmae and Anderson 1970). Subsequently it retreated as the climate, which then controlled its distribution in the north, became cooler in the late Holocene. In recent millenia white pine has expanded in the western part of its range, an apparent response to late-Holocene changes in climate in that region (Davis 1976).

A gradual northward extension of plant ranges could be interpreted as evidence for a gradual amelioration of climate at higher and higher latitudes as glacial climates gave way to interglacial climates. This interpretation cannot apply to the expansion of most tree species in North America because there are too many differences among species in rates and directions of movement. For example, white pine and hickory (*Carya*) (figure 2.4b) show contrasting patterns of migration. Still different patterns are exhibited by other species discussed in the literature (Davis 1976; Bernabo and Webb 1977; Kapp 1977). Each species moved outward in all directions from its refuges until climatic limits were reached. The rates of movement differed from one species to another, ranging between 100 and 400 m yr^{-1}. Similar Holocene migration rates are recorded in central Europe (Firbas 1949). The criss-crossing patterns of migration are in marked contrast to the regular, equiformal patterns of movement one would expect in response to a northward movement of climatic zones.

Even where climate does limit the geographical distribution of a plant species, the relationship between the edge of a species range and climate is complex, changing from one part of the species range to another (Iversen 1944). Marginal populations may also differ genetically from central populations (Lewontin 1974). Physiological responses, if they are studied, should be studied for relevant marginal populations, rather than generalized from individuals that may be more typical for the species as a whole. There is probably no way, however, that one could control for genetic differences of this magnitude in the past. For the palynologist, phenotypic responses may also create problems; marginal populations may pollinate less prolifically than populations growing under optimal conditions. The low pollen production may then add to difficulties in detecting the position of the species boundary. Nichols (1976) speculates, for example, that "relict" stands of spruce growing in the artic, beyond the limit of continuous forest, produce few or no pollen grains and have reproduced vegetatively by layering for hundreds or thousands of years.

Species boundaries can give climatic information only a) when the species boundary is detectable by pollen analysis, b) when the species boundary is in equilibrium with the climate and c) when it is controlled by a known climatic parameter. Such information is available in very few instances.

SPECIES ABUNDANCE AND CLIMATE

Pollen abundance is generally used to reconstruct entire communities of plants. Pollen abundances are frequently calculated as relative frequencies, and, consequently, the entire assemblage must be utilized in interpretation. Pollen grains in modern sediment demonstrate a correlation between pollen assemblages and vegetation (Webb 1974; Davis and Webb 1975). In general, fossil pollen percentages are easily interpreted when they duplicate modern assemblages.

A spatial shift through time along a geographical gradient in the position of a plant formation or its characteristic pollem assemblage suggests that the climate has changed. For instance, McAndrews' elegant study in Minnesota demonstrated an eastward mid-Holocene movement of a series of fossil pollen assemblages characteristic of modern vegetation ranging from prairie in the west through savanna to mixed forest in the east. The shift indicates an eastward movement of prairie climate during the mid-Holocene with decreased moisture and increased warmth at each site (McAndrews 1967). The reasoning is similar to the ideas expressed above for individual species. In this type of interpretation climatic values are frequently left unspecified, and investigators simply compare the paleoclimate to the geographical region where analogous pollen assemblages are found today. This seems appropriate to ecologists aware that critical factors have not been identified. However, the results are too vague to be useful to meteorologists constructing climatic models of the earth's climate or geologists estimating historical changes in physical processes.

Recognizing that pollen correlates with vegetation and that vegetation correlates with climate, Webb and Bryson (1972) have correlated modern pollen assemblages directly with the climatic parameters at the sites where the pollen samples were collected. These correlations, or transfer functions, are then applied to fossil pollen assemblages from the same general region, providing a climatic readout through time. The results are attractive because they express climatic interpretations in quantitative terms: as mm precipitation, degrees Celsius summer and winter, for example. The method leaves very little unexplained variance in data from mid- and late-Holocene sediments, especially from sites along the prairie-forest margin similar to those studied by McAndrews (1976). Webb and Bryson attribute the shift in prairie pollen assemblages to a 50 mm decrease in summer rainfall and a slight (ca. 0.5^{o}C) elevation in July temperature. In older sediments from these sites, however, the percentages for several pollen types, e.g. elm (*Ulmus*), exceed percentages observed in surface samples from any known sites. Although transfer functions can deal with mixtures of known assemblages, they cannot utilize assemblages different from the modern data base. Pollen types that occur in frequencies different from modern ones are therefore excluded from the analysis, as they cannot be translated into climate. Of course the plants themselves might have occurred at modern frequencies although the pollen percentages are higher than modern ones because the background pollen productivity provided by other taxa has changed. The interpretation Webb and Bryson (1972) give for the late-glacial and early Holocene climates has been based exclusively on the understood pollen types which make up about 70% of the assemblage. Spruce (*Picea*) is the major component in the interpretation. The degree to which omission of pollen types can cause errors is not really known, since there is as yet no independent way to check the accuracy of paleoclimatic interpretations of pollen based on transfer functions. Clearly omission of certain types does not interfere with correlations of modern material with climate, in fact the correlation may be enhanced. However, the unusual percentages of pollen observed in the late-glacial and early Holocene may have climatic significance, indicating a unique combination of climatic conditions; this information should be utilized in some way rather than discarded if a full picture of past environments is to be obtained.

Interpretation of fossil pollen frequencies by any method tends to break down when pollen assemblages represent a community different from any found today. Early Holocene communities in many parts of North America were unique. Forests at that time were species-poor due to the lag in the northward migration of forest dominants. Immature soils may have affected the success of some species. Another cause may have been unique climatic conditions. Several different methods have been used in attempts to interpret pollen assemblages that have no modern analogue. Where pollen dispersal is not a factor, e.g. very small sites beneath a forest canopy, corrections can be made for differences in pollen productivity (Andersen 1970, 1973). Very detailed interpretations have been possible in these instances. Most analyses, however, are from sediments in larger basins where pollen dispersal is important. Present data indicate that pollen dispersal is so variable and difficult to measure that it seems unlikely that correlations for pollen representation can be made accurately in these situations. Some authors have suggested that measurements of the absolute rate of accumulation of pollen grains (pollen influx) show promise for interpretating unique pollen assemblages (Davis et al. 1973). Influx in existing lakes can be calibrated against plant abundances in the modern forest and used as a very rough estimate of population size. The estimates for each genus can be added together to reconstruct unknown communities (Davis et al. 1973). The accuracy of the result is probably not very high because sedimentary processes produce large variance in pollen influx from site to site, and sometimes at the same site through time.

Despite the difficulty inherent in interpreting pollen assemblages produced by plant communities that have no modern analogue, attempts must be made to understand their meaning. They disprove the idea that communities merely shift position along geographical and climatic gradients whenever the climate changes. Striking changes in composition are documented for example by the fossil history of spruce-fir forests on the slopes of the White Mountains in northeastern United States. The spruce-fir forests today are one of a series of forest communities arranged along an elevation gradient. The spruce-fir zone is conspicuous, partly because the coniferous forests stand out in contrast to the deciduous forests on the lower slopes. Spruce (*Picea rubens* and *P. mariana*) and fir (*Abies balsamea*), with patchy admixtures of paper birch (*Betula papyrifera*) in disturbed sites, are dominant from about 750 m to 1500 m elevation. Above these elevations, alpine tundra extends to the summits at about 1800 m. Below the spruce-fir forests deciduous trees are dominant. Paleoecologists used to assume that the Boreal Forest of Canada had been pushed southward as the glaciers advanced, and that it had moved northward again as the glaciers retreated (e.g. Martin 1958). It was therefore natural to interpret the spruce-fir forest on the mountain slopes as a relict of the Boreal Forest that had moved through the area. The descriptive term "Boreal" is often applied to montane spruce-fir forest, despite some differences in species composition from the Canadian Boreal Forest (Whittaker 1956; Siccama 1974). However, fossil pollen data from the White Mountains shows that although both spruce and fir have been abundant at various times in the past in this region, they have had different histories and only recently have come to grow together in abundance as a distinctive community on the mountain slopes. The spruce-fir forest is therefore not a relict of a previous formation but is instead a new community that has existed as a conspicuous forest zone for only about 2000 years.

Although spruce and fir occupy a distinctive elevation belt, climate is not the only factor limiting their occurences. Competition with hardwood species seems to play a part in limiting red spruce at the lower limit of its elevation distribution (Bormann et al. 1970). Whittaker (1956) concluded that competition was occurring in the southern Appalachians, too, because deciduous trees grow there at much higher elevations on mountains where spruce is absent than where spruce is present. Factors controlling the upper limit for spruce are not so clear. The upper limit is diffuse, as patches of spruce become increasingly rare in fir forest

500 m or so below tree line. At these elevations fir competes successfully with spruce; fir is a short-lived tree and precedes spruce in biological succession. Apparently the lower limit for fir is determined by competition with spruce and at lower elevations in competition with deciduous trees. The upper limit for fir, however, is the upper limit for trees of any kind. Both fir and black spruce grow as krummholz at treeline. The upper limit of fir is doubtless controlled by the physical environment.

Spruce and fir pollen both occur in maximal abundance in surface sediments from the spruce-fir zone. Spruce pollen can be found in low percentages at sites at lower elevations, but fir pollen is quite local in its distribution (Davis et al. in preparation).

The history of spruce and fir is shown by the absolute accumulation rates of pollen from spruce and fir at a series of sites in northeastern United States (figure 2.5). Two of the localities are located several hundred km south of the White Mountains (figure 2.6). The other three localities are at various elevations in the mountains, from the valley floor (200 m) to the spruce-fir zone (1000 m).

As the ice left northeastern United States 13,000 to 14,000 years ago, tundra vegetation developed on the newly deglaciated landscape. After one or two thousand years spruce became established, first in the south and then moving northward. The absolute rate of spruce pollen deposition (probably not significantly different at the three low-elevation sites) was less than 5000 grains cm^{-2} yr^{-1}, suggesting an open spruce woodland extending from Connecticut northward to northern New Hampshire (Davis et al. 1973). Above 500 m in the mountains influx values are much lower, indicating that spruce grew only sparsely at these elevations. The presence of spruce needles at the 1000 m site (Linda Shane, University of Minnesota, personal communication) shows that spruce trees or krummholz did grow at this elevation, which must have been close to treeline at that time. About 10,000 years ago, spruce declined sharply in abundance at all elevations. Over the next millenium, it disappeared at low elevations, but spruce needles in the sediment show that it persisted near the higher sites at 650 and 1000 m elevation (Linda Shane, personal communication). The low pollen influx at these sites indicates that spruce was rare at these elevations and sites still higher on the slopes, now under study by Ray Spear, contain even fewer spruce pollen grains in late-glacial and mid-Holocene sediments (Ray Spear, University of Minnesota, personal communication).

Fir entered Connecticut about 11,000 years ago, becoming abundant at low elevations in both Connecticut and New Hampshire just as spruce declined. It remained abundant there a full 1000 years after spruce disappeared. It persisted somewhat longer near the 200 m site in the White Mountains. Fir is a poor pollen producer; hence, even though the influx rates in figure 2.5 are low, fir was more abundant than spruce. About 10,000 years ago fir arrived at higher elevations in the White Mountains including the site at 1000 m where it became abundant and has persisted as the dominant species to this day. Birch pollen and seeds indicate that paper birch arrived at about the same time.

Spruce became abundant again 2000 years ago. Its pollen increased sharply at all sites in the White Mountains. Macrofossil remains also increased in abundance at the sites at 650 m and 1000 m (Linda Shane, personal communication). Spruce grows today along the shore of the 650 m site, although the forests there as a whole are dominated by deciduous trees. Spruce is more common in forests just below the 1000 m site, although fir dominates the forest very near the lake. It should be noted that throughout the early history of this region, spruce and fir only briefly coincided; the peak in fir at low elevations persists after the decline of spruce. At high elevations fir remained as abundant as now throughout the mid-Holocene when spruce was rare. Only in the last two millenia has spruce become abundant at elevations where fir is common to form the spruce-fir forest. In this case, forest history does not involve a shift along the elevation gradient, but instead a complex series

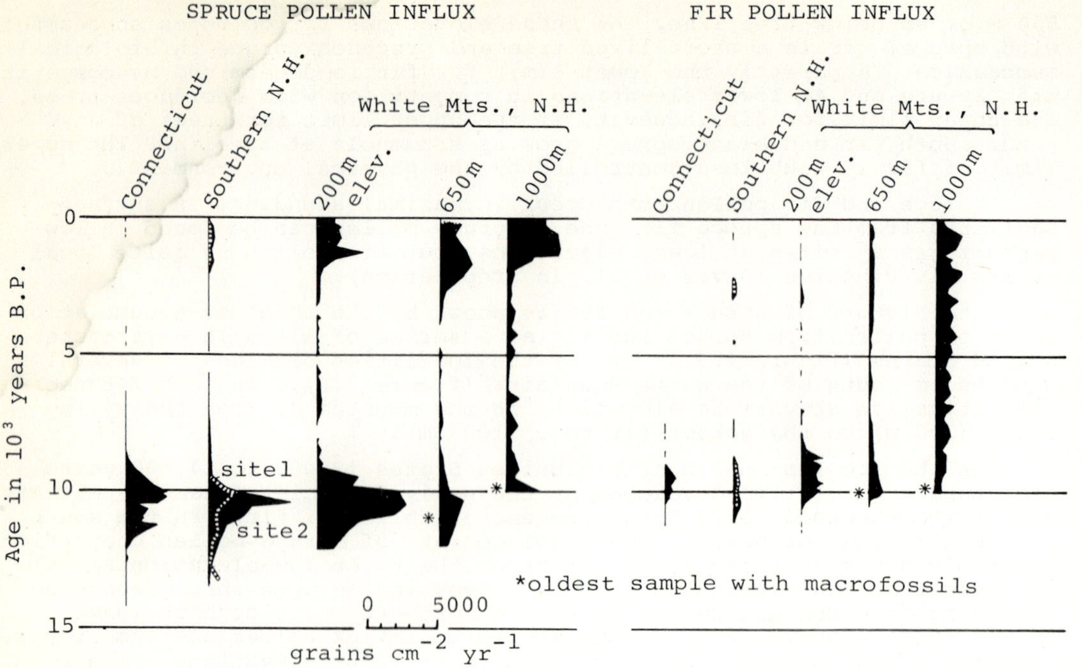

Figure 2.5 Absolute accumulation rate (influx) of spruce and fir pollen grains during the last 13,000 years at 5 localities in northeastern United States: Rogers Lake in Connecticut, Pout and Sandogardy Ponds in Southern New Hampshire, and Mirror Lake, Lost Pond and Carter Pond in White Mountains. The contrasting values at Pout and Sandogardy, which are only a few km apart, indicate the variability characteristic for pollen influx values.

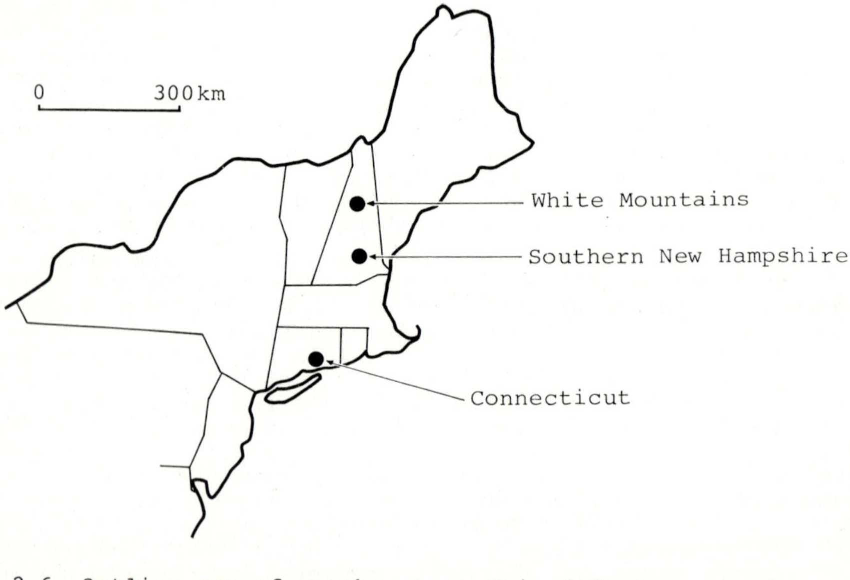

Figure 2.6 Outline map of northeastern United States, showing locations of sites in figure 2.5.

of changes in abundances, affecting each species differently. Climatic interpretations of these changes cannot utilize simple analogy with existing microclimate along the elevation gradient, but instead must depend on a deeper autecological understanding of each species. Any conclusion should also consider the interactions among species, since the community context for both spruce and fir has changed dramatically in the late-glacial and Holocene.

Species interactions, especially competition in forest communities, have doubtless been important in Quaternary history. Yet their role is seldom given emphasis because climatic interpretations tend to rely on autecological information derived from the behaviour of plants in isolation. A final example is included here to emphasize the importance of competition in regulating the abundance of forest trees. Throughout eastern North America, hemlock (*Tsuga*) pollen declines sharply at precisely 4800 radiocarbon years before present. Pollen profiles show the decline at all sites within what was then the range of hemlock from New Brunswick to Upper Michigan, a distance of 1700 km. The decline is rapid, involving a 90% reduction in pollen influx within 50 years. I have described this phenomenon in detail elsewhere (Davis 1976), attributing it to an outbreak of a pathogen or insect specific to hemlock. The populations were decimated and remained at low levels for about 2000 years before recovery to original abundances. The recovery presumably involved evolutionary changes that attenuated the virulence of the pathogen or developed resistance in the host. The importance of the hemlock decline in the present discussion relates to the accompanying release from competition of other forest trees. Percentages and influx rates for competing species of trees rose and remained high throughout the 2000 year interval when hemlock abundances were reduced, and then returned to previous levels. The response in each region was different, reflecting the diversity of competitors experienced by hemlock in different parts of its range. In the valleys of the White Mountains (Mirror Lake, 200 m elevation), the hemlock fall was followed by a brief increase in birch (a numerous seed producer) which, in turn, was followed by a long term rise in beech and maple, shade-tolerant climax species. Pine and oak also increased (figure 2.7a). In southern New Hampshire (Sandogardy Pond, 100 m), birch and beech responded at first, followed by pine and oak (*Quercus*) which occupied the space left vacant by hemlock (figure 2.7b). In Upper Michigan (Canyon Lake) birch was the first species to respond, followed by pine and other conifers (figure 2.7c).

The hemlock example shows how removal or partial removal of a species from the forest allows other species to increase. Although there is no proof that the competing species were expanding into new habitats from which they had been excluded previously by competition (i.e. expanding from their realized niche to their potential niche), the example does show that regional abundances of several species were suppressed by hemlock, and released by its demise. The variation in response from region to region hints at the complexity of the interactions involved in the ecological niche of a species in different parts of its geographical range. Because the removal of a species, once established, has affected these communities, it follows that the addition of new species through immigration (e.g. hemlock in the early Holocene) must also have had an effect. This effect may be difficult to distinguish from change in abundance caused by changes in climate. If more were known about the interactions among species in forest communities, it might be possible to model the impact of the addition of a species to a community, or changes in the physical environment could be considered, and the outcome of competition predicted under changed climate. Changes in the composition of communities over long stretches of time, for example from one interglacial to the next, could be among the most interesting changes from a biological point of view. Some of the differences may indicate chance differences in stocking, or order of immigration (Davis 1976); others may indicate evolutionary changes that have affected the competitive success of the dominant trees. Still others may reflect differences in

a.

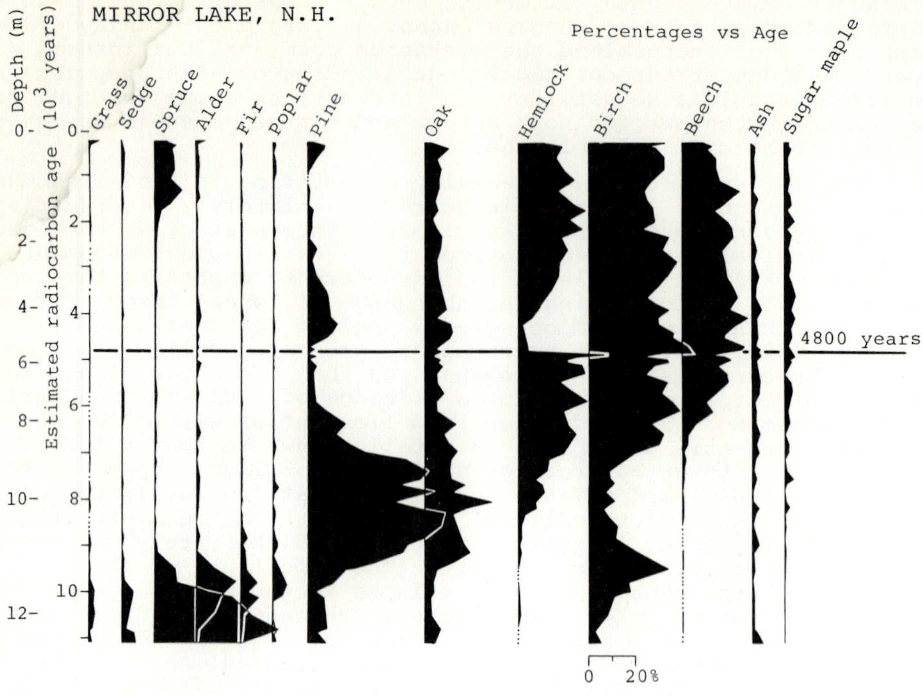
MIRROR LAKE, N.H.
Percentages vs Age
Depth (m)
Estimated radiocarbon age (10^3 years)
Grass
Sedge
Spruce
Alder
Fir
Poplar
Pine
Oak
Hemlock
Birch
Beech
Ash
Sugar maple
4800 years
0 20%

b.

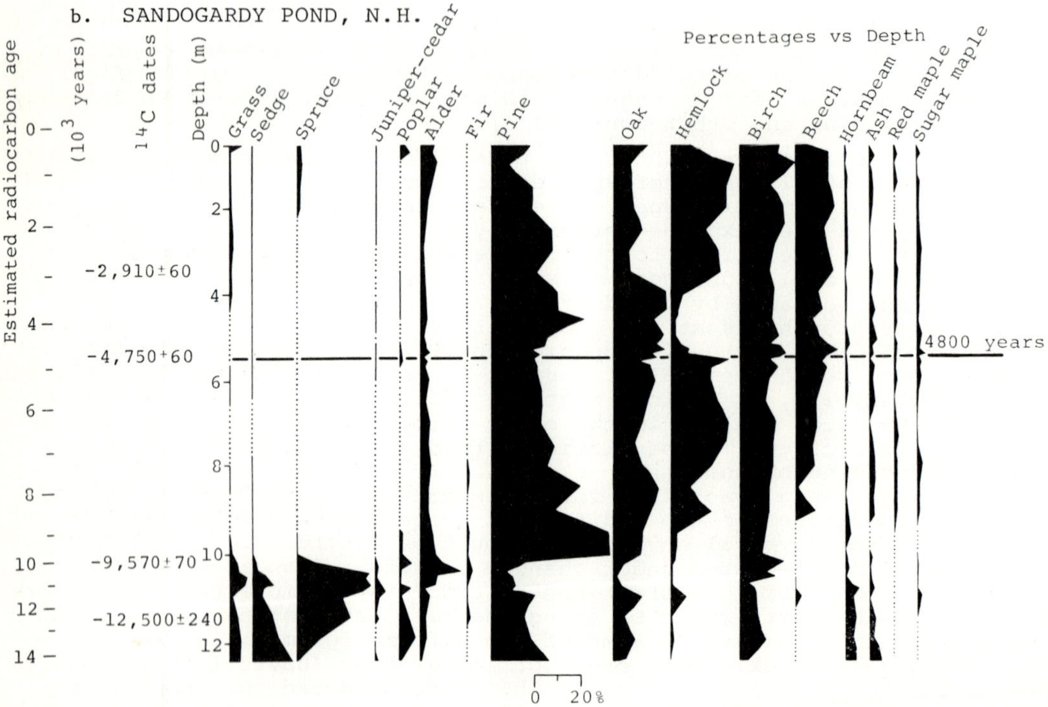
SANDOGARDY POND, N.H.
Percentages vs Depth
Estimated radiocarbon age (10^3 years)
14C dates
Depth (m)
-2,910±60
-4,750+60
-9,570±70
-12,500±240
Grass
Sedge
Spruce
Juniper-cedar
Poplar
Alder
Fir
Pine
Oak
Hemlock
Birch
Beech
Hornbeam
Ash
Red maple
Sugar maple
4800 years
0 20%

c. CANYON LAKE, MICH.

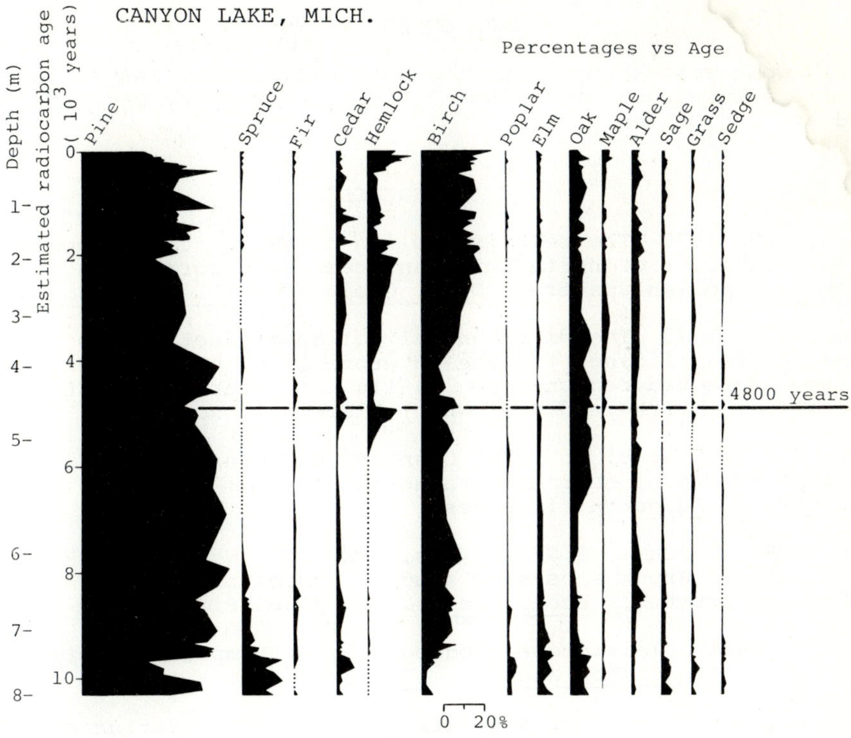

Figure 2.7 Pollen diagrams from three regions in northern United States showing the fall of hemlock pollen 4800 years ago, and the succession following it that is characteristic for each region. Pollen influx diagrams from these sites prove that a real increase in competing species occurred between 4800 and 3000 years ago when hemlock was rare. Geographical locations are shown in figures 2.2 and 2.6 (a) Mirror Lake, 200 m elevation, White Mountains in northern New Hampshire. (b) Sandogardy Pond, 100 m elevation in southern New Hampshire. (c) Canyon Lake, Upper Michigan, about 50 km west of the limit for beech.

climate from one interglacial to the next. In the long run, our best hope for valid estimates of paleoclimate will come from understanding these complex and difficult, but very real, interactions among species as they are affected by the physical environment.

ACKNOWLEDGEMENTS

This work was supported by the National Science Foundation. I gratefully acknowledge criticism of the manuscript by Ray Spear, Mary S. Ford and Jan Harlan.

REFERENCES

Andersen, S.T. 1970 The relative pollen productivity and pollen representation of north European trees, and correction factors for tree pollen spectra. Danm. Geol. Under. Ser II, No. 96, 1-99.

Andersen, S.T. 1973 The differential pollen productivity of trees and its significance for the interpretation of a pollen diagram from a forested region. pp 109-116 in Quaternary Plant Ecology (eds H.J.B. Birks and R.G. West) Blackwell, London.

Bernabo, J.C. and Webb, T. 1977 Changing patterns of the Holocene pollen record of northeastern North America: A mapped summary. Quaternary Research (in press).

Bormann, F.H., Siccama, T.G., Likens, G.E. and Whittaker, R.H. 1970. The Hubbard Brook ecosystem study: Composition and dynamics of the tree stratum. Ecol. Monogr. 40, 373-388.

Davis, M.B. 1976 Pleistocene biogeography of temperate deciduous forests. Geoscience and Man, 13, 13-26.

Davis, M.B., Brubaker, L.B. and Webb, T. 1973 Calibration of absolute pollen influx. pp 9-25 in Quaternary Plant Ecology (eds H.J.B. Birks and R.G. West), Blackwell, London.

Davis, M.B., Ogg, J.A. and Heathcote, I. Pollen percentages in surface sediments from the White Mountains, New Hampshire (in preparation).

Davis, R.B. and Webb, T. 1975 The contemporary distribution of pollen in eastern North America: a comparison with the vegetation. Quaternary Research, 5, 395-434.

Firbas, F. 1949 Spät- und nacheiszeitliche Waldegeschichte Mitteleuropas nördlich der Alpen: v. 1, Gustav Ficher, Jena.

Iversen, J. 1944 *Viscum*, *Hedera* and *Ilex* as climate indicators. Geol. Fören. Forhandl., 66, 463-483.

Kapp, R.O. 1977 Late Pleistocene and postglacial plant communities of the Great Lakes Region. pp 1-27 in Geobotany (ed R.C. Romans) Plenum Press, New York.

Lewontin, R.C. 1974 The Genetic Basis of Evolutionary Change. Columbia University Press, New York.

Maher, L.J. 1964. *Ephedra* pollen in sediments of the Great Lakes Region Ecology, 45, 391-395.

Martin, P.S. 1958 Pleistocene ecology and biography of North America: pp 375-420 in Zoogeography, Amer. Assn. Adv. Sci. Publ. 15, (ed C. Hubbs)

McAndrews, J.H. 1967 Pollen analysis and vegetational history of the Itasca region, Minnesota. pp 219-236. Quaternary Paleoecology (eds E.J. Cushing and H.E. Wright), Yale Press, New Haven.

Nichols, H. 1976 Historical aspects of the northern Canadian treeline. Arctic, 29, 38-47.

Porter, S.C. and Denton, G.H. 1967 Chronology of Neoglaciation in the North American Cordillera. Amer. Journ. Sci., 265, 177-210.

Siccama, T.G. 1974 Vegetation, soil and climate on the Green Mountains of Vermont. Ecol. Monogr. 44, 325-349.

Terasmae, J. and Anderson T.W. 1970 Hypsithermal range extension of White Pine (*Pinus strobus L.*) in Quebec, Canada. Can. Journ. Earth Sci., 7, 406-413.

Ward, R.T. 1961 Some aspects of the regeneration habits of the American Beech. Ecology, 42, 828-832.

Webb, T. III 1974 Corresponding patterns of pollen and vegetation in lower Michigan: a comparison of quantitative data. Ecology, 55, 17-28.

Webb, T. III and Bryson R.A. 1972 Late- and post-glacial climatic change in the Northern Midwest, U.S.A.: quantitative estimates derived from fossil pollen spectra by multivariate statistical analysis. Quaternary Research, 2, 70-115.

Whittaker, R.H. 1956 Vegetation of the Great Smoky Mountains. Ecol. Monogr., 26, 1-80.

3. PLANT MACROFOSSILS AND QUATERNARY PALEOECOLOGY

by W.A. Watts

SUMMARY

Plant macrofossils complement the information available to paleoecologists from pollen diagrams by enlarging the list of identifiable species, providing evidence of local presence, and defining the earliest certain time of local arrival for migrating species. Macrofossil studies are of greatest value if seeds and fruits occur abundantly and can be extracted quantitatively and with an exact stratigraphy from the same core as is used for pollen-analysis. Aspects of the extraction and identification of plant macrofossils, and the presentation of data as seed diagrams are discussed. Some features of the dispersal of plant macrofossils are poorly understood. The sequence of events and processes by which seeds and fruits are dispersed from the parental plant to the point of ultimate sedimentation in lakes is not known. A model is proposed that could be tested by simple experiments including seed trapping in air and water. Plant-macrofossil assemblages are often species-poor, with constant recurrence of the same species. An explanation in the dispersal and sedimentation process is sought. The unusual effectiveness in dispersal of pioneers, case histories of extinction, and the balance in reproductive strategy between fruiting and vegetative reproduction are seen as suitable subjects for experiment.

This account is divided into two sections. The first describes the procedures and principles involved in plant-macrofossil analysis and discusses the construction of seed diagrams. The second draws attention to problems arising in the interpretation of macrofossil records, particularly those that might be resolved by an experimental approach or by a more detailed knowledge of modern autecology.

PROCEDURES IN THE STUDY OF QUATERNARY PLANT MACROFOSSILS

Introduction.

The earliest studies of fossil flora from the Quaternary were concerned with plant macrofossils only. The classic works of Clement Reid in Britain, Hartz in Denmark and Weber in Germany showed that valuable results could be obtained by the method, particularly in demonstrating unexpected plant distributions in the past. These investigations were not quantitative, and the seeds found were not placed in an exact stratigraphic context. The development of the basic principles of pollen-analysis by von Post led to studies exemplified by the work of Jessen (1929), where technically rather primitive pollen-analysis provided a stratigraphic framework into which detailed information about fossil seeds, fruits and sediments was fitted. Jessen's classic studies in Ireland (Jessen 1949) provide a good example of his balanced use of both macro- and micro-fossils. During the late 1930's and 1940's Firbas and, more comprehensively, Iversen showed that it was possible to identify a very large range of pollen types by good microscopy, with careful analysis and description of pollen morphology backed by adequate pollen reference collections. The assessment of many workers at that time, especially in continental Europe and the United States, was that the problems in which they were interested could be resolved by technically sophisticated pollen analysis alone, and that sufficient additional information could not be obtained from plant macrofossils to make their study and the maintenance of large reference collections worthwhile. This viewpoint has proved transient and was never adopted generally.

At present there is a growing interest in plant macrofossils once more. The reasons may be explained in this way. Even with technically very advanced pollen analysis, pollen has limitations which can sometimes be overcome by the combined use of pollen and macrofossil studies. A few examples follow.

Because of its capacity to travel by air or water, pollen may be sedimented at a considerable distance from its source. Macrofossils may travel some distance before sedimentation, but probably not very far, and their presence provides better evidence for local presence than pollen can. A good example is provided by the Late-glacial vegetation of the northeastern United States. In this region pollen of pine, spruce and other trees predominates in relative pollen counts. Davis (1967) shows that the pollen influx early in the Late-glacial is so low that it is unlikely that trees were present and that the pollen must have originated at a considerable distance. Macrofossil studies in Pennsylvania (Watts, in press) show that tundra plants predominate in sediments equivalent in age and pollen assemblage to those studied by Davis.

Studies of tree migration rates have attracted attention recently (Watts 1973; Davis 1976). It is often of critical interest to know when a migrating tree species has arrived in a new location for the first time. This is difficult to establish from pollen diagrams because a rising pollen curve has first a component of pollen that has travelled from a distance, then an additional and ultimately more abundant locally produced component. The moment at which the local component takes over from the long-distance component cannot be distinguished by pollen analysis but the occurrence of macrofossils provides evidence of local presence. The migration history of white pine (*Pinus strobus*) in North America illustrates the point. This species is a very large pollen producer, with morphologically distinctive pollen. Fragments of its needles are very common as macrofossils and can be used effectively to establish a minimum date for local arrival (Watts 1973).

Wind-pollinated species predominate in pollen counts, with a very much smaller contribution from insect-pollinated species, many of which are rarely or never seen by pollen analysts. A macrofossil assemblage is not subject to the same limitations and may contain many species not encountered in pollen counts of the same sediments. Data from the two sources combined greatly enlarge the basis of primary information on which interpretations can be based. *Dryas* (mountain avens) species illustrate the point well. In Late-glacial floras in Europe and North America this insect-pollinated rosaceous genus is rarely identified from pollen counts, but its leaves are among the most ubiquitous macrofossils. At Kirchner Marsh in the northern United States (Watts and Winter 1966) some 80 taxa were determined to species by macrofossils. Of these, fewer than 20 were also determined to genus or family level by pollen analysis.

In some families or genera pollen is either morphologically homogeneous or not sufficiently distinctive to be readily determined to species. In such cases the progress that can be made in distinguishing species by pollen morphology is time-consuming and may ultimately still leave many uncertainties. The Ericaceae (heaths), Cyperaceae (sedges), *Pinus* (pine) and *Betula* (birch) belong in this category in the northern hemisphere. Well preserved macrofossils of all these taxa can be referred to genus or species relatively easily. For example, the distinction between the diploxylon pines *Pinus banksiana* (jack pine) and *Pinus resinosa* (red pine) in the eastern United States can be made immediately on inspection of quite small needle fragments. This leads to the important conclusion that, of these two northern species, *P. banksiana* alone was present in the Late-glacial and *P. resinosa* is known as a rare macrofossil only from the postglacial period in the Middle West. The two species have pollen very closely similar in size and morphology and, although they might be separated on pollen morphology, macrofossil separations can be made with much more certainty in far less time.

Ideally, then, macrofossil studies complement pollen analysis by adding greatly to the lists of species determined and so to the exactness and resolution of the fossil record. It is possible to determine the time of arrival of important tree species and the disappearance of others, and the rates of migration of important species. A catalogue of the advantages of macrofossil studies should not obscure the fact that, as with pollen-analysis, critical examination of the results reveals that there are a number of important aspects where our understanding is inadequate and where further observation or research are necessary if the significance of the data is to become clear.

Some practical considerations.

The material studied and choice of site. Plant macrofossils may be seeds, fruits, stamens, sporangia, leaves, buds, cuticle fragments, rhizomes, or pieces of wood of any size. Specimens may be complete or fragmentary and they may be carbonised by natural or man-made fires. Among lower plants mosses preserve very well as macrofossils but liverworts and lichens are hardly ever found, and seem not to be fossilised except under very exceptional conditions not understood. The relative ease of determination of different kinds of macrofossil and the commitment of time and resources to building up reference collections and acquiring skill and experience with each category are very important limitations on what workers with macrofossils can hope to achieve. Seeds and fruits may be identified by comparison with reference collections and, after some experience with a flora, most can be determined and counted rather quickly. Mosses commonly preserve well and can be identified by the same procedures as live specimens. Buds and small fragments of leaf, cuticle, rhizome or bark are more difficult because reference manuals or collections scarcely exist, although usually a small number of species are so distinctive in one or more characteristics that they are easy to identify. Among leaves, conifer needles are often abundant as macrofossils, as are leaves of dwarf shrubs in deposits from tundra environments. Leaves of deciduous forest trees are rarely preserved except as small fragments. They appear to be destroyed in lakes mechanically or by decomposition. Perfect leaves are occasionally found in quantity in fine-grained alluvial deposits where rapid burial in still back-waters ensures preservation (Clendening et al. 1967). Wood, except in the case of some very obvious species, must be either macerated or seen in thin section to be determined. Reference collections usually exist only for commercially valuable species. The identification of wood is not necessarily technically difficult but it is time-consuming. The same applies to charcoal and both tend to be left to specialists.

All of these considerations show that one cannot hope to identify all the material that is, in fact, identifiable. In practice macrofossil analysis in most of the circumstances encountered by Quaternary paleo-ecologists is limited to the study of seeds, fruits and, in glacial floras, mosses, supplemented by information from a limited number of other easily identified plant parts such as conifer needles and certain leaves. Most investigators will wish to obtain the largest possible amount of information for the smallest expenditure of time. For this reason, macro-fossils may be worth studying only if they are abundant, well-preserved and easily extracted from the sediments in which they occur. Abundance may be defined as 50 or more seeds per 100 ml of sediment. If there are fewer than 10 seeds per 100 ml the study may not be sufficiently rewarding to justify the time spent. Occasionally, but not often, it will be worth while to study a site where macrofossils are infrequent because they may help to solve a particularly difficult problem encountered in pollen analysis.

If one of the objectives in macrofossil studies is to find very numerous fossils, the type of site that should be studied needs to be considered. Most commonly plant macrofossils are studied as part of a wider paleoecological study. They are usually extracted from a core that is also used for pollen analysis, for radiocarbon dating, and perhaps also for paleomagnetic studies, chemical profiles, and fossil pigments, and for algae, Cladocera, Coleoptera, or other fossil organisms. The advantage of using a single core is the very exact stratigraphical and quantitative control that is obtained. The general objective of paleoecological studies of the natural environment is to establish the character, species composition and productivity of upland and aquatic ecosystems through time, and to make inferences about the climate and environment of the past. There are some conflicts between the different techniques used. Pollen influx studies of lakes are best carried out in fine off-shore sediments and especially in sediments with annual banding where the chronological control is particularly good. But such sediments are unlikely to contain macrofossils, which tend to accumulate near-shore in the zone of floating-leaved aquatics where the sediments may be coarser. If pollen and macrofossil studies are to be obtained on the same core a compromise is necessary, and a core must be taken near-shore. Alternatively, a practice of taking and correlating an in-shore and a deep-water core may be adopted.

One very important objective of paleoecological study is the nature of the upland plant cover in the past. Because of this, the most desirable sites for study lack broad fringing belts of sedge-mat, swamp or bog, because they act as traps that prevent seeds and nutrients alike from reaching the lake from upland communities. The author's experience in the eastern United States is that small well-vegetated sinkhole ponds in forested regions where the forest grows to the pond margin are excellent seed-traps for upland and aquatic species alike.

It follows that the best site for combined macrofossil and other paleoecological studies is a small pond or lake, without major inflows or outflows that might cause irregularities in sedimentation and thus make pollen influx counts impossible. A relatively stable water table is desirable to avoid destruction of seeds by oxidation in shallow-water sediments. Upland vegetation should immediately adjoin the pond. In practice, it is impossible to do more than indicate the kind of site that is likely to be good for macrofossil studies, because sites may vary very greatly in the numbers of macrofossils they contain for reasons that may defy analysis.

Although lakes probably provide the best sites for paleoecological work because so many types of study can take place on the same material, plant macrofossils may occur abundantly in several other environments.

Acid peat bogs preserve a wide range of plant material excellently. However, macrofossils are likely to reveal the local succession at the bog surface only and to yield little or no information about the upland. The local sequences may themselves be of great ecological interest in-so-far as they illuminate the difficult question of the dynamics of bog surface growth (Walker and Walker 1961; Grosse-Brauckmann 1968). Acid peats, if one excludes species of *Sphagnum*, may have low densities of identifiable plant macrofossils, and, if fibrous, may be troublesome to disaggregate. These considerations also apply to fen peats, where oxidative destruction of plant debris because of water-level fluctuations may be a serious problem.

Backwaters of rivers, ox-bow lakes, terrace sediments and channel fills of streams are often rich in macrofossils, particularly large fruits such as acorns and walnuts, which are rarely found in lakes. They appear to be floated to their point of sedimentation by flood-water and to survive because of rapid burial by alluvium. Some very important species-rich early-Quaternary floras in Europe are of this type, such as the interglacial deposit at Tegelen (Reid and Reid 1915; Zagwijn 1960). Very interesting assemblages of large-fruited deciduous forest trees have also been demonstrated in river-terrace alluvium in the southeastern United States (Delcourt and Delcourt 1976). Unfortunately such sediments by definition are laid down irregularly and are subject to erosion, so that they are not so suitable for pollen analysis and radiocarbon dating as pond and lake sediments. They fulfil a very valuable role complementary to lake studies in-so-far as they may have different species assemblages not normally found in lakes.

Sediments or soil from archaeological sites may contain very species-rich assemblages of cultivated plants, weeds of cultivation, and uncultivated species collected for food. The determination of fossil cereals and weeds, often on material swollen by carbonisation and difficult for this reason to compare with reference material, is a specialised skill developed to a very high degree by some workers (Jessen and Helbaek 1944; van Zeist 1968). There has also been considerable interest in recent years in remains of cultivated plants, plants of trade, and weeds in town excavations from mediaeval and earlier times (Behre 1969, 1972). As this paper is concerned with natural environments uninfluenced by man, these very interesting specialised studies are not considered further here.

Extraction and identification of seeds and fruits. Seeds may be extracted by washing small volumes of lake sediment through sieves with the aid of a strong jet of water. It is usually convenient to divide sediment into a coarse fraction for residues held on a 40-mesh sieve and a fine fraction held on a 140 mesh sieve (approximately 0.4 and 0.1 mm apertures respectively). The smallest seeds, about 0.3 mm in length, remain in the fine fraction. Residues are dispersed in water and examined on a white plate under a low-power stereo-microscope. Identifiable fossils are picked out with a small artist's brush from which most bristles have been removed and the remainder shortened. Fossils are stored in an equal mixture of glycerol and 2% formalin, an effective preservative for long periods, preferable to dry storage, where surfaces are damaged for abrasion and flaking. When it is necessary to examine specimens again they can be washed in water and dried on filter paper for observation. The author has not found a system preferable to this very simple one. The various chemical treatments of sediment and flotation procedures referred to in the literature do not have advantages that compensate sufficiently for the expense and handling difficulties.

As with pollen, identifications of fossils are best made by comparison of the fossil with reference material, and descriptions, drawings or photographs are of lesser value. However, manuals for seed identification can be of considerable value in a first approximation to identifying unknowns. Beijerinck (1947) and Katz et al. (1965) are useful for this purpose. Recently, a number of descriptions of seeds and fruits of individual families or genera have been published. These are extremely useful in giving critical morphological analyses of the features by which identifications may be made. Aalto (1970) on the difficult genus *Potamogeton*, Koerber-Grohne (1967) on wild grasses, and Berggren (1969) on sedges may be mentioned.

Scanning electron microscopy has not yet been fully exploited as a technique for identification of fossil seeds. Small seeds with strongly expressed and complex sculpturing features and with small but clear differences among species are especially suitable for study. Studies are available, amongst others, for fossil seeds of *Erica* and *Calluna* (Huckerby et al. 1972). The technique is of greatest value where material is abundant and it is unnecessary to worry about the possibility of destroying specimens during photomicrography.

Presentation of data. A weakness of older studies of plant macrofossils was that they were non-quantitative and inexact stratigraphically, so that it is often impossible now to say exactly where interesting fossil records should be placed in relation to pollen diagrams. There is uncertainty, for example, about the position of a number of species presently of markedly southern distribution in deposits of the Irish Late-glacial period (Watts 1977). More recently it has become common practice to work with cores obtained by piston corers such as the Livingstone sampler (Cushing and Wright 1965). This makes it possible to work with standard volumes of sediment, usually half of a core segment 10 cm long and 5 cm in diameter. The sediment sample is related exactly to pollen samples from the same core. All the seeds are extracted, identified, counted and the results presented graphically as a seed diagram. Examples are available from

Kirchner Marsh (Watts and Winter 1966), Jacobson Lake (Watts and Wright 1969) and Wolf Creek (Birks 1977), all in Minnesota.

The idea of a 'seed sum' and of presentation of data as percentages of a seed sum (Watts and Winter 1966) is no longer held to have any advantage over a presentation of absolute figures per unit volume of sediment. The quantitative occurrence of seeds is not as significant as it is for pollen and the greatest interest lies in knowing presence or absence at a particular stratigraphic horizon and, as a lesser consideration, whether present as a single individual or more frequently. As the relationship of seeds to pollen diagrams is particularly interesting and important in demonstrating local presence, the author has experimented recently with diagrams in which pollen and macrofossil data are drafted together (Watts, in press). This enables one to highlight the most important relationships between occurrence as pollen and as macrofossils for the main tree genera.

PROBLEMS IN THE INTERPRETATION OF PLANT MACROFOSSIL ASSEMBLAGES

Dispersal and sedimentation.

The species composition of fossil macrofloras commonly shows some unexplained characteristics: (a) they are often species-poor, even where a rather diverse flora might have been expected and an increase in the volume of material examined adds very few additional species, (b) the same assemblage of species is found over and over again at many sites within one region, (c) the assemblage does not appear to derive from any single, readily identified, plant community. These considerations suggest that the fossils found have a common mechanism for becoming incorporated in lake sediments that is not shared by most other species. If the mechanism were understood it might explain the puzzling selectivity by which only a minority of species become fossils.

Tauber (1965) pointed out that pollen analysts really know very little about how pollen grains travel from the point of dispersal to their final position in lake sediments. Tauber's proposed model for pollen transport aroused much interest in experimental studies of pollen dispersal. Subsequent investigations (Davis and Brubaker 1973; Bonny 1976; Tauber 1977) have gone far to show that pollen recruitment to lakes does indeed involve previously unsuspected pathways of transport and sedimentation. The considerations that apply to pollen cannot apply equally to seeds because of their greater unit weight and high diversity of size and morphology. Nevertheless the success of the experimental approach initiated by Tauber suggests that a similar attempt to establish an experimentally testable model for seed transport and sedimentation would be of value.

As with pollen, productivity, dispersal mechanism, mean distance of dispersal and mode of sedimentation must all be taken into consideration. Seed production varies enormously between species and between individuals of the same species depending on their reproductive strategies and their vegetative plasticity in response to the environment (Harper 1977). Certain trees and annual weedy plants, especially those of mud surfaces of lakes with fluctuating water-levels (Salisbury 1942; Watts and Winter 1966) may have very large seed production. At the other end of the scale, woodland shrubs and herbs tend to produce few, often large, seeds that have a very small chance of becoming fossils. Fowells (1965) records that predation by small mammals and attacks by animal and fungal parasites may destroy a large proportion of the total seed production of some North American tree species. The seed population available for dispersal is what remains after such predation.

Ridley (1930) and van der Pijl (1969) detail mechanisms, many of them elegant and bizarre, by which seeds are thought to be dispersed. Webb (1966) pointed out that many widespread and abundant species have no apparent dispersal mechanism of any kind, and that few experimental observations have been made to establish whether presumed dispersal mechanisms always play the role in the survival strategies of species that has been inferred from their morphology. Little is known about dispersal distance of seeds but the few available experimental observations show that most seeds fall close to their parent and that seed density falls off steeply with distance from the dispersal point (Harper 1977). It might be noted that seeds which become fossilised in lake sediments are losses to the living population, so that their fossilisation is accidental and cannot be part of the reproductive strategy of the species.

Macroflora commonly contain both aquatic and land plants. A distinction must be made between aquatic genera that are pollinated and produce seeds under water, such as *Naias* and *Ceratophyllum*, and those that flower above water and are therefore more comparable to land plants in their dispersal. For land plants the three possible basic dispersal mechanisms are wind, water and animals.

Wind, especially on stormy days, may be very important in moving quite large seeds for considerable distances. The writer has observed the effect of wind associated with a thunderstorm at Lake Annie in the Florida Peninsula. Before the storm, the surface of the water was clear of debris, afterwards pine needle fragments and other plant material floated in some abundance over the whole surface of the lake which is several hundred metres wide. As this was an ordinary summer storm, the less frequent severe storms or hurricanes must produce much greater effects and may disperse seeds for unusually great distances. Rather rare episodes of violent weather may have a greater significance than is appreciated in dispersal. Even relatively weak winds are sufficient to carry seeds from plants growing on lake shores on to the water surface, particularly from trees where the seed is launched from some height and, if winged or relatively light, may take some time to come to ground by gravity. Seeds of low herbaceous plants, especially if growing under a woodland canopy clearly cannot be dispersed as effectively by wind, though turbulent winds may lift plant material from an unprotected ground surface and carry it for some distance.

Rainfall, particularly if intense enough to cause water to saturate the soil and flow over the surface as well as cause strong flows in small streams or water courses, may be responsible for carrying seeds from the herb layer and from stream-side plants into lakes either by flotation or rolling of waterlogged seeds. Streams can be observed to carry numerous whole leaves in deciduous forest areas but, as these are rarely found as fossils, they must be destroyed mechanically or by decomposition in lakes. Birds or animals may carry seeds to lakes either in or on their bodies and excrete or drop them there.

Once plant material has arrived at a lake surface it may become waterlogged and sink readily or it may remain afloat for varying periods of time depending on its anatomy. Praeger's pioneer work (Praeger 1913) showed that many species can remain afloat for several to many days in still water. Although an open body of water with wave action is not comparable with a laboratory experiment, Praeger's work suggests that there is an important difference between species that stay afloat long enough to be carried to the lake margin by wind and wave action where they will be ejected on to the shore and those that sink quickly. Different capacities to remain afloat would explain why some unexpected species occur very commonly as macrofossils while others which might be expected are never found at all. Birks (1973) notes that some aquatic species depend on flotation for dispersal and are supported in water for some time by corky tissue. We do not know whether seeds, once they have become waterlogged, are moved about on lake bottoms by redeposition processes such as are known to be important in the case of pollen (Davis 1968; Davis and Brubaker 1973; Bonny 1976). There seems to be no reason why this would not be so but the size and density of seeds must determine to a great extent how much movement can take place before permanent sedimentation occurs while the large size of seeds, as compared with other sedimentary particles, must increase their chance of sedimenting out early particularly if their movement is hampered by obstacles such as the stems of rooted aquatics.

Taking all these factors into consideration, the following model for seed recruitment to lake sediments is proposed. (a) After predation and parasitism have been accounted for, the surviving seed population is dispersed. (b) Wind, especially strong or violent winds, carry seeds to lake surfaces, most effectively in the case of trees where the seeds are launched from a height. Alternatively some seeds are carried to lakes by streams and surface flooding during periods of high rainfall. Birds and animals play a minor role in transporting soft or hooked fruits. (c) Seeds that are not already waterlogged remain afloat and are carried into shallow water by wind and wave action. (d) Seeds of some species become waterlogged preferentially and sink, others succeed in resisting waterlogging completely and are washed ashore. (e) The waterlogged seeds are moved along the lake bottom by turbulence until they settle out by gravity in the coarser marginal sediments of the lake or are trapped near the margin by submerged parts of aquatic plants.

The model is speculative and it is desirable to design trapping experiments similar to those carried out for pollen (Bonny 1976) to establish what pathways are actually followed by seeds on their way to sediments. The amount of experimental work available in this field is very limited. Ryvarden (1971) studied the dispersal of macrofossils on to recently deglaciated ground in Norway making use of traps. He observed the role of both terrestrial and water transport and estimated the annual influx of macrofossils. Birks (1973) extracted plant macrofossils from surface samples of mud from a series of small lakes in Minnesota and demonstrated that macrofossils were deposited almost entirely within the zone of rooted aquatics. She also demonstrated that assemblages of aquatic macrofossils were correlated with conductivity and that aquatic plants present in lake basins tended to be represented also as macrofossils whereas the occurrence of upland plants as macrofossils was accidental and unpredictable. Further studies of this type, including an elaboration of Praeger's (1913) studies on flotation will be necessary to provide a remedy for the "desperate poverty of hard quantitative information" (Harper 1977) that still exists in this field.

Dispersal in Arctic Regions.

In a 'tundra' flora dating from the time of maximum extent of the Last Glaciation at Longswamp, Pennsylvania (Watts, in press), there is a poor correlation between the pollen and a species-poor macroflora. The pollen flora is dominated by grass, with relatively diverse herbs. The macroflora is composed largely of bryophytes, with leaves of dwarf shrubs including *Dryas integrifolia*, *Vaccinium vitis-idaea* var. *alpina*, *Empetrum* cf. *nigrum*, and *Salix* cf. *herbacea*, and leaves and fruits of *Betula glandulosa*. Grass fruits are sparse and the few other species are represented by only a small number of specimens, although a substantial volume of sediment was examined. Leaves of *Dryas*, *Vaccinium*, *Salix* and *Betula glandulosa* are found at virtually all 'tundra' sites examined (Watts 1967). This suggests a preservation or sedimentation mechanism that favours these species and rejects others.

It seems possible that selective dispersal of leaves and bryophyte debris (Miller and Ambrose 1976) in arctic regions may depend on wind, especially in winter. It can be observed that late snowbeds in the Arctic may be discoloured by the presence of plant debris blown across snow surfaces in winter from snow-free areas such as exposed ridges. Snowbeds in hollows by lakes may act as traps for wind-blown debris. When the spring thaw comes, plant remains fall into lakes where some sink and some float on the surface until carried ashore by wind. P. Glaser (personal communication), who has collected plant debris from snowbeds in Alaska, has observed that the species diversity in the snowbeds is much greater than that of the pond sediments adjoining them. He attributes this to floating off of seeds and other remains of species that do not sink readily. Thus the selective accumulation of the leaves of some species can be explained by their persistence in exposed places during the winter period, their resistance to mechanical damage during wind transport, and the ease with which they become waterlogged and sink. A smooth snow surface in winter provides no obstacle to trap wind-blown plant debris and material may travel for considerable distances on it. The characteristic abundance of conifer needles in lake sediments may also be explained by winter dispersal over snow. Glaser has also observed that snowbed debris may contain species including conifer needles that do not occur locally and have been blown from other communities at lower altitudes, thus limiting the general principle that seeds are likely to be of strictly local origin. As has been discussed above, the macroflora of sediments tends to be less diverse than might be expected, apparently because propagules of many species float. It is interesting that in northwest Europe the macrofloras of Late-glacial age, especially of the Younger *Dryas* Period, may be rich in leaf debris but also relatively species-rich. There is much evidence of solifluction from the sediments of this period. It is possible that the species diversity may result from solifluction masses slumping into lakes as the ice cover melts in Spring, trapping seeds that would normally have floated.

Aquatic Plants.

For obvious reasons aquatic plants occur frequently as macrofossils and their seeds and fruits may greatly outnumber those of upland species. As well as preserving the plant-sociological record of local hydroseres, they may yield information about the trophic status of lakes in the past and about paleoclimate. However, the significance of the occurrence of the seeds of certain aquatic species may not be clear, because of aspects of their biology that are not sufficiently understood. Some of these are discussed below. They include the remarkable effectiveness of many aquatic plants as pioneering migrants, the near-extinction of some species in the later Postglacial and the varying balance between vegetative reproduction and fruiting in many species.

Aquatics as pioneers. It has long been observed that aquatic plants may respond to climatic change with great speed and vigour. This interesting problem was discussed at length by Iversen (1954) who observed that, in Denmark, many aquatics showed sudden large expansions in population size immediately after the cold Younger *Dryas* Period before land plants demonstrated a comparable response. The same phenomenon has been noted recently for sites in Ireland (Watts 1977) and appears to be general in northwest Europe. What enables the aquatics to expand? The variables include climate, with its separate components of mean temperature of the growing season and its duration, nutrient status of the water, and decreasing turbidity of the water as slopes began to be stabilised. It is possible that many aquatics survived the Younger *Dryas* Period vegetatively with no or infrequent flowering and fruiting, but the largely negative evidence suggests that they were new immigrants at most sites. After a very sudden and vigorous expansion marked by high peaks in the pollen curves for aquatics, they either disappear from the record or fall to much lower frequencies. The phenomenon is puzzling. If the aquatics were absent during the Younger *Dryas* Period, even if they survived in a few favoured sites, the short-term population expansion, which calls for the invasion of all available sites in a very brief period, poses these questions: what was the dispersal mechanism and how big a parental stock was necessary to sustain an expansion of such a scale over a period of time perhaps as brief as a century? It appears that nutrition has something to do with the expansion, the collapse in the pollen curves after very brief high peaks being due to the exhaustion of a critical nutrient. It is possible that there was a brief period of very vigorous flowering triggered by warmer, decreasingly turbid water rich in nutrients followed by a period in which the biomass was not very different but flowering was less frequent. The species involved include *Myriophyllum alterniflorum*, *M. spicatum*, several *Potamogeton* species and *Ceratophyllum demersum*. Although the expansion and collapse after the Younger *Dryas* Period is especially strongly expressed, probably because the change in climate was very marked and took place very quickly, similar invasions by aquatics are also known for other periods. It seems probable that the behaviour of the aquatics in the past could be understood better if additional observations were available on possibly analgous present-day behaviour, in particular the performance of aquatics in flowering and fruiting in a variety of field conditions and the conditions affecting balances in their flowering and vegetative growth. Newly-created reservoirs from which the soil is removed before filling may provide valid analogues in which to study this invasion of aquatics and their flowering performance and population size as succession takes place. The reproductive behaviour of aquatics in waters thermally polluted by power-station discharges may indicate potential response to a warming climate. The history of the introduction of *Anacharis canadensis* to the British Isles was initially one of a very aggressive invasion, with growth on such a scale that canals and watercourses were obstructed. Subsequently, the population fell and attained stability at a much lower level. This seems to provide an unevaluated but, unfortunately, poorly documented analogy for the behaviour of, for example, *Myriophyllum alterniflorum* at the end of the Younger *Dryas* Period.

Case histories of extinction. Species of the genus *Naias* are annuals and must produce seed regularly if they are to survive. No doubt their seeds remain viable for some time so that they can survive intermittent drying of small lakes or temporary burial by sediment. Even if a species were to become extinct for a time at one site, it might re-immigrate from a surviving population nearby. *Naias flexilis* has suffered a great loss of range in the later post-glacial period in Europe. It is known as a fossil from mnay sites in Britain, Ireland and Scandinavia (Godwin 1975; Hulten 1950). It survives in a very much reduced number of localities throughout its former range. In North America it is widespread in the Great Lakes Region and in New England. Although the species is easily overlooked and may well be under-recorded, it seems certain that it has disappeared from most of its former localities because of natural causes. What might these be? It is difficult to believe that water-temperature can be involved. As it overwinters by seed, *N. flexilis* should not be affected by winter temperatures. It now occurs in oceanic areas of western Ireland with many cloudy days, limited sunshine and low average summer temperatures and in warmer summer areas of Scandinavia. It seems unlikely that high mean summer temperatures (unavailable in Ireland) or long growing season (unavailable in Finland) can determine its occurrence. *N. flexilis* is found in crystalline rock regions in Ireland in oligotrophic lakes, some of which are brown because of humic materials derived from surrounding peatlands. Many fossil occurrences are in what must have been mesotrophic or eutrophic lakes, judging by the sediments and the associated flora. A hypothesis that extinction may be explained by decreasing availability of nutrients as a result of progressive leaching of soils and bog development in the catchment area is not reconcilable with the oligotrophic lakes where the species now characteristically occurs. The same consideration applies to the possibility that reduced light penetration and photosynthesis in brown-water lakes would account for extinction, because the species survives in just such lakes. The possibility cannot be discounted that *Naias flexilis* is very variable genetically and that part of its population was vulnerable to environmental pressures that did not affect the remainder. If so, it is not at all clear why the surviving population is unable to re-invade the many oligotrophic lakes that appear ecologically suitable for it in Ireland. The fact is that very little is known about the autecology of this inconspicuous species and much could be found out by observation and experiment.

Naias marina presents a more definable problem than *N. flexilis*. In the earlier Postglacial it occurred widely in Ireland, Britain and inland Scandinavia. It now has one occurrence in a brackish coastal lake in Britain and it occurs in shallow coastal waters of the Baltic Sea in Scandinavia. Farther south in Europe it has scattered localities in inland lakes. In North America it has also suffered loss of range (Watts and Bright 1968). *N. marina* is limited to brackish, alkaline or eutrophic, usually carbonate-sedimenting, waters. Marl lakes are still common in limestone regions of Ireland, so that suitable habitats are available. A possible cause of extinction may have been a reduction in mean temperature during the growing season, or in its length, resulting in a shorter period with the high average temperatures necessary for growth and fruiting. At Pickerel Lake, South

Dakota (Watts and Bright 1968), one of the North American sites at which *N. marina* has become extinct, there is direct evidence (Stuiver 1970), from the $^{18}O/^{16}O$ isotope ratio in mollusc shells, that the climate was warmer than now in the period 9000 – 4000 years ago. However, even now, mean summer temperatures in South Dakota are probably much higher than they were in Ireland in the earlier Postglacial. Genetic variation may also be involved, because the Irish populations survived at a temperature probably lower than that at which the South Dakota population became extinct. The possibility that *N. marina* is also dependent on a specific nutrient that may have fallen below a critical threshold of availability due to progressive leaching must also be considered.

The reason for speculating about these two species in such detail is that they occur abundantly as macrofossils at many sites and both seeds and spiny leaf tips, especially of *N. marina*, can be recovered. As *Naias* species, being annuals, cannot survive vegetatively their presence or absence can be determined with some certainty. If either species could be used as a 'thermometer' or indicator of some other exactly definable environmental variable, it would be possible to use the seed diagram procedure to bracket exactly the period of time concerned at any site and to identify variation in its duration from place to place. It would be particularly valuable to be able to delimit the elusive 'climatic optimum' of the Postglacial in this way.

It is possible that the conditions governing the occurrence of the *Naias* species are a combintaion of genetic, climatic, nutritional and other environmental factors that vary too much to be capable of analysis. Nevertheless, a thorough knowledge of the autecology of these species, based on laboratory and field experiment and on observation, may make it possible to reconstruct, for example, paleotemperatures on a more satisfactory scientific basis than has been available up to the present. Some other species have similar extinction histories, for example the large aquatic sedge *Cladium mariscus*, now rare in Scandinavia but not in Ireland, and *Trapa natans* (water chestnut), for which extinction has been progressive in Scandinvia and north-central Europe into recent historically recorded time.

<u>The balance between vegetative growth and fruiting in aquatics</u>. A number of perennial aquatic plants may reproduce vegetatively or by seeds and fruits. The circumstances in which either vegetative or sexual reproduction predominates are of interest to paleoecologists. Fruiting may require higher mean summer temperatures than vegetative growth whereas a species growing purely vegetatively may have been an important part of the aquatic biomass but escapes detection as a fossil.

West (1958) has drawn attention to the interest of the aquatic macrofossils of the Ipswichian (Last Interglacial) flora of Britain. An assemblage which includes fruits of *Stratiotes aloides* and *Hydrocharis morsus-ranae* (both Hydrocharitaceae) is found at several Ipswichian sites. *Stratiotes* is scattered throughout Britain now, mainly in eastern England. Only female plants occur normally in Britain, so that reproduction is purely vegetative. Further south, both male and female plants are found and fruiting occurs. At the southern limit of its range, populations are predominantly male. *Hydrocharis* flowers but rarely or never fruits in Britain and Ireland where it is more widespread than *Stratiotes*. It reproduces by turions. There is a clear hypothesis to be made that summer temperatures were higher in the Ipswichian than ever in the Postglacial, but the circumstances in which fruiting takes place have not been quantified by observation or experiment.

Lemna species are found erratically as fossil seeds or pollen. They are also characteristic of the Ipswichian, but are known to have occurred in quantity occasionally in the Postglacial (Walker and Lambert 1955). They probably rely predominantly on vegetative reproduction and the circumstances in which mass flowering and fruiting may take place are not known.

Ceratophyllum demersum does not fruit freely under Irish conditions at present. It is possible to collect very large amounts of vegetative

Ceratophyllum without finding fruits. Its irregular fruiting pattern has been investigated in Scandinavia (Julin and Luther 1959) where fruiting is said to take place in summers warmer than average. Overwintering may be by vegetative buds (Muenscher 1944). Fruits and spiny leaf-tips both occur frequently as fossils, the leaf-tips in preparations made for pollen-analysis. It is possible, therefore, to separate the histories of vegetative growth and of reproduction by fruiting in a sediment profile. As Muenscher noted, *Ceratophyllum* is a vigorous invader of "newly made lakes in which water is rich in organic matter and materials in solution". It is an effective pioneering invader, appearing in Ireland as fruits in very early Late-glacial deposits (Craig 1973). It is also an early pioneer in North America where its fruits are found in a spruce/dwarf birch assemblage in New Jersey in the first sediments formed in a small pond near the glacial limit (Watts, in press). Terasmae and Craig (1958) have recorded its occurrence as fossil leaves north of its present range in Boreal Canada.

Ceratophyllum represents an extreme case of a species that may consistently be underrepresented in fossil fruit counts. It is possible to find lakes in which it is very abundant but from the surface sediments of which its seeds are absent, whereas those of *Naias* species are frequent although living plants of *Naias* cannot be found. Birks (1973) also notes occurrences of *Naias* of this type. Thus, seed diagrams may give an incorrect indication of the species that are major contributors to the lake biomass. It seems possible that *Ceratophyllum* fruits readily when pioneering new lakes rich in mineral nutrients, either man-made or resulting from glacial retreat, and in lakes made eutrophic by sewage. In a stable ecosystem, competing with other macrophytes, reversion to predominantly vegetative growth may occur.

CONCLUSION

The purpose of this paper has been to review the status of plant macrofossil studies in Quaternary sediments and to identify fields where experimental investigations with living plants and ecosystems might illuminate problems identified by fossil studies. To some extent the result has been to reveal very large areas of ignorance and rather limited fields in which consultation between paleoecologists and ecologists of living vegetation is immediately profitable. Nevertheless, it would now be common ground among Quaternary paleoecologists that their findings are not merely of stratigraphic and floristic interest, but contribute directly to fundamental ideas in ecology in fields such as dispersal, migration, community stability and extinction, all essentially in the field of population ecology. The production of seed diagrams now enables paleoecologists to present their data with such quantitative stratigraphic accuracy that further questions naturally arise about the significance of macrofossil finds. There are three subjects in which knowledge is seriously defective at present and where modern experimental work would be of great value. The first is the extension to seeds of pollen trapping and surface sampling procedures (Bonny 1976; Tauber 1977) as has already been done to a limited extent (Ryvarden 1971; Birks 1973). It is essential to establish what is the mechanism for the high selectivity of occurrence of upland species in lake sediments. The second area is the biology of aquatic plants, particularly its relationship between vegetative and sexual reproduction; even anecdotal literature is sparse and there is very little experimental work. Because aquatic plants occur commonly as macrofossils they have the advantage that the widespread fossil occurrences of species such as *Naias marina*, now extinct over most of their former ranges, provide the necessary basis for a detailed case history of extinction. Finally, Harper (1977) has reviewed our present knowledge of seed dispersal and described the very limited amount of experimental information that is available. In this field substantial progress by ecologists would be of great value to paleoecologists who are capable of establishing migration rates for many tree species by combined pollen and macrofossil studies, but who lack an adequate concept of how migration takes place in the field.

REFERENCES

Aalto, M. 1970 Potamogetonaceae fruits. 1. Recent and subfossil endocarps of the Fennoscandian species. Acta Botanica Fennica, 88, 1-85.

Behre, K-E. 1969 Untersuchungen des botanischen materials der frühmittelalterlichen Siedlung Haithabu. Berichte über die Ausgrabungen in Haithabu 2, 7-55.

Behre, K-E. 1972 Kultur- und Wildpflanzenreste aus der marschgrabung Jemgumkloster/Ems (um Christi Geburt). Neue Ausgrabungen und Forschungen in Niedersachsen, 7, 164-184.

Beijerinck, W. 1947 Zadenatlas der nederlandsche Flora. Veenman, Wageningen.

Berggren, G. 1969 Atlas of Seeds. Part 2. Cyperaceae. Berlingska Boktryckeriet, Lund.

Birks, H.H. 1973 Modern macrofossil assemblages in lake sediments in Minnesota. pp 173-190 in Quaternary Plant Ecology (eds H.J.B. Birks and R.G. West), Blackwell, Oxford.

Birks, H.J.B. 1976 Late-Wisconsinan vegetational history at Wolf Creek, Central Minnesota. Ecol. Monogr., 46, 395-429.

Clendening, J.A., Renton, J.J. and Parsons, B.M. 1967 Preliminary palynological and mineralogical analyses of a Lake Monongahela (Pleistocene) Terrace Deposit at Morgantown, West Virginia. West Virginia Geological and Economic Survey, Circular Series, 4.

Craig, A.J. 1973 Studies on the ecological history of south-east Ireland, using pollen influx analysis and other methods. Unpublished Ph.D. thesis, Trinity College, Dublin.

Cushing, E.J. and Wright, H.E. Jr. 1965 Hand-operated piston corers for lake sediments. Ecology, 46, 380-384.

Davis, M.B. 1967 Pollen accumulation rates at Rogers Lake, Connecticut, during late- and postglacial time. Rev. Palaeobotan. Palynol., 2, 219-230.

Davis, M.B. 1968 Pollen grains in lake sediments: redeposition caused by seasonal water circulation. Science, 162, 796-799.

Davis, M.B. 1976 Pleistocene biogeography of temperate deciduous forests. Geoscience and Man, 13, 13-26.

Davis, M.B. and Brubaker, L.B. 1973 Differential sedimentation of pollen grains in lakes. Limnol. and Oceanog., 18, 635-646.

Delcourt, P.A. and Delcourt, H.R. 1977 The Tunica Hills, Louisiana-Mississippi: Late Glacial locality for spruce and deciduous forest species. Quat. Res. 7, 218-237.

Fowells, H.A. 1965 Silvics of Forest Trees of the United States. U.S. Dept. of Agriculture Handbook 271.

Godwin, H. 1975 The History of the British Flora (2nd edn) University Press, Cambridge.

Grosse-Brauckmann, G. 1968 Einige Ergebnisse einer vegetationskundlichen Auswertung botanischer Torfuntersuchungen, besonders im Hinblick auf Sukzessionsfragen. Acta Bot. Neerl., 17, 59-69.

Harper, J.L. 1977 Population biology of plants. Academic Press, London 892 p.

Huckerby (née Vokes) E., Marchant, R. and Oldfield, F. 1972 Identification of fossil seeds of *Erica* and *Calluna* by scanning electron microscopy. New Phytol., 71, 387-392.

Hultén, E. 1950 Atlas of the distribution of vascular plants in north-west Europe. Generalstabens Litografiska Anstalts Forlag, Stockholm.

Iversen, J. 1954 The late-glacial flora of Denmark and its relation to climate and soil. Danm. Geol. Unders. Ser. 2., 80, 87-119.

Jessen, K. 1949 Studies in Late Quaternary deposits and flora-history of Ireland. Proc. R. Irish Acad., 52B, 61-85.

Jessen, K. and Milthers, V. 1928 Interglacial fresh-water deposits in Jutland and north-west Germany. Damn.Geol. Unders. Ser. 2., 48, 1-379.

Jessen, K. and Helbaek, H. 1944 Cereals in Great Britain and Ireland in prehistoric and early historic time. K. danske vidensk. Selsk., 3, 1-100.

Julin, E. and Luther, H. 1959 Om Blomning och fruktsättning hos *Ceratophyllum demersum* i Fennoskandien. Bot. Notiser, 112, 321-338.

Katz, N.J., Katz, S.V. and Kipiani, M.G. 1965 Atlas and keys of fruits and seeds occurring in the Quaternary deposits of the USSR (in Russian). Publishing House Nauka, Moscow.

Koerber-Grohne, V. 1967 Geobotanische Untersuchungen auf der Feddersen Wierde. Franz Steiner, Wiesbaden.

Muenscher, W.C. 1944 Aquatic plants of the United States. Cornell University Press, Ithaca, N.Y.

Miller, N.G. and Ambrose, L.S.H. 1976 Growth in culture of wind-blown bryophyte gametophyte fragments from Arctic Canada. Bryologist, 79, 55-63.

van der Pijl, L. 1969 Principles of dispersal in higher plants. Springer, Berlin.

Praeger, R.L. 1913 On the buoyancy of the seeds of some Britannic plants. Sci.Proc. R. Dublin Soc., 14 (N.S.), 13-62.

Reid, C. and Reid, E.M. 1915 The Pliocene floras of the Dutch-Prussian border. Meded. van de Rijksopsporing van Delfstoffen, 6.

Ridley, H.N. 1930 The dispersal of plants throughout the world. Reeve, Ashford, Kent.

Ryvarden, L. 1971 Studies in seed dispersal, 1. Trapping of diaspores in the alpine zone of Finse, Norway. Norw. J. Bot., 18, 215-226.

Salisbury, E.J. 1942 The reproductive capacity of plants. Bell, London.

Samuelsson, G. 1934 Die Verbreitung der höheren Wasserpflanzen in Nordeuropa (fennoskandien und Dänemark). Acta Phytogeographica Suecica, 6.

Stuiver, M. 1970 Oxygen and carbon isotope ratios of fresh-water carbonates as climatic indicators. J. Geophys. Res., 75, 5247-5257.

Tauber, H. 1965 Differential pollen dispersion and the interpretation of pollen diagrams. Damn. Geol. Unders. Ser. 2., 89, 1-69.

Tauber, H. 1977 Investigations of aerial pollen transport in a forested area. Dansk Botanisk Arkiv, 32, No. 1.

Terasmae, J. and Craig, B.G. 1958 Discovery of fossil *Ceratophyllum demersum* L. in Northwest Territories, Canada. Can. J. Botany, 36, 567-569.

Walker, D. and Lambert, C.A. 1955 Boreal deposits at Kirkby Thore, Westmorland, New Phytol., 54, 209-215.

Walker, D. and Walker, P.M. 1961 Stratigraphic evidence of regeneration in some Irish bogs. J. Ecol., 49, 169-185.

Watts, W.A. 1967 Late-glacial plant macrofossils from Minnesota. pp 89-97 in Quaternary Paleoecology (eds E.J. Cushing and H.E. Wright), Yale University Press, New Haven.

Watts, W.A. 1973 Rates of changes and stability in vegetation in the perspective of long periods of time. pp 195-206 in Quaternary Plant Ecology (eds H.J.B. Birks and R.G. West), Blackwell, Oxford.

Watts, W.A. 1977 The Late Devensian vegetation of Ireland. Phil. Trans. R. Soc. London, B, 100, 171-191.

Watts, W.A. 1977 The vegetation of northern Appalachia in the Late Wisconsin and Holocene periods (in press).

Watts, W.A. and Bright, R.C. 1968 Pollen, seed and mollusk analysis of a sediment core from Pickerel Lake, northeastern South Dakota. Geol. Soc. America Bull., 79, 858-876.

Watts, W.A. and Winter, T.C. 1966 Plant macrofossils from Kirchner Marsh, Minnesota - a paleoecological study. Geol. Soc. America Bull., 77, 1339-1360.

Webb, D.A. 1966 Dispersal and establishment: what do we really know? pp 93-102 in Reproductive biology and taxonomy of vascular plants (ed J.G. Hawkes), Pergamon Press, London.

West, R.G. 1958 Interglacial site at Bobbitshole. Phil. Trans. R. Soc. London, B, 241, 1-31.

Wright, H.E. Jr., Watts, W.A. and others 1969 Glacial and vegetational history of northeastern Minnesota. Minn. Geol. Surv. Spec. Publ. 11.

Zagwijn, W.H. 1960 Aspects of the Pliocene and Early Pleistocene vegetation in the Netherlands. Mededel. Geol. Sticht., C3.

van Zeist, W. 1970 Prehistoric and early historic food plants in the Netherlands. Palaeohistoria, 14, 41-173.

4. RECONSTRUCTION OF PAST LACUSTRINE ENVIRONMENTS BASED ON THE REMAINS OF AQUATIC INVERTEBRATES*

by T.L. Crisman

SUMMARY

The five invertebrate groups offering the greatest potential for reconstructing past lacustrine environments are: Cladocera, midges, rotifers, ostracods and molluscs. Of these, the Cladocera have been the most thoroughly studied. Individual species display differential sensitivity to variations in water chemistry and lake tropic state. In addition, variations in community structure, size-selective predation, and the importance of littoral versus planktonic Cladocera have been recorded during periods of environmental fluctuation. Midges are sensitive indicators of oxygen concentration, sediment composition and water depth but the value of midges as a paleolimnological tool is diminished by taxonomic and statistical problems. Rotifers are represented in sediments by their resting eggs, but inadequate taxonomic study of these remains and the lack of detailed ecological investigations hinder paleolimnological reconstructions based on this group. Ostracods are particularly sensitive to variations in water chemistry, oxygen concentration, sediment composition, and lake trophic state. Nevertheless, differential preservation of the carbonate-rich shells complicates interpretations of past environments. Finally, few quantitative investigations on molluscs from lake cores are available. Molluscs are especially sensitive to changes in the lacustrine environment, but large volumes of sediment must be processed in order to obtain a statistically significant sample.

INTRODUCTION

The exoskeletal remains of benthic and planktonic aquatic invertebrates are common in lake sediments. As the exoskeleton of aquatic invertebrates completely disarticulates into its several component parts upon death of the organism, the researcher is faced with two important problems: (a) identification of each exoskeletal fragment to species and (b) reconstruction of population levels of individual species from identified exoskeletal components. In addition, complications arise from differential preservation of exoskeletal components as well as changes in mechanical fragmentation of remains, as determined by sediment composition, and biological fragmentation by detritus feeders.

Stratigraphic changes in species composition and community structure of fossil invertebrate assemblages (Cladocera, midges, ostracods, rotifers, molluscs) provide valuable information regarding intra-basin fluctuations in water chemistry, trophic state, oxygen concentration, and lake level associated with lake aging and basin infilling. Rather than relying on a few "indicator species", the preferred paleolimnological reconstructions combine analyses of several invertebrate groups with data on diatoms, sediment chemistry, pollen, and chlorophyll degradation products (Cowgill et al. 1966; Hutchinson et al. 1970; Whitehead et al. 1973).

A lake is not a closed ecosystem, for it depends for its nutrient supply on atmospheric precipitation and on chemical release from its catchment. Whereas the available nutrient pool in the catchment is ultimately controlled by bedrock type and climatically related weathering processes, the quality and quantity of nutrient export from the catchment is strongly controlled by the composition and age structure of local vegetation. For example, increased chemical release from mature coniferous, compared to deciduous, forests is a reflection of differences in primary production rates, acidity of forest litter, and rainfall interception. In addition, Vitousek and Reiners (1975) report that chemical release from forested catchments decreases sharply as forest biomass increases

*Contribution 167 of the Limnological Research Centre, University of Minnesota, Minneapolis.

in successional systems. Variations in chemical export from catchments accompanying late- and post-glacial changes in regional vegetation can result in major oscillations in lake trophic state and the community structure of invertebrates (Whitehead et al. 1973; Crisman 1977; Whitehead and Crisman 1977).

In order for an invertebrate group to be useful in paleolimnological studies, the following criteria must be met. First, statistically significant numbers of remains must be obtained from relatively small sediment volumes. Second, fossil remains must be identifiable to species. Third, detailed ecological information on the living species must exist. Finally, the organism must be sensitive to environmental change. An excellent review of the faunal groups encountered in lake sediments is given by Frey (1964, 1976), but only five groups (Cladocera, midges, ostracods, molluscs, and possibly rotifers) approach the above-mentioned criteria.

The following discussion is a survey of the variety of paleolimnological questions that can be approached from an examination of the remains of these five invertebrate groups; it is not intended to serve as a historical review of paleolimnology. Whenever possible multidisciplinary investigations that compare late- and post-glacial lacustrine environmental changes with contemporaneous changes in regional vegetation and climate are discussed.

CLADOCERA

Cladoceran exoskeletons completely disarticulate upon death of the organism; and the components are subject to differential preservation. Fossil cladoceran assemblages isolated from lacustrine sediments are composed of species from two distinct cladoceran communities, littoral and planktonic. The littoral community (principally Chydoridae) is well preserved and may be identified to species from a variety of exoskeletal components (shell, head shield, postabdomen, postabdominal claws, antennal segments, mandibles). With the exception of *Bosmina*, and to a lesser extent *Daphnia*, *Ceriodaphnia* and *Diaphanasoma*, planktonic cladoceran species are rarely encountered in lake sediments. The remains of *Bosmina* (whells, head shield, postabdominal claws, mandibles) are often the most common cladoceran remains in sediments, while identification of other planktonic cladocerans must be based on isolated ephippia, postabdominal claws, antennal segments, and mandibles. This generally poor preservation of planktonic species explains the frequent inability to identify remains of planktonic cladocerans beyond the generic level.

Cladocerans reach sexual maturity after passing through a determinate number of growth stages called instars. Each instar is terminated by a moulting of the exoskeleton, and the number of instars required to reach sexual maturity is species-specific. Once sexual maturity is reached, growth becomes indeterminate and is accompanied by a variable number of moults throughout the remaining life of the organisms. Each exoskeletal moult quickly disarticulates and, with the exception of the earliest instars, has an equal chance of preservation in the sediments. The cladoceran assemblage isolated from lake sediments is, therefore, based on remains from several exoskeletal moultings.

An important problem that must be faced by each investigator of fossil cladocerans is the reconstruction of population levels of individual species based on exoskeletal fragments found in sediments. The most common practice for reconstruction of cladoceran populations is to assume that each identified fragment represents an individual cladoceran (Frey 1958, 1960, 1961; Harmsworth 1968; Whiteside 1970). This method is not favoured because it does not account for differential fragmentation of exoskeletal components or for major stratigraphic variations in the intensity of biological or mechanical fragmentation processes.

Other workers have relied on specific exoskeletal remains as estimates of past cladoceran populations. Alhonen (1967) counted only whole head shields, even though other remains could be identified. Goulden (1964, 1966a, b) and Tsukada (1972) used both head shields and shells for their analyses but also counted fragmented remains selectively. The implied assumption that all species are equally represented by the exoskeletal remain chosen for tabulation is in error, because species that are not represented at all by a specific remain can be well represented by other readily identifiable parts.

An alternative method for construction of population levels for fossil cladocerans has recently been proposed by Crisman (1977). This method utilizes both fragmented and whole remains in order to estimate cladoceran individuals. Each remain encountered for a given species is tabulated under the appropriate exoskeletal category, and the degree of fragmentation is noted. For purposes of calculation, two fragmented half-remains are equated to one whole remain. The maximum number of individuals that could possibly be represented by each category of fragmented and whole remain are calculated, and the largest estimate of number of individuals is selected as the population level for that sample. Rather than reliance upon one or two remain categories, which may grossly underestimate the cladoceran population level, it is felt that this technique will provide a more valid estimate of population size. This method does not exclude species that are never represented in the fossil record by a particular remain category that has been arbitrarily chosen as the standard for determination of cladoceran abundance.

Fossil assemblages from deep-water sediments integrate the patchy species distributions and seasonal population fluctuations of individual species of littoral cladocerans, thereby serving as the best index of the littoral community for the entire lake. Through a comparison of detailed collections of modern populations of littoral cladocerans with subfossil assemblages isolated from surface sediments, Frey (1960) not only was able to demonstrate that the species composition of the sediment assemblage of the lakes at Madison, Wisconsin, was similar to the living population, but he encountered six littoral species not previously reported in modern collections from those lakes.

Mueller (1964) examined assemblages of littoral and planktonic cladocerans isolated from surface sediments along transects of increasing water depth for three Indiana lakes (figure 4.1). His results indicate that the percentage composition of sediment assemblages of littoral cladocerans displays only minor variation with increasing water depth. On the other hand, the quantity of remains of littoral cladocerans (chydorids) decreases steadily with increasing distance from shore. Sediments from the profundal zone display less lateral variation in number of remains than sediments of intermediate water depth. The remains of planktonic cladocerans exhibit the same profundal distribution as the littoral remains but they display a concentration zone of remains in sediments approximating the upper portion of the hypolimnion. While the percentage composition of the littoral assemblage is little affected by fluctuations in lake level, reconstructions of past population levels for both littoral and planktonic cladocerans are strongly influenced by water depth. In particular, the downward displacement of the concentration zone of planktonic remains accompanying a reduction in water level could falsely be interpreted as representing a real increase in planktonic cladoceran population. It is essential, then, that all cladoceran reconstructions take into account possible water-level fluctuations.

The littoral cladoceran community can be divided further into macrophytic and sediment assemblages, according to detailed examination of functional morphology (Fryer 1968, 1974) and collections of modern populations (Flössner 1964). Quade (1969, 1971) attempted to relate species composition of the macrophytic assemblage to individual species of macrophyte but later he rejected the feasibility of this approach. It is suggested that Cladocera respond to the availability of a particular substrate type (e.g. leaf form) rather than to the presence of particular species of macrophyte.

Stratigraphic variation in the importance of macrophytic, relative to substrate, Cladocera has potential for determining the extent of macrophyte development within a basin. Whiteside (1970) observed a drastic reduction in the abundance of macrophytic cladocerans associated with increasing productivity in Danish lakes. Furthermore, the importance of sediment cladocerans remains essentially independent of lake trophic state. Periods of increased productivity in the trophic history of Berry Pond, Massachusetts, are accompanied by an increase in the importance of sediment versus macrophytic cladocerans, but the relationship at nearby North Pond is not so distinct (Crisman and Whitehead 1977). North Pond has always been less productive than Berry Pond and displays only minor increases in macrophytic cladocerans during periods of increased productivity. Although it appears that major increases in productivity are accompanied by a reduction in macrophytes due to algal shading, the relationship must be interpreted with caution. The influence of sediment composition, water chemistry and basin morphology in controlling the areal extent and species composition of macrophyte assemblages must also be considered in any paleolimnological interpretation.

Information regarding the response of individual cladoceran species to various parameters of water chemistry and lake trophic state stem mainly from the work of Whiteside (1970) in Denmark. Cladocera assemblages were isolated from surface sediments of 77 lakes for which adequate physical and chemical data exist. The distributions of the 35 littoral species encountered were then correlated with inter-basin fluctuations in seven physical and chemical parameters in addition to productivity data where available. These data indicate that chydorid cladocerans prefer lakes of low productivity, as demonstrated by the general reduction in species number and diversity with increasing productivity.

In addition, Whiteside studied the community response of littoral cladocerans to major differences in lake chemistry and trophic state. The Danish lakes were divided into three arbitrary groups by a discriminant analysis of seven physical and chemical parameters: (a) clear-water unpolluted lakes, (b) clear-water polluted lakes and (c) ponds and bogs. Utilizing the 22 most common chydorid species, he grouped cladoceran assemblages from these lakes by discriminant analysis. In 83% of the cases the lake type predicted by the cladoceran assemblage for a particular lake agreed with the lake type designation based on chemical and physical parameters. These results demonstrate the sensitivity of the littoral cladoceran community to variations in lake chemistry and trophic state.

While this information is extremely helpful in European investigations, strict application elsewhere is difficult. European species designations have been applied to morphologically similar forms in North America without detailed taxonomic analysis. Taxonomic uncertainty in addition to possible differential responses of regional cladoceran populations to various lacustrine parameters demonstrate the need for similar ecological studies in North America. Synerholm (1974) relates the distribution of littoral cladoceran species isolated from surface sediments of 32 Minnesota lakes to changes in lake conductivity. She divides the lakes into three arbitrary groups on the basis of increasing conductivity. By means of discriminant analysis 81% of the lakes examined were assigned to the proper conductivity group on the basis of the percent composition of four cladoceran species: *Chydorus sphaericus*, *Acroperus harpae*, *Leydigia leydigi*, and *Alona intermedia*.

Crisman (unpublished) has recently expanded the work of Synerholm (1974) to include an analysis of the response of littoral cladocerans to 16 physical and chemical parameters for 150 lakes in Minnesota and Florida. Initial results derived from multiple linear regression and principal component analysis indicate that, while lake conductivity is important in controlling the distribution of cladoceran species, lake alkalinity and trophic state are also important controlling factors.

Individual species of littoral Cladocera are rarely used as indicator species in paleolimnological investigations. The notable exceptions are *Chydorus sphaericus* and *Alona rectangula*, a species not present in North

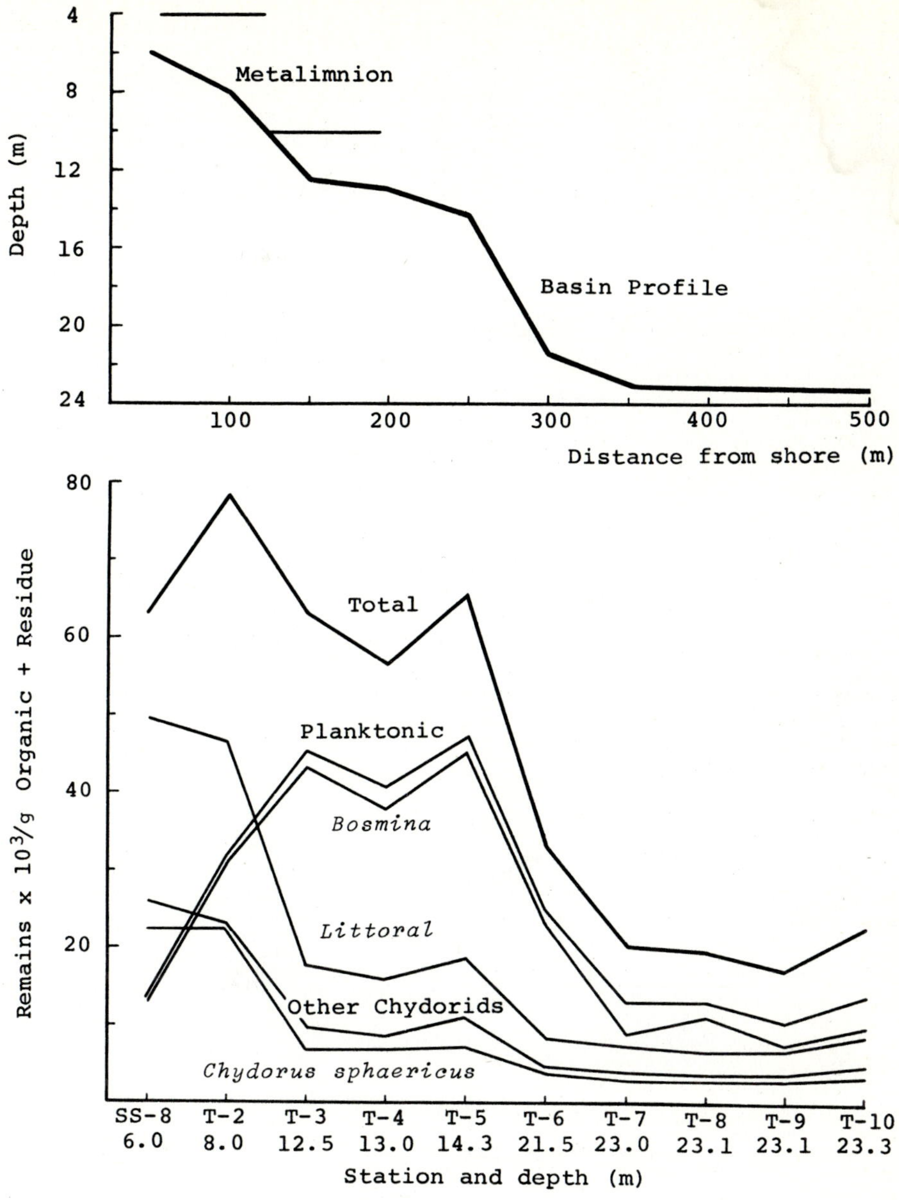

Figure 4.1 Abundance of cladoceran remains in surface sediments collected along a transect from shore to profundal waters at Winona Lake, Indiana. Concentration of littoral cladoceran remains decreases with increasing distance from shore, whereas planktonic cladoceran remains display a concentration peak at the top of the hypolimnion. (From Mueller 1964).

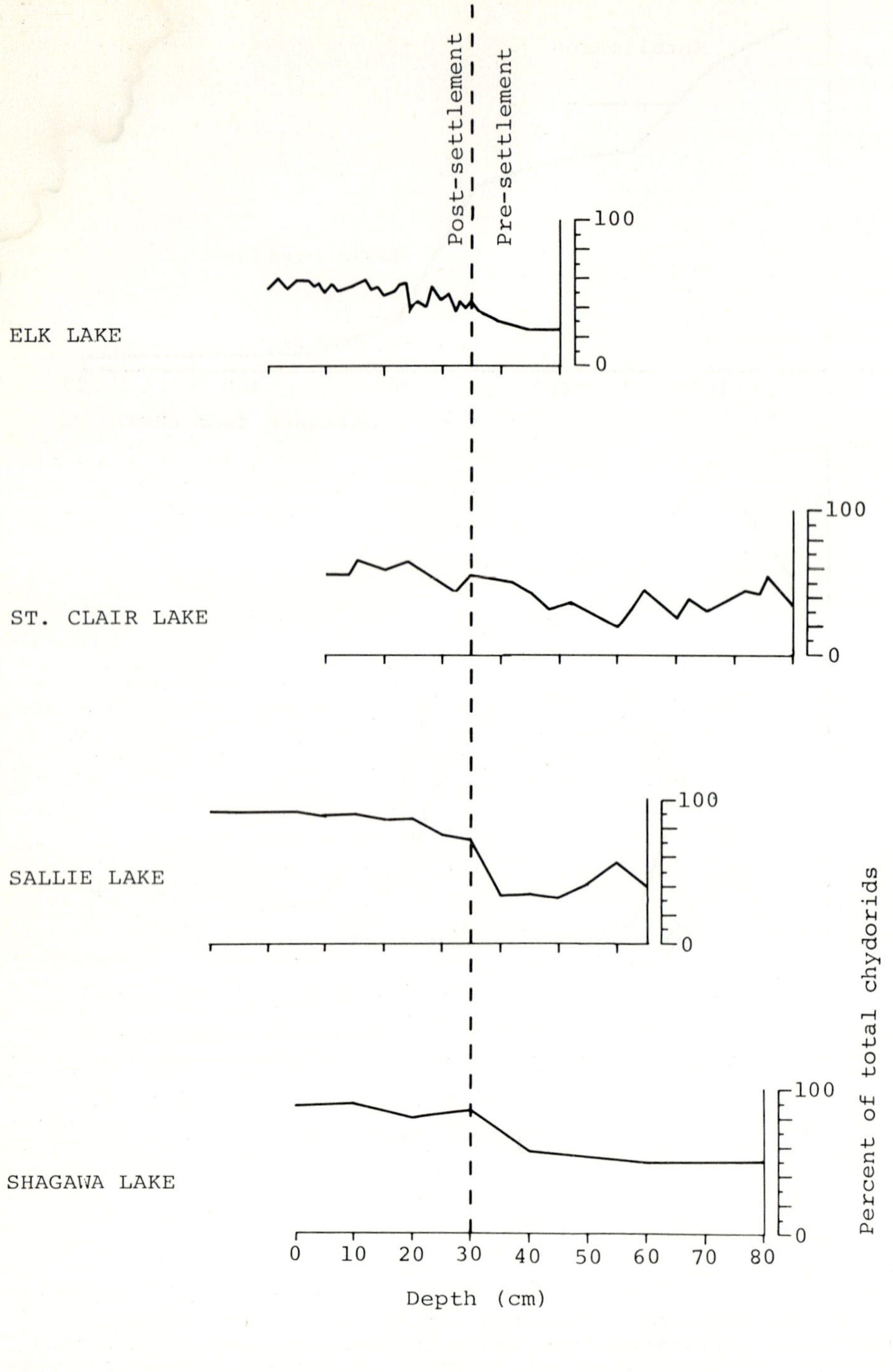

Figure 4.2 The response of *Chydorus sphaericus* to cultural disturbance in four Minnesota lakes. The settlement horizon is defined by the dramatic increase in pollen of *Ambrosia* a common plant of disturbed ground. (Data from Bradbury and Megard 1972; Birks et al. 1976).

America. Both of these normally littoral species become planktonic and display peak abundance in highly polluted European lakes (Whiteside 1970; Hogan 1977). In particular, *Chydorus sphaericus* is known to become planktonic in productive lakes by attachment to rafts of blue-green algae.

The recent cultural eutrophication of temperate lakes presents a unique opportunity to test the sensitivity of individual species of littoral Cladocera to changing lake productivity and to see if such responses can be detected from the sediment record. Shagawa Lake in northeastern Minnesota is extremely eutrophic as a result of logging, iron mining, and most recently an expanded tourist trade. The sedimentary record for the past 100 years at Shagawa Lake (Bradbury and Megard 1972) reveals a dramatic percentage increase of *Chydorus sphaericus* concurrent with a general reduction in littoral Cladocera during the postsettlement period (figure 4.2). The stratigraphic increase of *Chydorus sphaericus* is consistent with the known history of the basin and is supported by sediment profiles for diatoms (Bradbury and Megard 1972) and chlorophyll pigments (Gorham and Sanger 1976). Birks et al. (1976) note a similar increase of *Chydorus sphaericus* associated with land clearance and municipal sewage (St. Clair Lake and Lake Sallie) and catchment logging (Elk Lake) in northwestern Minnesota (Figure 4.2). These investigations demonstrate the sensitivity of lake ecosystems to catchment disturbances and add credibility to the use of littoral Cladocera in delineating late- and post-glacial trends in lacustrine productivity.

In addition to the response of individual species, changes in cladoceran community structure are valuable in reconstructing the response of littoral Cladocera to environmental fluctuations. Goulden (1966b) reported a transition in the community structure of littoral cladocerans from the "Type IV" distribution (superabundance of a few species, with all others rare) in the initial colonizing assemblage to a MacArthur "Type I" distribution (no one species excessively abundant) during the later developmental history of a small Guatemalan pond (figure 4.3). Similar establishment of cladoceran community equilibrium following an unstable colonization phase is recorded for a number of lakes in Europe (Goulden 1964, 1969), Japan (Tsukada 1972) and North America (Crisman and Whitehead 1977).

Once an equilibrium community structure is established, major deviations from a Type-I distribution may be associated with alterations of physical habitat or lake trophic state. Tsukada (1967, 1972) recorded a sudden but transitory decline in cladoceran community equitability and diversity associated with habitat alteration from two volcanic ash layers at Lake Nojiri, Japan (figure 4.4). A significant negative correlation (-.932) between species diversity of littoral Cladocera and lake productivity in Danish lakes is attributed in part to a limitation of macrophyte growth by algal reduction of light transmission associated with increased productivity (Whiteside and Harmsworth 1967). Similar relationships may be identified from the paleolimnological record at Berry Pond, Massachusetts (Whitehead et al. 1973; Crisman and Whitehead 1977). Following an early postglacial change from "Type IV" to "Type 1" community structure (8800 BP), chydorid diversity remains relatively constant until about 4800 BP (pollen zone C1-C2 transition)(figure 4.5). Subsequently chydorid diversity declines sharply and remains low for the remainder of the postglacial. This fluctuation in community structure coincides with an increase in lacustrine productivity initiated by removal of hemlock as a major catchment species (4800 BP) and sustained as the hypolimnion was obliterated by basin shallowing (3500 BP) and as macrophytes increased (3000 BP). Although productivity peaks cannot be quantified by cladoceran diversity indices, parameters of community structure may be more sensitive to environmental stress than individual species.

Given the sensitivity of littoral cladoceran communities to fluctuations in catchment vegetation, water chemistry and lake trophic state, the next question concerns the value of Cladocera as climatic indicators. The most detailed cladoceran biogeographical survey is that of DeCosta (1964), who examined littoral cladoceran assemblages of fifty-six lakes situated along a 2000 km transect in the Mississippi River Valley

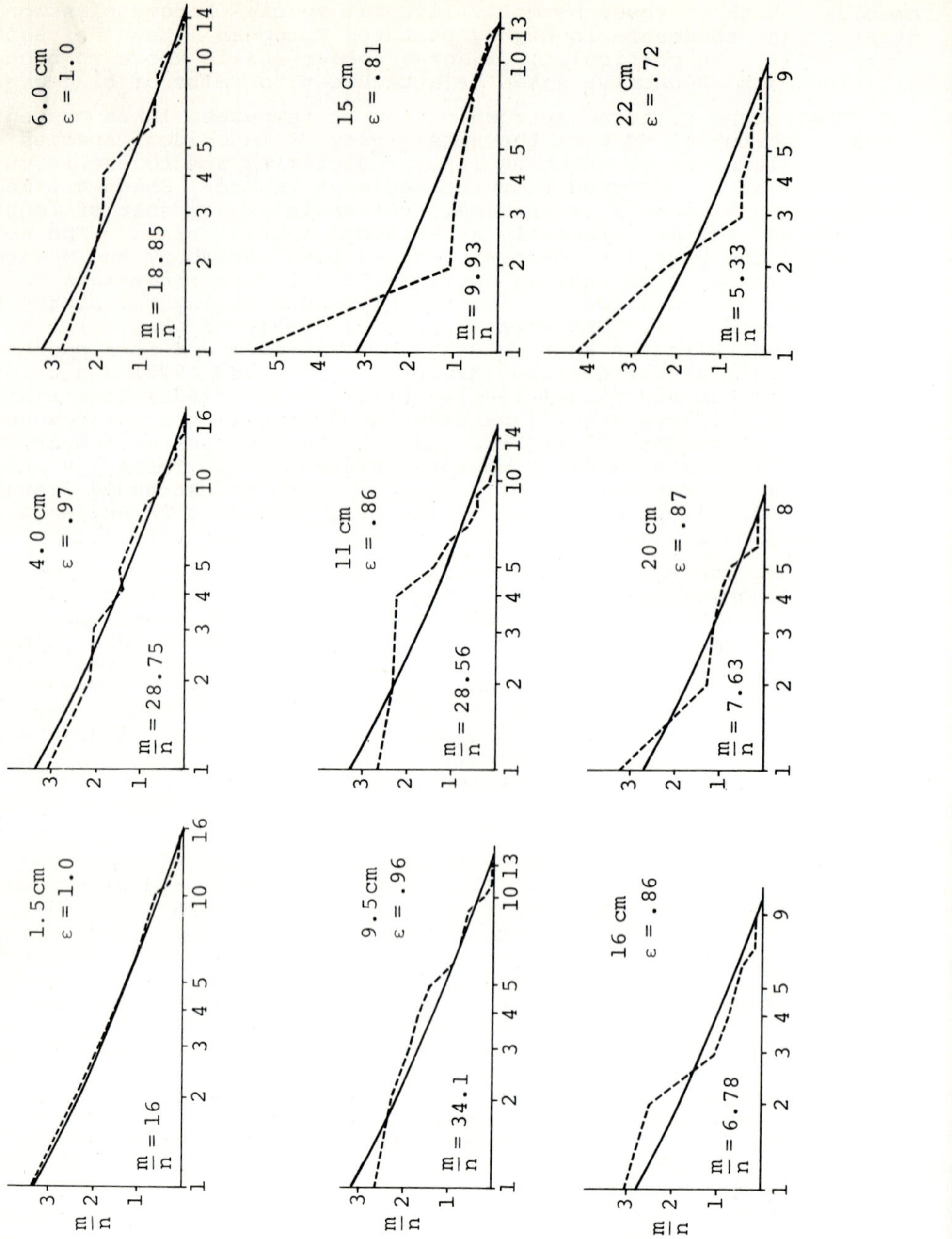

Figure 4.3 Comparison of observed abundances of fossils of littoral Cladocera (dashed line) from select levels of a core from Santa Anna Vieja, Guatemala with the theoretical relative abundance (solid line) predicted by the MacArthur Type I distribution. The establishment of Mayan agriculture in the catchment (15 cm) momentarily interrupted the trend of the cladoceran assemblage to gradually approach the theoretical population distribution of Macarthur. (m = number of individuals; n = number of species; e = equitability component of diversity) (From Goulden 1966).

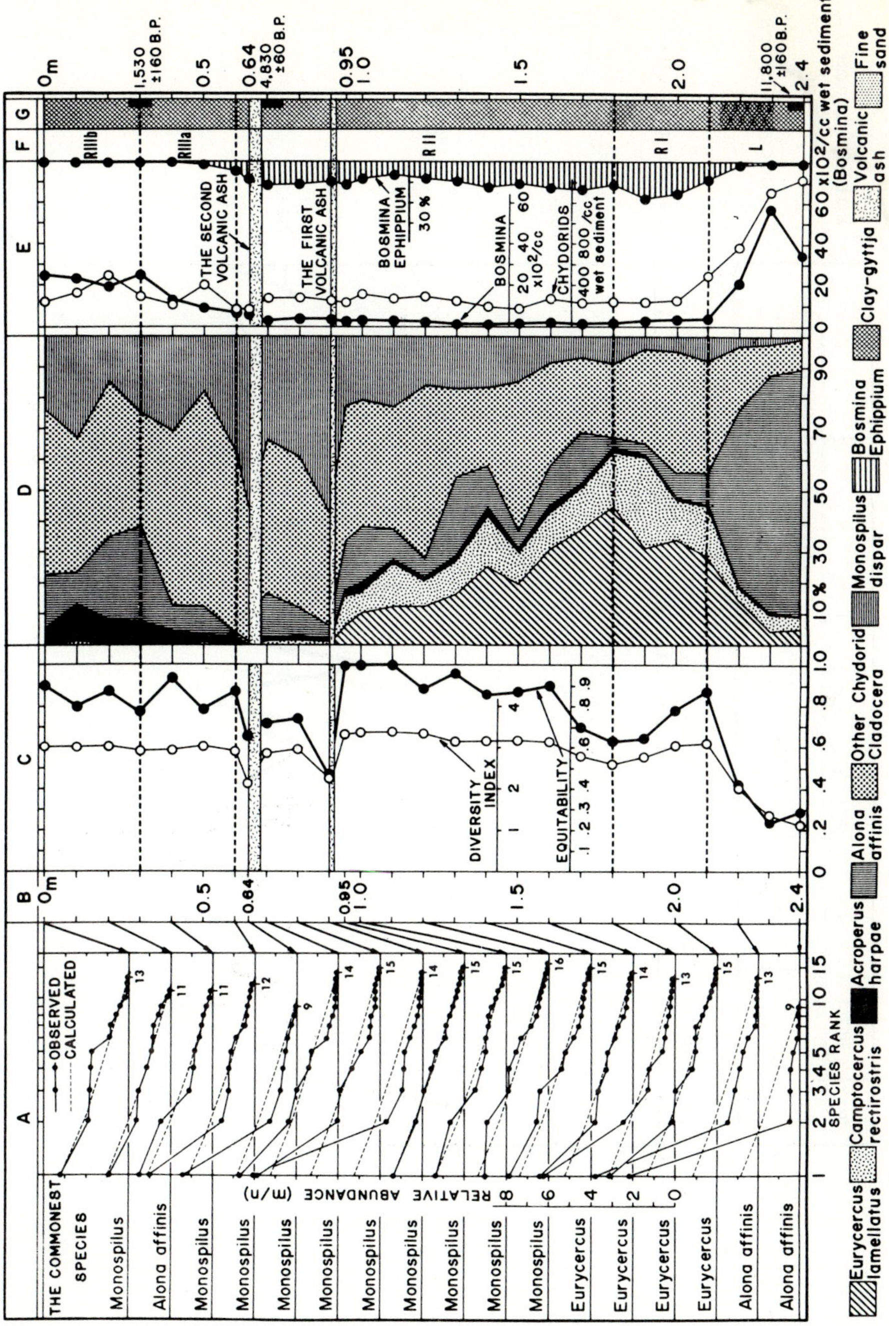

Figure 4.4 Summary diagram of the history of Lake Nojiri, Japan indicating the relative abundance and rank order of chydorid species (A): depth (B); species diversity and equitability (C); percentage diagram for chydorids (D); absolute abundance of chydorids, *Bosmina* and *Bosmina* ephippia (E); pollen zones (F); and sediment composition (G). Note the major deviation of observed community structure from the theoretical and the sharp decline in both diversity and equitability associated with the two volcanic ash falls. (From Tsukada 1972).

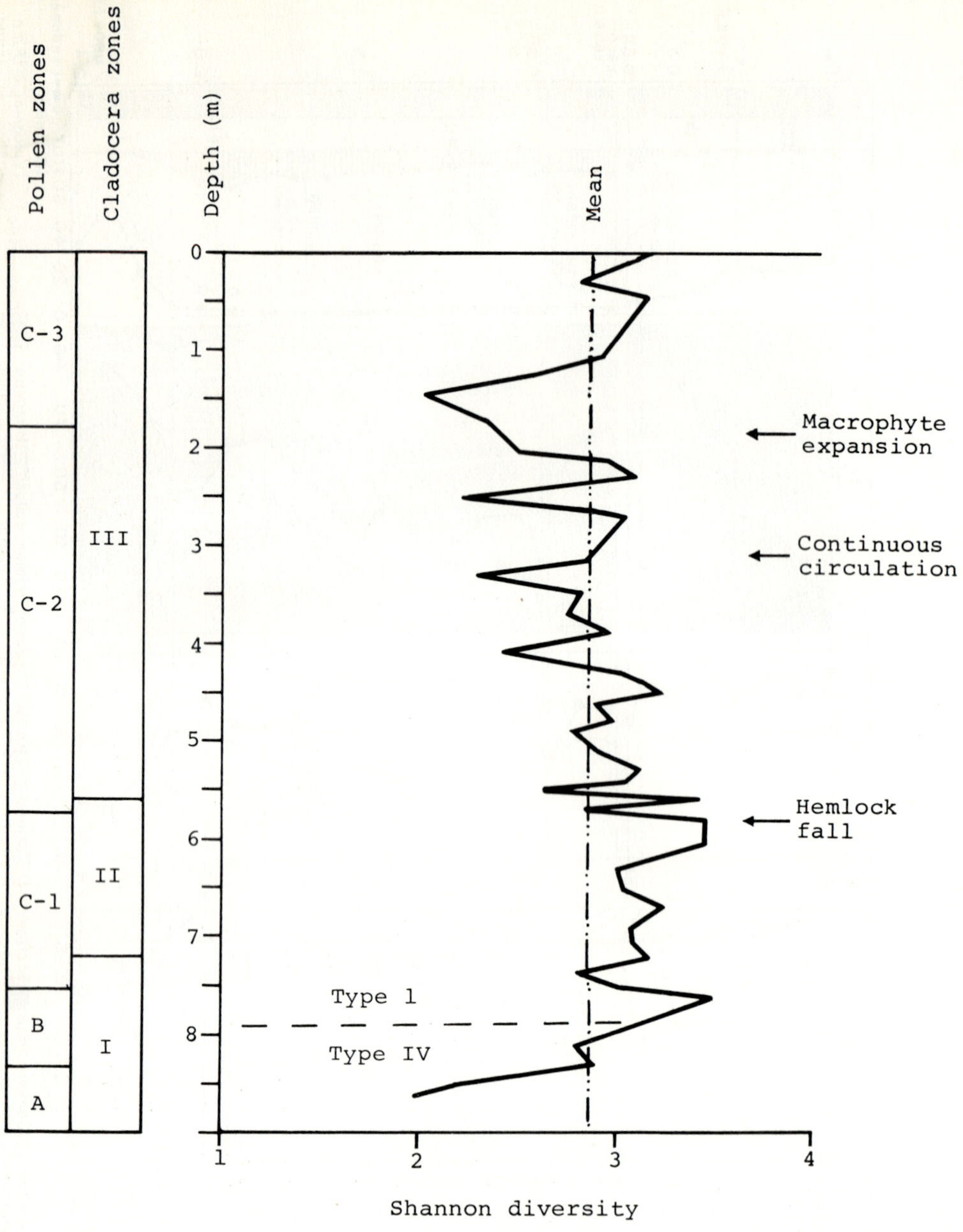

Figure 4.5 Late- and postglacial fluctuations in the diversity of the chydorid cladoceran assemblage at Berry Pond, Massachusetts, and the transition from a Type IV to Type I distribution. Postglacial productivity-increase associated with the removal of hemlock as an important catchment species. Subsequent destruction of the hypolimnion from basin infilling helped sustain high production levels and contributed to a general expansion of aquatic macrophytes. (Data from Whitehead and Crisman 1978).

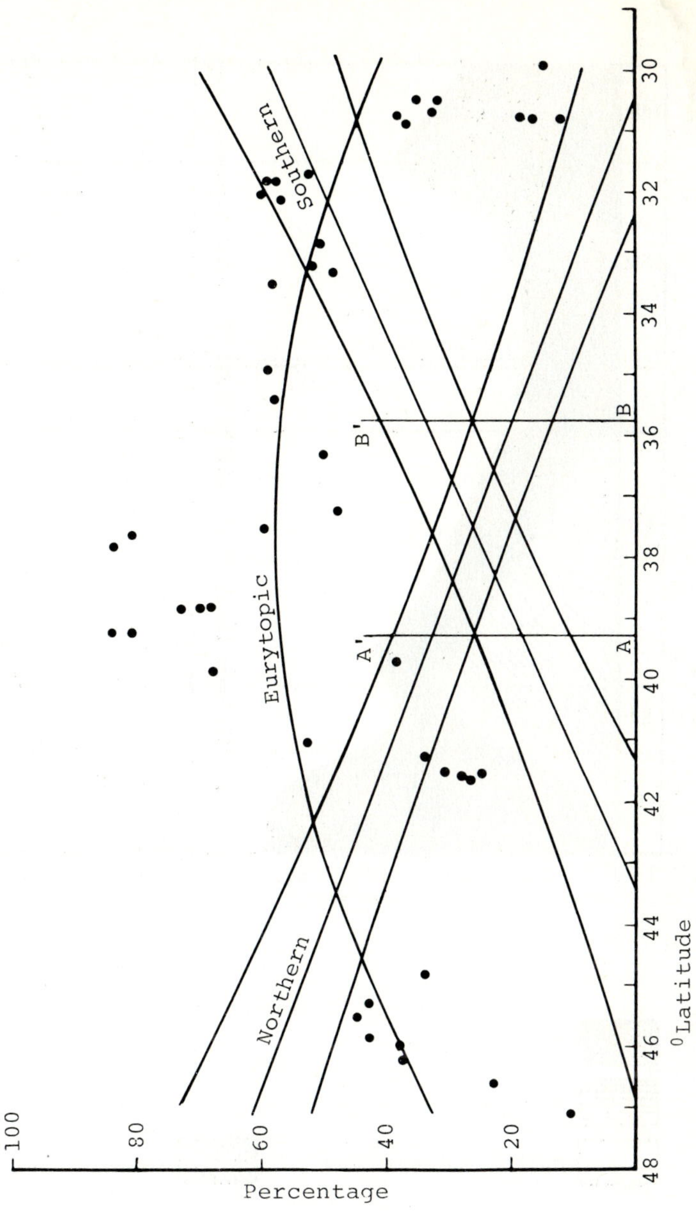

Figure 4.6 The latitudinal distribution within the Mississippi River Valley of northern, southern, and eurytopic assemblages of chydorids. Regressions with 95% confidence intervals are given for the northern and southern groups. (From DeCosta 1964).

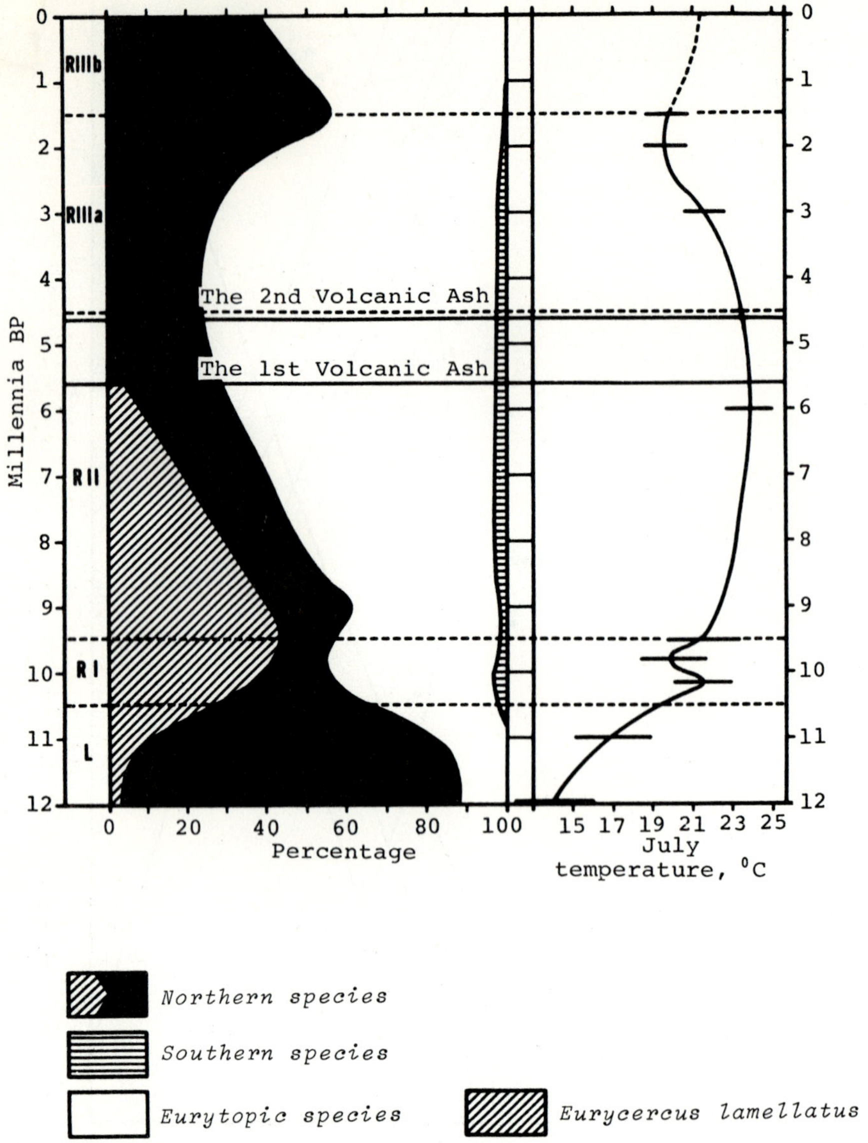

Figure 4.7 Variations of northern, southern, and eurytopic chydorid species during the history of Lake Nojiri, Japan. Estimates of fluctuations in mean July temperature, ^0C are based on interpretations of pollen data from several sites in central Japan. (From Tsukada 1972).

from Minnesota to Louisiana. DeCosta assigned cladoceran species to affinity groups and was able to delineate a succession of community dominance from northern to eurytopic to southern species with decreasing latitude (figure 4.6). Following the classification of DeCosta, Tsukada (1972) attributed the postglacial establishment of southern species and associated reduction in northern species at Lake Nodjiri, Japan, to climatic warming during the hypsithermal (figure 4.7). Similarly, Megard (1967), in an investigation of the cladoceran history of Lake Zeribar, Iran, documented a replacement of northern species by southern and cosmopolitan elements associated with climatic amelioration between 12,000 and 5000 years ago.

Species reported by Poulsen (1940) as characterizing arctic and sub-arctic ponds (*Acroperus harpae, Alona affinis, Alonella excisa* and *Chydorus sphaericus*) dominate late-glacial cladoceran assemblages in temperate Europe (Frey 1962; Goulden 1964; Harmsworth 1968) and North America (Crisman 1977). The use of these assemblages in direct climatic interpretations, however, is complicated by (a) the continued importance of these species in modern temperate cladoceran assemblages and (b) the apparent inability to separate climatic responses from those resulting from changes in lake conditions (productivity, chemistry, substrate). Paleolimnological investigations at the temperate-tropical zonal transition offer greater potential for delineating cladoceran climatic responses than more temperate sites. This problem is being approached by examination of the cladoceran history of Lake Annie, a site near the temperate-tropical transition in southern Florida (Crisman unpublished).

Although relatively few species are preserved in sediments, planktonic cladocerans should not be disregarded in paleolimnological investigations. The ratio of total abundance of planktonic-to-littoral cladocerans may be useful in detecting past fluctuations in lake level. Alhonen (1970) correlates increased importance of littoral relative to planktonic Cladocera in cores from four Finnish lakes with a regional lowering of lake level during the drier Subboreal period (figure 4.8). A similar interpretation has been applied to Subboreal changes in cladoceran assemblages in two Danish lakes (Whiteside 1970). According to this interpretation a lowering of water level would result in a habitat reduction for planktonic Cladocera through a diminishing of limnetic area. Concurrently, expansion of the areal extent of macrophytes would favour expansion of the littoral cladoceran community.

Additional factors may have a pronounced influence on observered planktonic-to-littoral ratios. As previously mentioned, Mueller (1964) documents a concentration zone for planktonic cladoceran remains corresponding to the uppermost hypolimnion. Downward displacement of this concentration zone would be expected during a lowering of lake level. If the concentration zone reached the coring site, the increase in the remains of planktonic cladocerans could be falsely interpreted as representing a true population increase rather than a general reduction in limnetic area.

Increased importance of planktonic relative to littoral cladocerans is characteristic of periods of increased productivity at North Pond and Berry Pond, Massachusetts (Crisman and Whitehead 1977). The positive response of planktonic Cladocera to increasing productivity is attributed to (a) a decline in littoral cladoceran habitat (macrophytes) as a result of photic-zone reduction by algal shading and (b) an increase of algae as a food source (figure 4.9). These investigations demonstrate the necessity of dealing with alternative explanations before interpreting stratigraphic variations in planktonic-to-littoral ratios strictly on the basis of water-level fluctuations.

Major adjustments within the planktonic cladoceran assemblage associated with increased productivity can be recognized in the sediment record. Recent shifts in planktonic Cladocera from larger *Daphnia* species to small *Daphnia* and *Bosmina longirostris* associated with recent cultural eutrophication have been documented for Lake Ontario (McNaught and Buzzard 1973) and Lake Michigan (Roth and Stewart 1973) and in paleolimnological investigations at Frain's Lake, Michigan (Kerfoot 1974) and Berry Pond,

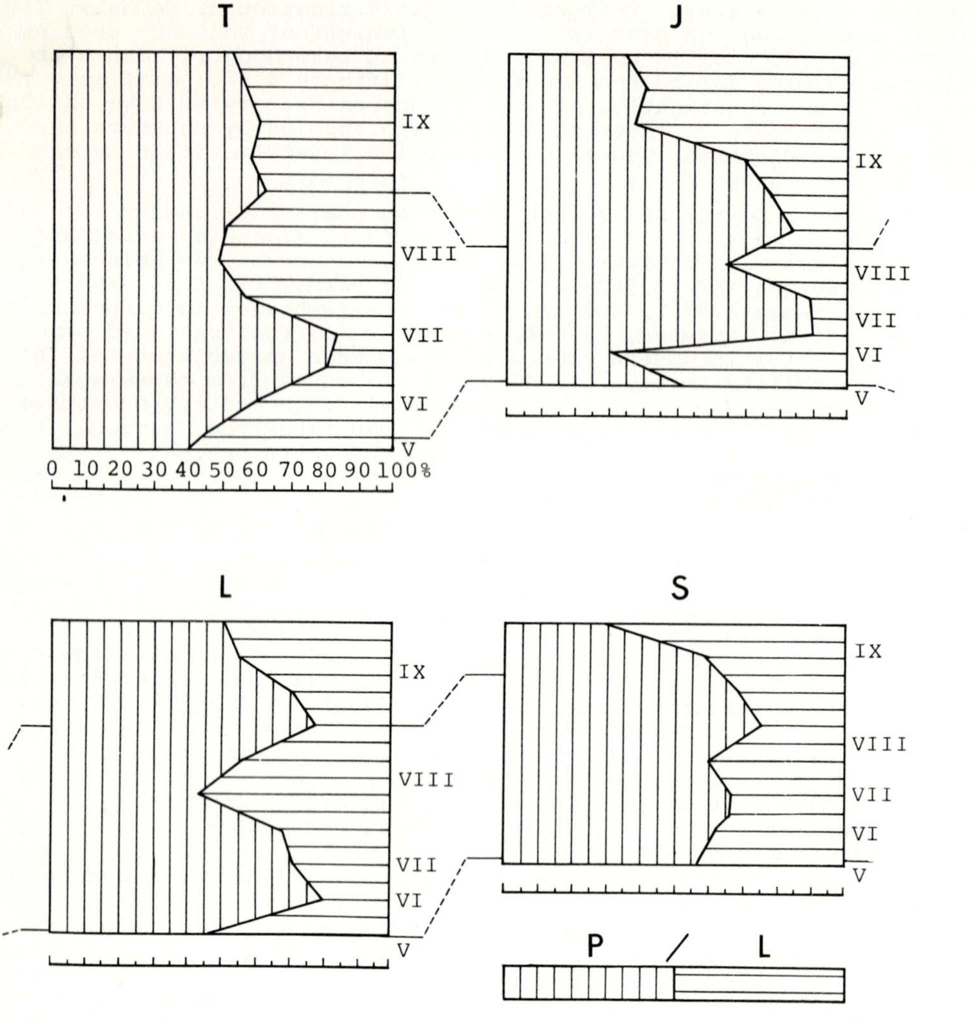

Figure 4.8 Stratigraphic variations in the importance of planktonic (P) and littoral (L) Cladocera in cores from four Finnish lakes: Telkko (T); Järvenhylänjärvi (J); Leppäsjärvi (L); Särkijärvi (S). All lakes display an increase in littoral Cladocera during the drier Sub-Boreal period (VIII). (From Alhonen 1970).

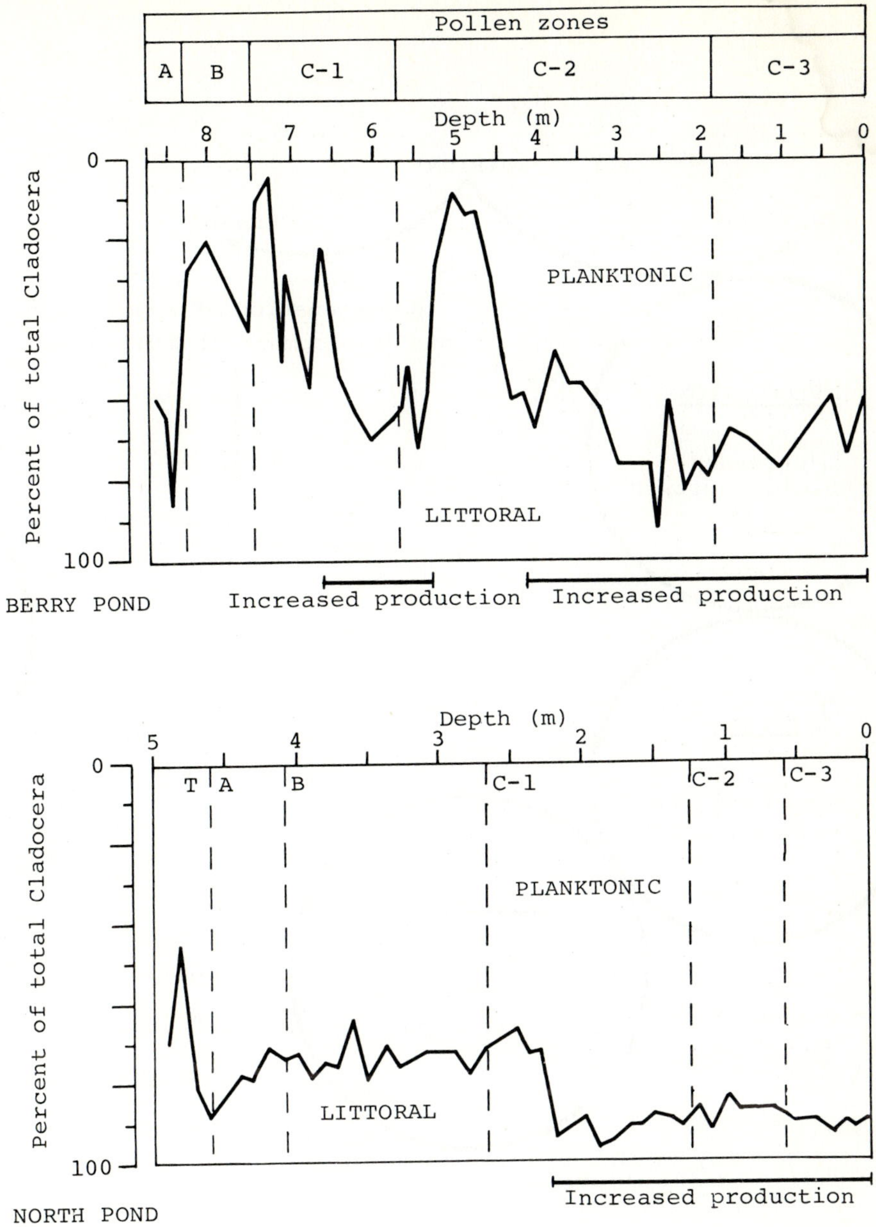

Figure 4.9 Late- and postglacial variations in the importance of planktonic and littoral Cladocera for Berry Pond and North Pond, Massachusetts. Both sites display maximal representation of the planktonic assemblage during periods of increased production. (Data from Crisman and Whitehead 1978).

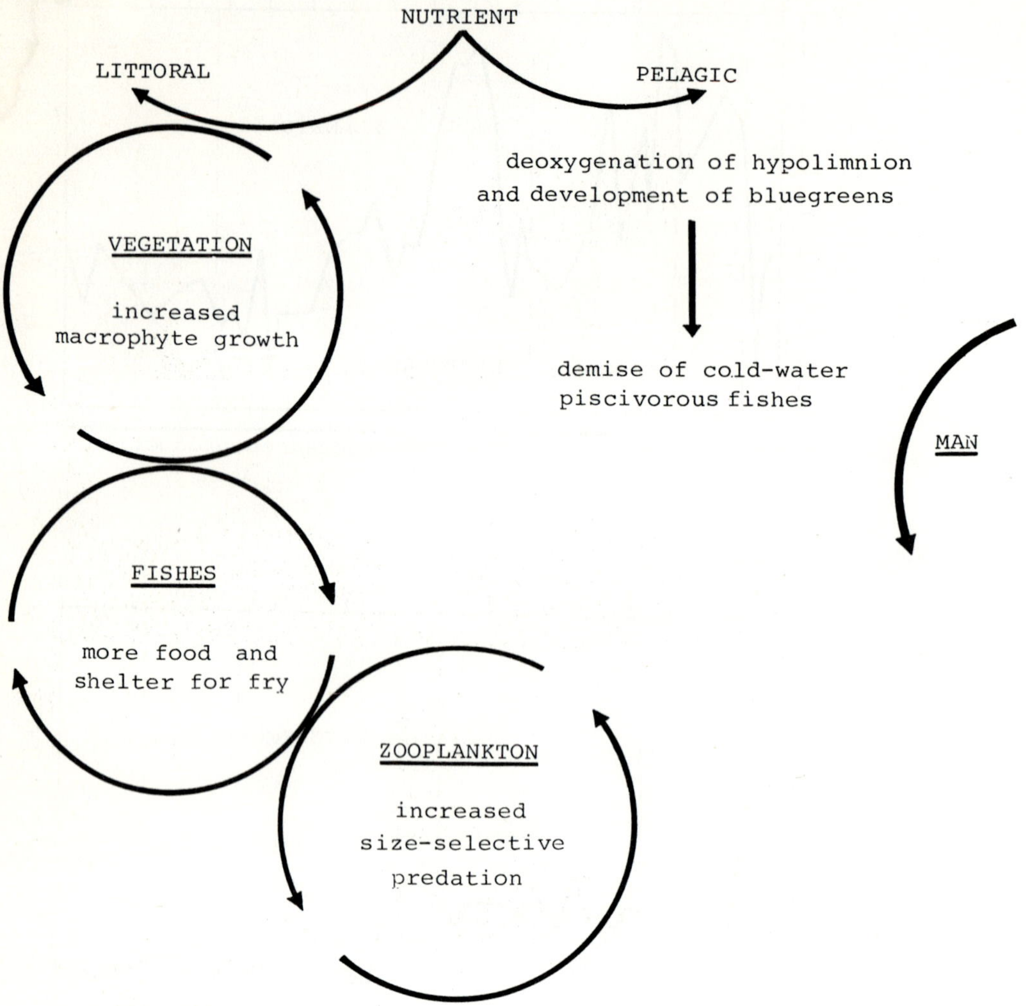

Figure 4.10 Summary diagram of the proposed effects of cultural eutrophication at Frain's Lake, Michigan. (From Kerfoot 1974).

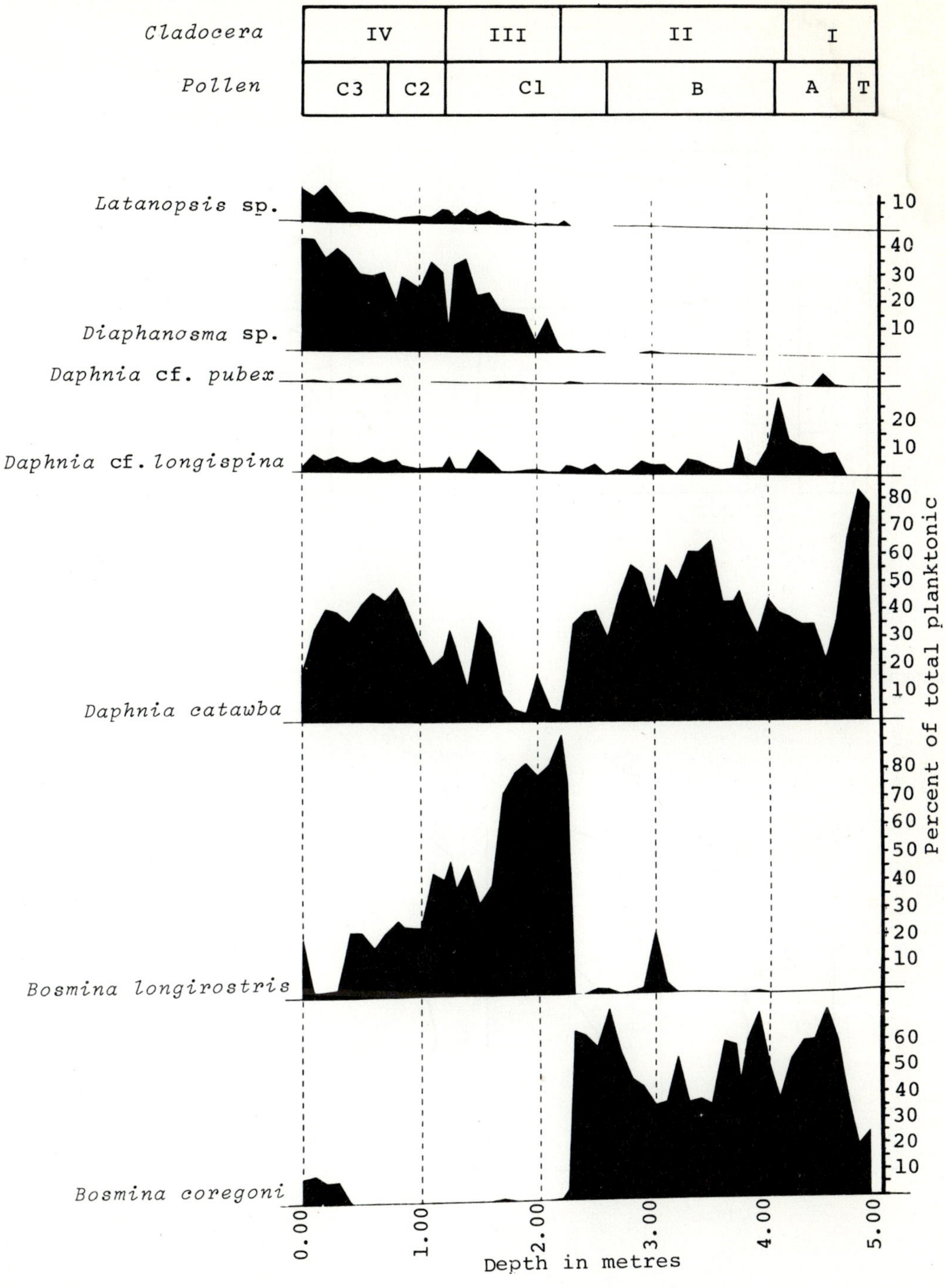

Figure 4.11 Percentage diagram depicting late- and postglacial fluctuations within the planktonic cladoceran assemblage at North Pond, Massachusetts. Cladoceran zones III and IV represent a period of increased productivity. (From Crisman 1977).

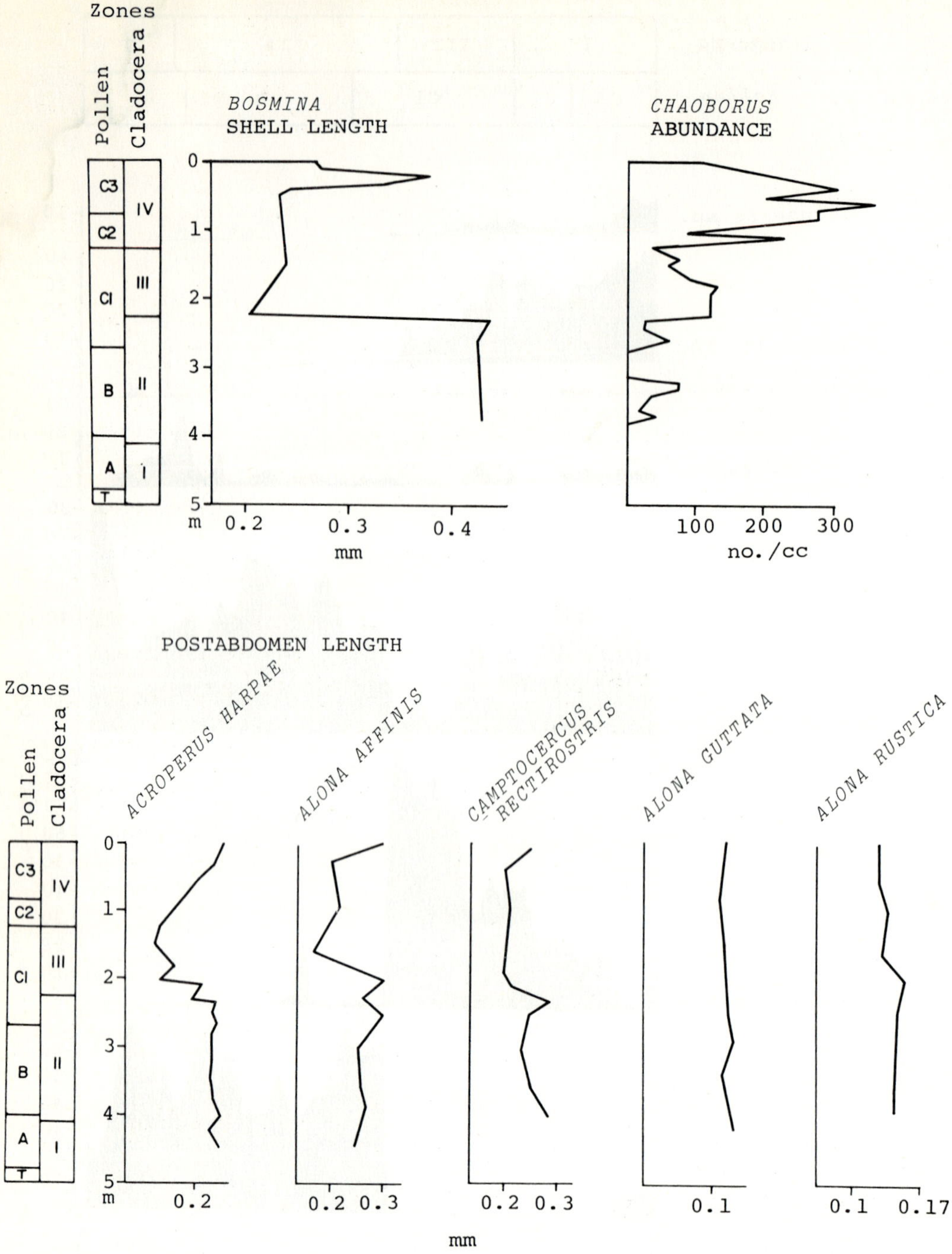

Figure 4.12 Stratigraphic variations in the size of shells of the planktonic cladoceran *Bosmina* and postabdomens of five littoral cladoceran species at North Pond, Massachusetts. In addition, fluctuations in the abundance of *Chaoborus*, a major invertebrate predator on Cladocera, are given. Cladoceran zones III and IV represent a period of increased producitivty. (From Crisman and Whitehead 1978).

Massachusetts (Crisman unpublished). Similarly the replacement of *Bosmina coregoni* by the smaller *B. longirostris* during periods of increased lacustrine productivity commonly has been observed in paleolimnological investigations (Deevey 1942, 1969; Goulden 1964; Harmsworth 1968; Geyh et al. 1974; Crisman 1977; Hogan 1977; Hofman 1977) and is substantiated by modern eutrophication studies (Minter 1938; Edmonson et al. 1956; Wells 1960).

Recent investigations suggest that changes in predation pressure may be the major factor controlling the response of planktonic cladocerans to increasing lacustrine productivity. Deevey (1969) postulates that an increase in the abundance of the predatory midge *Chaoborus* was responsible in part for the postglacial replacement of *Bosmina coregoni longispina* and *B. tubiscen* by the smaller *B. longirostris* at Roger's Lake, Connecticut. *Bosmina tubicen* joins the fauna with *B. longirostris* only following the recent cultural eutrophication of the basin. Deevey also suggests that the observed bosminid succession could be controlled by population fluctuations of the alewife (*Alosa pseudoharengus*) an important fish predator on zooplankton. Unfortunately, this is difficult to document because of the poor preservation of fish remains in lake sediments.

Strongly influenced by the arguments for size-selective predation (Brooks and Dodson 1965), Kerfoot (1974) attributes the post-settlement shift from *Daphnia* to *Bosmina* at Frain's Lake, Michigan, to increased fish predation on larger zooplankton (*Daphnia*) (figure 4.10). According to this hypothesis, increased nutrient input to Frain's Lake following watershed deforestation by Europeans resulted in an expansion of aquatic macrophytes. Macrophyte expansion favoured the survival of fry of planktivorous fishes, the principal vertebrate predators on Cladocera.

The survival rate of planktivorous fishes was further enhanced by concurrent events in the hypolimnion. Cultural eutrophication also resulted in a major phytoplankton increase and the establishment of an anoxic hypolimnion. The population of cold-water piscivorous fishes, which previously lived in profundal waters, was restricted following hypolimnetic deoxygenation, thereby reducing predation pressure on the fry of planktivorous fishes. Finally, the resulting increased abundance of planktivorous fishes intensified size-selective predation on larger zooplankton and eliminated the larger species of *Daphnia* from the lake.

Vertebrate and invertebrate predation may also strongly influence the species composition, size-class distribution, and population levels of littoral cladoceran assemblages. Goulden (1971) demonstrates convincingly that the midsummer population reduction for chydorid Cladocera at Lake Lacawac, Pennsylvania, is largely the result of increased predation pressure from tanypodine midges. The influence of fish predation on cladoceran populations was shown to be of minor importance.

Recently, the influence of predation on the littoral community has been documented from the paleolimnological record (Crisman and Whitehead 1977). The postglacial replacement of the planktonic cladoceran *Bosmina coregoni* by *B. longirostris* at North Pond, Massachusetts (figure 4.11), is associated with a major decline in *Bosmina* size and an increase in the predatory midge *Chaoborus* (figure 4.12). *Bosmina* shell size increases in the most recent sediments together with the reappearance of *B. coregoni* and a reduction in *Chaoborus*. Within the littoral cladoceran assemblage, the mean size of three large species (*Acroperus harpae, Alona affinis, Camptocercus rectirostris*) parallels the trend observed for *Bosmina*, while the mean size of smaller cladoceran species (*Alona guttata, A. rustica*) displays only minor variation (figure 4.12). The size reduction of large littoral species and the elimination of *B. coregoni* from the planktonic assemblage are attributed to an increase in size-selective predation during periods of increased productivity. These investigations stress the importance of considering the possible effects of predation before environmental reconstructions based solely on cladoceran (littoral and planktonic) assemblages are made.

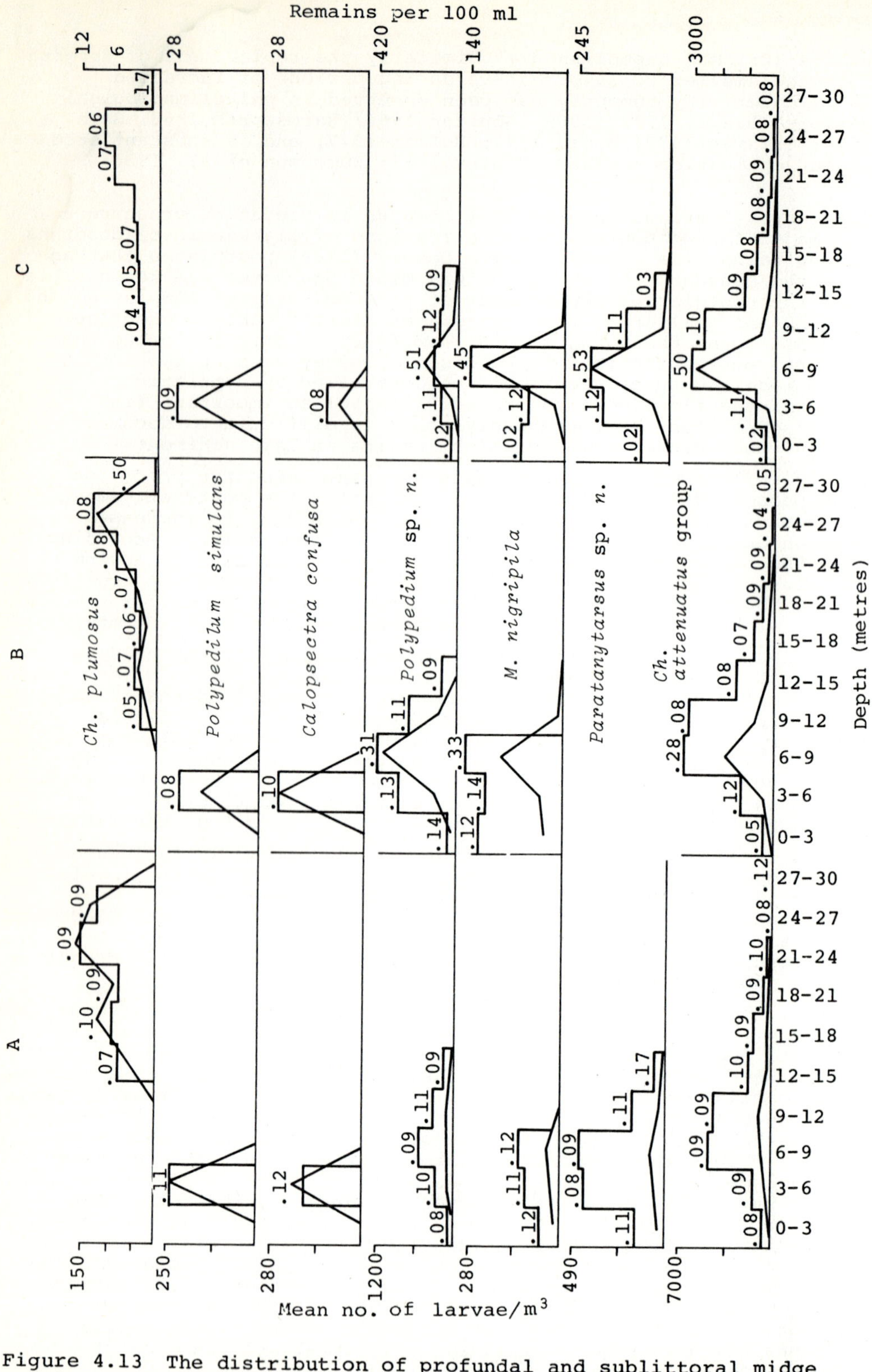

Figure 4.13 The distribution of profundal and sublittoral midge species along three transects (A, B, C) from shore to deep-water at Pretty Lake, Indiana showing: (1) annual means of larval numbers (histograms); (2) number of remains per 100 ml of sediment (line); and (3) the ratio between 1 and 2 (decimals). (From Iovino, 1975).

MIDGES

Head capsules, mandibles and labia of midges (larvae of aquatic Diptera) are the most numerous insect remains in lake sediments. Of the taxa frequently encountered (Chironomidae, Ceratopogonidae, Orthocladinae, Diamesinae, Tanypodinae, Chaoboridae), remains of the Chironomidae tribes (Tanytarsini and Chironomini) are the most common. With the exception of the planktonic Chaoboridae, the majority of midges are associated with macrophytes on sediments. Excellent reviews of paleolimnological analyses of midge remains are presented by Frey (1964, 1976), Hofmann (1971a) and Stahl (1969).

Although midges are widely distributed within lakes, individual species often display restricted depth distributions. Iovino (1975) investigated the importance of intra-basin lateral transport of midge remains by comparing living and subfossil assemblages isolated from surficial sediments along depth profiles at Pretty Lake, Indiana. His results indicate that lateral transport of remains may be important in that portion of the littoral zone above the base of the epilimnion (effective wave base). Remains of littoral midges are often concentrated at the epilimnion-metalimnion boundary, but below this point subfossil assemblages reflect the distribution of living species, with little evidence of contamination from littoral midge remains (figure 4.13). The recent midge investigations of Stark (1976) at Elk Lake, Minnesota, lead to the same conclusions.

Midge species display marked variation in their oxygen requirements. This fact together with general hypolimnetic deoxygenation with increased productivity serves as the basis for a trophic classification of European lakes based on profundal midge assemblages (Thienemann 1913, 1922). According to this sytem low productivity lakes (oligotrophic) are dominated by "*Tanytarsus*", high productivity lakes (eutrophic) by *Chironomus*, and lakes of intermediate productivity by *Stictochironomus* and *Sergentia*. Brundin (1949, 1956) is responsible for later refinements of this typology. Expansion of the European classification to include North American midge species is based largely on the work of Deevey (1941) and Saether (1975).

The first quantitative paleolimnological midge investigation was that of Deevey (1942), who records a *Tanytarsus*-to-*Chironomus* succession during the postglacial eutrophication and hypolimnetic deoxygenation of Linsley Pond, Connecticut.

Stahl (1959) documents a postglacial succession in the profundal midge assemblage at Myers Lake, Indiana, from *Sergentia* to *Chironomus*. In addition, the planktonic midge *Chaoborus* became important only after the decline in *Sergentia*. These midge changes are attributed to gradual morphometric eutrophication of an initially mesotrophic lake. Reduction of hypolimnetic volume (40%) by basin infilling gradually reduced hypolimnion oxygen concentrations (moderately low initially) to such low levels that *Sergentia* was replaced by *Chironomus*.

In contrast to interpretations of midge remains in stratified lakes discussed above, reconstruction of trophic histories of shallow unstratified lakes based solely on midge oxygen requirements are questionable. The midge fauna of Dead Man Lake, a shallow (maximum depth 1 m) eutrophic lake in New Mexico, is presently dominated by Tanytarsini (Megard, 1964). These normally oligotrophic indicators are able to survive because the absence of stratification permits replenishment of bottom oxygen concentrations. Even though the earliest lacustrine sediments were deposited 16 m below a possible outlet, Megard (1964) contends that Dead Man Lake has always been shallow, because Tanytarsini dominate the midge fauna throughout most of the postglacial, and planktonic cladocera (*Bosmina* and *Daphnia*) are uncommon.

In addition to changes in oxygen concentrations, midges are sensitive to sediment changes, especially in organic content. Bryce (1962) attributes the gradual reduction and eventual replacement of *Chironomus* by Tanytarsini during the Boreal period at Malham Tarn Moss, England, to a

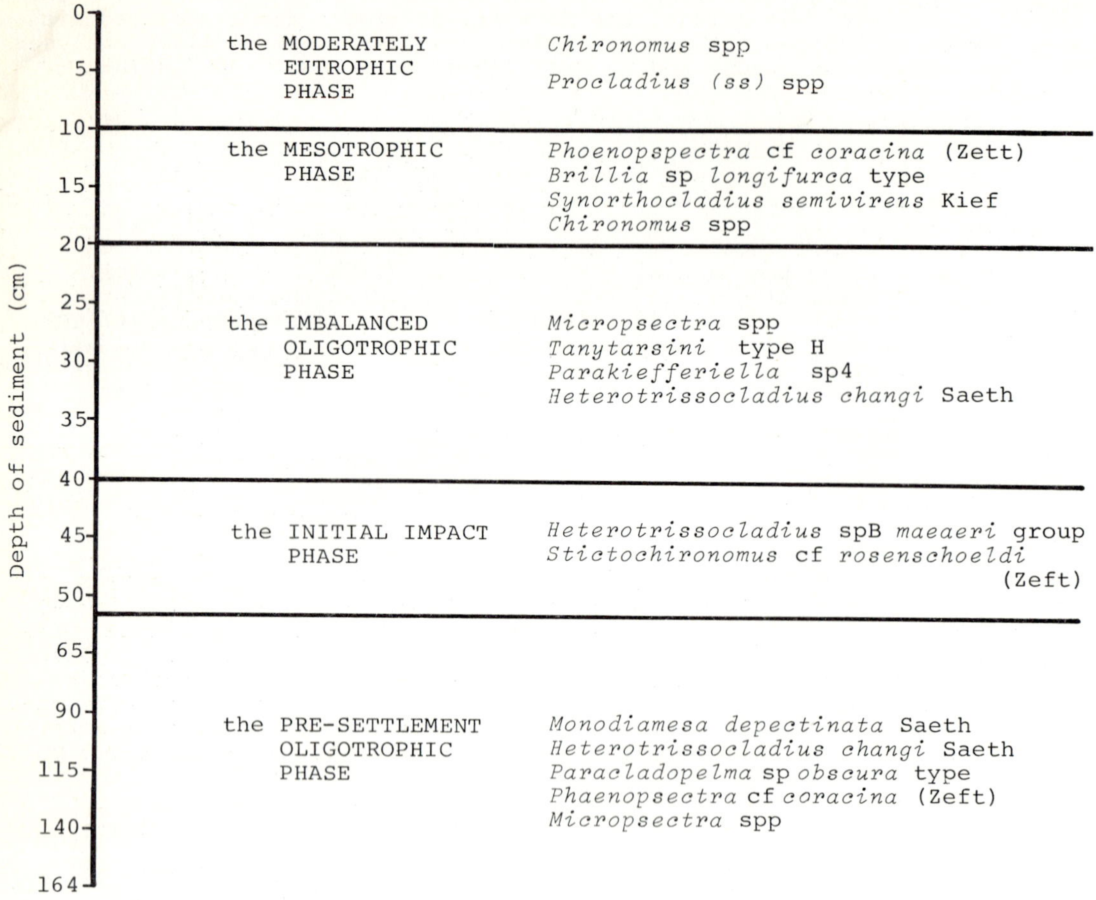

Figure 4.14 Response of the midge community of the Bay of Quinte, Lake Ontario to watershed disturbance by European settlers and associated basin eutrophication. (From Warwick 1975).

substrate change from clay to calcareous sediments during a reduction in water depth. Similarly, reduction in *Chironomus* in shallow-water cores from Elk Lake, Minnesota, is attributed to increased silt and sand in the substrate (Stark 1976). Although no midge taxa were eliminated following deposition of two volcanic ash layers at Lake Nojiri, Japan, Tsukada (1972) does note a pronounced reduction in the accumulation rate of midge remains.

Warwick (1975) documents four phases of midge succession in the Bay of Quinte, Lake Ontario, following European colonization in the catchment (figure 4.14). The earliest post-settlement period is marked by the deposition of increased clay and decreased organic content concurrent with the establishment of a midge fauna more strongly oligotrophic than the pre-settlement oligotrophic community. The remainder of the profile records a midge succession from an imbalanced oligotrophic to a mesotrophic to an eutrophic assemblage dominated by *Chironomus*. Sediment organic content gradually increased throughout this period of cultural eutrophication, even though inorganic influx into the basin remained high. The establishment of the early post-settlement oligotrophic midge assemblage is attributed to a major reduction in sediment organic content resulting from increased watershed erosion following land clearance, whereas later midge changes are a response to progressive hypolimnetic deoxygenation associated with increased cultural eutrophication.

Two additional investigations by Roback (1970) and Lawrenz (1975) have demonstrated major midge responses to lake-level fluctuations. The complete disappearance of the chironomid fauna during part of the history of Lago di Monterosi, Italy, is ascribed to a major reduction in lake level (Roback 1970), but the midge succession described by Lawrenz (1975) at Green Lake, Michigan, is more complicated. *Tanytarsus* and *Heterotrissocladius* dominated the oligotrophic midge assemblage colonizing Green Lake during the late-glacial. *Phaenopsectra* gradually replaced *Heterotrissocladius* in the profundal assemblage during the late-glacial as sediment organic content increased, but the transition was interrupted by a lowering of lake level associated with the Valders glacial advance (3 km to the west). The sublittoral assemblage (*Chironomus anthracinus*, *Glyptotendipes*) was short-lived and was soon replaced by a more profundal oligotrophic midge assemblage. The subsequent establishment of a zone dominated by sublittoral midges (*Chironomus anthracinus*, *Microtendipes*) coincides with increased oak and the replacement of spruce by pine as the dominant pollen taxa and is attributed to a lake lowering during a climatically warmer and drier period. Lawrenz (1975) speculates that alternatively the shallowing could be the result of a lowering of the ground-water table during a low phase of nearby (20 km) Lake Michigan (9000 to 4000 BP). The final 4000 years of lake history is characterized by an oligotrophic profundal midge assemblage.

Perhaps the most detailed investigation of fossil midges to date is that of Hofmann (1971b) on two shallow-water and two deep-water cores from Schohsee, Germany. The profundal midge assembles from the Younger Subarctic to the Subboreal was dominated by an oligotrophic *Tanytarsus lugens* community. Reduced profundal oxygen concentrations associated with hypolimnetic volume restriction accompanied a decline in lake level during the early Subboreal. During this period *Chironomus anthracinus* migrated into the profundal, and by the Subatlantic the present eutrophic *Chironomus* assemblage had completely replaced the oligotrophic *Tanytarsus lugens* assemblage.

In addition to productivity-related changes, Hofmann (1971b) was able to attribute a midge response to climatic change. The midge taxa recognized from the Younger Subarctic period all display northern or "boreoalpine" distribution today, but postglacial faunal additions have more temperate modern ranges. Finally the migrations of *Lauterbornia* into deeper cooler water during the Atlantic and of *Sergentia* during the Subboreal period are attributed to post-glacial climatic amelioration.

Unlike the Cladocera, structural adjustments within the midge community associated with environmental changes are poorly known. A notable exception is the investigation of Clair and Paterson (1976) on Front Lake, a coastal pond in New Brunswick. The midge community at this pond was destroyed by a salt-water intrusion during a storm in 1869. Re-colonization by midges was rapid. Twenty-two of the twenty-three midge species present before the storm returned. In addition, midge diversity displays a rapid rise of pre-disturbance values (figure 4.15). This rapid community adjustment to environmental stress is not unexpected given the dispersal advantage offered by a winged adult.

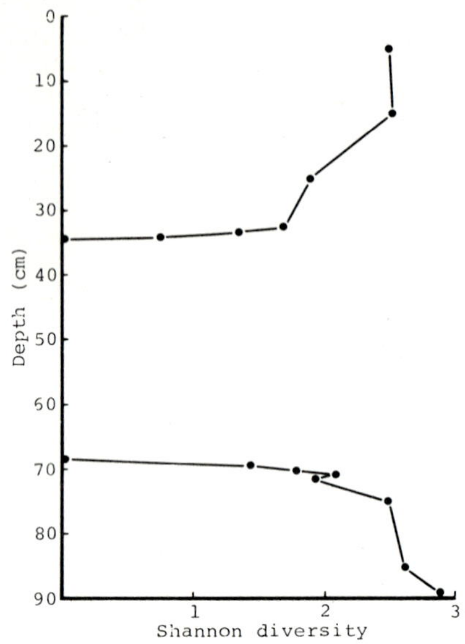

Figure 4.15 Fluctuations in the species diversity of the midge assemblage of Front Lake, New Brunswick in response to a salt-water intrusion during an 1869 storm. Rapid sedimentation associated with the storm is indicated by the interval between 67 and 34 cm. (From Clair and Paterson 1976).

The absence of data on midge community structure for long cores is a reflection of two major interpretational problems. The first problem concerns the difficulty in obtaining a statistically significant quantitative fossil sample. Because midge remains are often sparse in lake sediments, the usual procedure is to isolate and mount each remain individually. Consequently the time and sediment volume required to obtain a statistically significant sample are often too great. The second problem is the inability to identify fossil remains beyond genus.

The taxonomy of midges is based principally on adult characters. Association of particular larval types with individual species of adult is possible only after extensive life-history investigations. In addition, the lack of intra-generic morphological diversity of larvae hinders separation of individual species. The resulting inability to distinguish between a potentially large number of ecologically diverse species within a genus complicates reconstruction of past environmental fluctuations.

In spite of taxonomic problems, reconstruction of lacustrine trophic history based on midges corresponds closely with cladoceran interpretations for the same basin (figure 4.16). Goulden (1964) documents the replacement of *Bosmina coregoni* by *B. longirostris* as the dominant bosminid in the planktonic cladoceran assemblage at Esthwaite Water, England, following a productivity increase associated with human disturbance. Midge data

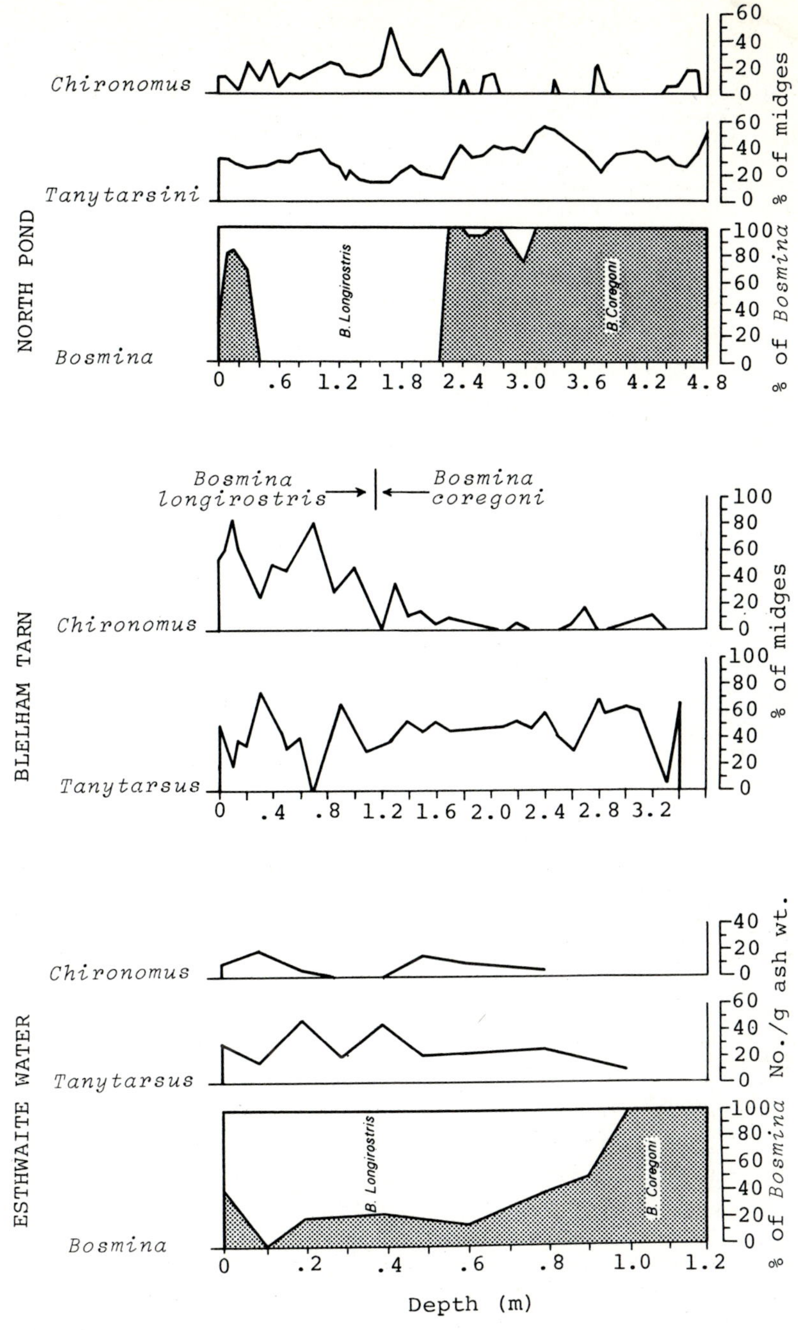

Figure 4.16 Midge fluctuations coincident with the replacement of *Bosmina coregoni* by *B. longirostris* during periods of increased productivity at Esthwaite Water and Blelham Tarn, England, and North Pond, Massachusetts. Note how the distribution of *Chironomus* parallels that of *B. longirostris*. (Data from Goulden 1964; Harmsworth 1968; Crisman 1977 and unpublished).

from the same core suggest a similar interpretation. The importance of *Chironomus* relative to *Tanytarsus* increases during basin eutrophication concurrent with the increased importance of *Bosmina longirostris*. Both cladoceran and midge data suggest that cultural eutrophication of the basin was not severe enough to destroy completely the pre-existing oligotrophic faunal assemblages. Additional investigations from Blelham Tarn, England (Harmsworth 1968), and North Pond, Massachusetts (Crisman 1977), demonstrate further the close agreement between cladoceran and midge sensitivity to environmental fluctuations (figure 4.16).

ROTIFERS

Rotifers are represented in lake sediments predominantly by their abundant resting eggs. The taxonomic identification of rotifer eggs is based on the work of Sudzuki (1964), Müller (1970, Sebestyén (1974) and especially Nipkow (1961), who identified sixteen rotifer species from sediments. The first rotifer stratrigraphic profile was for six species at Otterstedter Sees, Germany (Müller 1970). The upper Atlantic period at this site was characterized by maximal abundance of *Filinia longiseta, Trichocerca cylindrica* and *Conochiloides natans*, but their representation declined during the lower Subboreal concurrent with an increase in blue-green algae. The only other rotifer profiles to date are from Schleinsee, Germany (Geyh et al. 1974), North Pond, Massachusetts (Crisman unpublished) and Linsley Pond, Connecticut (Brugam 1975). Cultural eutrophication at Linsley Pond was accompanied by increased yearly influx for *Filinia* and *Anureopsis fissa* and the near elimination of *Polyarthra* from the rotifer fauna.

The limited number of profiles available demonstrate the potential value of rotifers in paleolimnology. The response of rotifer populations to periods of increased lacustrine productivity and to pronounced size-selective predation on cladocerans would be especially interesting. Interpretation of rotifer remains at present is hindered both by the absence of a detailed taxonomic survey of rotifer resting eggs and the need for modern ecological information on individual rotifer species.

OSTRACODS

The preservation of the principal ostracod remains (mandibles and shells) displays marked inter-basin variation. Dissolution of the carbonate-rich shells is accelerated at low pH values, and differential preservation may be related to shell thickness and carbonate content. The taxonomy of living ostracods is based on unpreserved characters including antennae, thoracic legs and spermatic glands, whereas fossil identification must rely on variations in shell morphology. Taxonomy of the fossils is complicated further by the fact that a species' shell morphology may display pronounced instar-specific variation as a response to changing appendage number and form (Hoff 1942).

Ostracods may be divided broadly into macrophytic communities and benthic (littoral and profundal) communities. Stark (1976) compared living with subfossil ostracod assemblages from surficial sediments along a depth profile at Elk Lake, Minnesota. She found that the profundal benthic ostracod assemblage is little affected by postmortem transport processes and therefore is reasonably representative of the local population. Remains of littoral species (benthic and macrophytic) are thoroughly integrated by wave action, with a concentration zone of littoral remains commonly observed at the depth of the effective wave base (epilimnion-metalimnion transition). In addition, the contamination of profundal sediments by remains of macrophytic species is ascribed to the periodic inclusion of these active swimmers in the plankton.

Paleolimnological interpretation of fossil ostracod assemblages is hindered by the lack of adequate modern ecological information for individual species. Delorme (1971a) relates the distribution of 102 ostracod species in 4300 samples from Canadian lakes to 18 physical and chemical parameters. The distribution of a given species can thereby be defined on the basis of minimal, maximal, and mean values of each lacustrine parameter considered. Delorme then used these data to reconstruct postglacial variations in lake chemistry for two sites in Saskatchewan (1971a) and Manitoba (1971b), making estimates of past chemical parameters based on a weighted average of the observed modern mean values of that parameter for each species in the ostracod assemblage.

Few quantative ostracod paleolimnological studies exist, but two recent investigations are of note. Klassen et al. (1967) interpreted the ostracod stratigraphy of an extinct Pleistocene lake in Manitoba on the basis of climatically related changes in lake cation concentration, water level and trophic state. In addition, the occurrence of two ostracod species is used to define two periods of subarctic climate (*Cytherissa lacustria*) and a period of humid continental climate (*Candona obtusa*). Stark (1976) declines to make climatic inferences based on postglacial ostracod assemblages in profundal and littoral cores from Elk Lake, Minnesota, but feels that the deep-water assemblage reacted to hypolimnetic oxygen reduction related to productivity, where the littoral assemblage is more strongly controlled by substrate composition and macrophyte development.

MOLLUSCS

The paleoecological literature for molluscs is more extensive than that for any other faunal group (see Frey 1964). Carbonate shells of both terrestrial and aquatic molluscs have been identified from lake sediments and recently larval glochidia of two genera have been described from Polish lake sediments (Brodniewicz 1969). As with ostracods, the preservation of mollusc shells is a reflection of both shell structure and lake pH. In addition, postmortem transport of littoral-zone shells and subsequent concentration at a depth corresponding to the wave base may complicate fossil interpretation. The following brief remarks are not intended as a complete literature review. Instead I have chosen a few recent papers that consider molluscs as part of multi-disciplinary investigations designed to correlate lacustrine with terrestrial events.

Valuable information regarding past fluctuations in such lake parameters as water level, cation concentration, trophic state, substrate and oxygen concentration has been obtained from fossil mollusc assemblages. Watts and Bright (1968) combine mollusc with plant-macrofossil and pollen studies in order to interpret the effect of climatic change on the development of Pickerel Lake, South Dakota. Their results indicate that Pickerel Lake became shallower, with increased alkalinity, hardness, and carbonate content concurrent with the postglacial development of prairie in the area. More recently Stark (1976) used a decline in the oxygen sensitive genus *Pisidium* to define in part a post-glacial increase in productivity at Elk Lake, Minnesota. Finally, Ohlhorst and Hutchinson (1977) were able to interpret major molluscan changes at Lake Huleh, Israel, with climatic and lake level fluctuations.

Cultural eutrophication and associated anoxia at St. Clair Lake, Minnesota, were accompanied by a total replacement of clams and proso-branch snails by pulmonate snails (principally *Gyraulus parvus*), a group able to utilize atmospheric oxygen (Birks et al. 1976). Similarly the dominance of two pulmonate genera (*Biomphularia*, *Stenophysa*) at Lago de Peten, Guatemala, is attributed to cultural eutrophication resulting from increased nutrient input following Mayan land clearance for farming (Covich 1976). Covich further suggests that the increased abundance of a smooth-shelled prosobranch (*Pyrgophorus*) during cultural eutro-phication may have resulted from the recent human exploitation of major predators such as turtles.

At present the number of quantitative paleolimnological investigations of molluscs from lake cores remains small. The main hindrance to core investigations is the large volume of sediment that must be processed in order to obtain a quantitative mollusc sample. The amount of geologic time represented by a lake-core sample necessary for detailed mollusc analysis is therefore often too great to be sensitive to minor environmental fluctuations. It is for this reason that a majority of mollusc studies have concentrated on exposed lacustrine deposits where large volumes of sediment can be processed while still maintaining close-interval stratigraphic sampling.

CONCLUSIONS

Aquatic invertebrates are extremely sensitive to disturbances of lacustrine physical, chemical and biological processes. Past water-level fluctuations may be defined from the ratio of planktonic to littoral remains (Cladocera) or the habitat preferences of benthic species (molluscs, ostracods, midges). In addition, estimates of past water chemistry, lake trophic state, hypolimnetic oxygen concentration and substrate texture and organic content have been based on the requirements of individual invertebrate species. Consequently the precision of any paleolimnological interpretation is determined by the quality and type of modern ecological data that exist for the invertebrate group in question. Because of the paucity of ecological information for many groups and the possibility of regional variation, it is best to collect modern regional ecological data as part of any paleolimnological investigation.

A lake is not a closed system but is closely linked with its catchment, especially for its nutrient supply. Although the chemical base for a lake is ultimately established by the composition of catchment bedrock, catchment chemical release is strongly controlled by the composition and age structure of the local vegetation. When combined with stratigraphic studies of pollen, plant macrofossils, and sediment chemistry, fluctuations in lacustrine parameters recorded by invertebrates may be related to changes in watershed vegetation and associated patterns of chemical release (e.g. Whitehead and Crisman 1977). Aquatic invertebrates may therefore display an indirect response to climatic change.

Making direct climatic interpretations from invertebrate fossil assemblages is often difficult. Climatic inferences are hindered both by the fact that a majority of the taxa encountered display wide geographical ranges and because of the difficulty in distinguishing the response to climate from that of the often accompanying oscillations in lake trophic structure. Investigation of sites near the transition of major biomes offers the best chance of direct climatic interpretation based on invertebrate remains.

Finally, as frequently stressed in this discussion, reconstruction of past lacustrine environments based on a single invertebrate group may be weakened by a paucity of modern ecological information, differential preservation, insufficient sample size and inadequate taxonomy. Therefore, the most readily acceptable paleolimnological interpretations are those based on several invertebrate groups in combination with other independent paleolimnological data sets (pigments, chemistry, pollen). Such multi-disciplinary investigations offer the best chance of delineating the trophic history of a lake and relating it to concurrent fluctuations in climate and catchment processes.

ACKNOWLEDGMENTS

I wish to thank Drs H.E. Wright Jr. and D.G. Frey for their helpful criticism of this manuscript.

REFERENCES

Alhonen, P. 1967 Palaeolimnological investigations of three inland lakes in south-western Finland. Acta. Bot. Fenn., 76, 3-59.

Alhonen, P. 1970 On the significance of the planktonic/littoral ratio in the cladoceran stratigraphy of lake sediments. Commentationes Biologicae, 35, 3-9.

Birks, H.H., Whiteside, M.C., Stark, C.M. and Bright, R.C. 1976 Recent paleolimnology of three lakes in northwestern Minnesota. Quaternary Research, 6, 249-272.

Bradbury, J.P. and Megard, R.O. 1972 Stratigraphic record of pollution in Shagawa Lake, northeastern Minnesota. Geol. Soc. Amer. Bull., 83, 2639-2648.

Brodniewicz, I. 1969 Les glochidia et leur importance pour l'etude des depots quaternaires d'eau douce. Mitt. Internat. Verein. Limnol., 17, 315-318.

Brooks, J.L. and Dodson, S.I. 1965 Predation, body size, and composition of plankton. Science, 150, 28-35.

Brugam, R.B. 1975 The human disturbance history of Linsley Pond, North Branford, Connecticut. Ph.D. Thesis, Yale University, New Haven. 171 p.

Brundin, L. 1949 Chironomiden und andere Bodentiere der sudschwedischen Urgebirgsseen. Ein Beitrag zur Kenntnis der bodenfaunistischen Charakterzuge schwedischer oligotropher Seen. Rept. Inst. Freshwat. Res. Drottningholm, 30, 1-914.

Brundin, L. 1956 Die bodenfaunistischen Seetypen und ihre Anwendbarkeit auf die Sudhalbkugel. Zugleich eine Theorie der produktionsbiologischen Bedeutung der glazialen Erosion. Rep. Inst. Freshwat. Res. Drottningholm, 37, 185-235.

Bryce, D. 1962 Chironomidae (Diptera) from freshwater sediments with special reference to Malham Tarn (Yorks.). Trans. Soc. Brit. Ent., 15, 41-54.

Clair, T. and Paterson, C.G. 1976 Effect of a salt water intrusion on a freshwater Chironomidae community: a paleolimnological study. Hydrobiologia, 48, 131-139.

Covich, A. 1976 Recent changes in molluscan species diversity of a large tropical lake (Lago de Peten, Guatemala). Limnol. Oceanogr., 21, 51-59.

Cowgill, U.M., Hutchinson, G.E., Racek, A.A., Goulden, C.E., Patrick, R. and Tsukada, M. 1966 The history of Laguna de Petenxil, a small lake in northern Guatemala. Conn. Acad. Arts Sci., 17, 1-126.

Crisman, T.L. 1977 North Pond, Massachusetts: post-glacial variations in lacustrine productivity as a reflection of changing watershed-lake interactions. Ph.D. Thesis, Indiana University, Bloomington, Indiana.

Crisman, T.L. and Whitehead, D.R. 1977 Paleolimnological studies of small New England (U.S.A.) Ponds. II: Cladoceran community responses to trophic oscillations. Pol. Arch. Hydrobiol.,(in press).

DeCosta, J.J. 1964 Latitudinal distribution of chydorid cladocera in the Mississippi Valley, based on their remains in surficial lake sediments. Invest. Ind. Lks. and Streams, 6(2), 65-101.

Deevey, E.S., Jr. 1941 Limnological studies in Connecticut. IV. The quantity and composition of the bottom fauna of thirty-six Connecticut and New York lakes. Ecol. Monogr., 11, 413-455.

Deevey, E.S., Jr. 1942 Studies on Connecticut lake sediments: III. The biostratonomy of Linsley Pond. Am. J. Sci., 240 (4), 233-264.

Deevey, E.S., Jr. 1969 Cladoceran populations of Rogers Lake, Connecticut during late- and postglacial time. Mitt. Internat. Verein. Limnology, 17, 56-63.

Delorme, L.D. 1971a Palaeoecology of Holocene sediments from Manitoba using freshwater ostracodes. Geol. Assoc. Can., Special Paper 9, 301-304.

Delorme, L.D. 1971b Paleoecological determinations using Pleistocene freshwater ostracodes. Bull. Centre Rech. Pau-SNPA, 5, 341-347.

Edmondson, W.T., Anderson, G.C. and Peterson, D.R. 1956 Artificial eutrophication of Lake Washington. Limnol. Oceanogr., 1, 47-53.

Flossner, D. 1964 Zur Cladocerenfauna des Stechlinsee-Gebietes II. Ökologische Untersuchungen über die litoralen Arten. Limnologica(Berl.), 2, 35-103.

Frey, D.G. 1958 The late-glacial cladoceran fauna of a small lake. Arch. fur Hydrobiol., 54, 209-275.

Frey, D.G. 1960 The ecological significance of cladoceran remains in lake sediments. Ecology, 41, 684-699.

Frey, D.G. 1961 Developmental history of Schleinsee. Verh. Internat. Verein. Limnol., 14, 271-278.

Frey, D.G. 1962 Cladocera from the Eemian interglacial of Denmark. J. of Paleont., 36, 1113-1154.

Frey, D.G. 1964 Remains of animals in Quaternary lake and bog sediments and their interpretations. Arch. fur Hydrobiologie Eigebn. Limnol., 2, 1-114.

Frey, D.G. 1976 Interpretations of Quaternary paleoecology from Cladocera and midges, and prognosis regarding usability of other organisms. Can. J. Zool., 54, 2208-2226.

Fryer, G. 1968 Evolution and adaptive radiation in the Chydoridae (Crustacea:Cladocera): A study in comparative functional morphology and ecology. Phil. Trans. Roy. Soc. London B, 254, 221-385.

Fryer, G. 1974 Evolution and adaptive radiation in the Macrothricidae (Crustacea:Cladocera): a study in comparative functional morphology and ecology. Phil. Tran. Roy. Soc. London B, 269, 137-274.

Geyh, M.A., Merkt, J., Muller, H. and Strief, H. 1974 Reconstitutions paleoclimatiques et paleoecologiques a partir de l'etude des sediments lacustres de l'Allemagne Meridionale. Journées de l'Hydraulique, 13, 1-17.

Gorman, E. and Sanger, J. 1976 Fossilized pigments as stratigraphic indicators of cultural eutrophication in Shagawa Lake, northeastern Minnesota. Geol. Soc. Amer. Bull., 87, 1638-1642.

Goulden, C.E. 1964 The history of the cladoceran fauna of Esthwaite Water (England) and its limnological significance. Arch. Hydrobiol., 60, 1-52.

Goulden, C.E. 1966a The animal microfossils. Conn. Acad. Arts Sci., 17, 84-120.

Goulden, C.E. 1966b La Aquada de Santa Ana Vieja: an interpretative study of the cladoceran microfossils. Arch. Hydrobiol., 62, 373-404.

Goulden, C.E. 1969 Developmental phases of the biocoenosis. Proc. Nat. Acad. Sci., 62, 1066-1073.

Goulden, C.E. 1971 Environmental control of the abundance and distribution of the chydorid cladocera. Limnol. Oceanogr., 16, 320-331.

Harmsworth, R.V. 1968 The developmental history of Blelham Tarn (England) as shown by animal microfossils, with special reference to the cladocera. Ecol. Monogr., 38, 223-241.

Hoff, C.C. 1942 The ostracods of Illinois. III. Biol. Monogr. 19, Nos. 1-2. Univ. Illinois, Urbana, Ill.

Hofmann, W. 1971a Zur Taxonomie und Palökologie subfossiler Chironomiden (Dipt.) in Seesedimenten. Arch. Hydrobiol. Beih. Ergebn. Limnol., 6, 1-50.

Hofmann, W. 1971b Die postglaziale Entwicklung der Chironomiden und *Chaoborus* - Fauna (Dipt.) des Schöhsees., Arch. Hydrobiol. Suppl. 41, (1/2), 1-47.

Hofmann, W. 1977 *Bosmina (Eubosmina)* populations of the Grosser Segeberger See during late glacial and postglacial times. Arch. Hydrobiol., 81, 349-359.

Hogan, M.E. 1977 Studies of cladoceran microfossils from short cores of Lough Leane, Killarney. B.Sc. Thesis, University College, Dublin, Ireland. 68p.

Hutchinson, G.E., Bonatti, E., Cowgill, U.M., Goulden, C.E., Leventhal, E.A., Mallett, M.E., Margaritora, F., Patrick R., Racek, A., Roback, S.A., Stella, E., Ward-Perkins, J.B. and Wellman, T. 1970 Ianula: an account of the history and development of the Lago di Monterosi, Latium, Italy. Trans. Amer. Philosoph. Soc., 60, 1-178.

Iovino, A.J. 1975 Extant chironomid larval populations and the representativeness and nature of their remains in lake sediments. Ph.D. thesis, Indiana University, 54p.

Kerfoot, W.C. 1974 Net accumulation rates and the history of cladoceran communities. Ecology, 55, 51-61.

Klassen, R.W., Delorme, L.D., and Mott, R.J. 1967 Geology and paleontology of Pleistocene deposits in southwestern Manitoba. Can. J. Earth Sci., 4, 433-447.

Lawrenz, R.W. 1975 The developmental palaeoecology of Green Lake, Antrim County, Michigan. M.S. thesis, Central Michigan University. 78p.

McNaught, D.C. and Buzzard, M. 1973 Changes in zooplankton populations of Lake Ontario (1939-1972). Proc. 16th Conf. Great Lake Res., 76-86.

Megard, R.O. 1964 Biostratigraphic history of Dead Man Lake, Chuska Mountains, New Mexico. Ecology, 45, 529-546.

Megard, R.O. 1967 Late Quaternary cladocera of Lake Zeribar, western Iran. Ecology, 48, 179-189.

Minter, L. 1938 Der Zurichsee als Eutrophierunsphanomen. Geol. Meere Binnegewasser, 2, 284-299.

Mueller, W.P. 1964 The distribution of cladoceran remains in surficial sediments from three northern Indiana lakes. Invest. Indiana Lakes and Streams, 6, 1-63.

Muller, H. 1970 Oekologische Veranderungen in Otterstedter See im Laufe Nacheiszeit. Ber. Naturhist. Ges., 114, 33-47.

Nipkow, F. 1961 Die Radertiere im Plankton des Zurichsees und ihre Entwicklungsphasen. Z. Hydrol., 23, 398-461.

Ohlhorst, S. and Hutchinson, G.E. 1977 The waters of Merom: a study of Lake Huleh. V. Temporal changes in the molluscan fauna. Arch. Hydrobiol., 80, 1-19.

Poulsen, E.M. 1940 Freshwater Entomostraca. The zoology of East Greenland. Meddel. om Gronland, 121, 1-72.

Quade, H.W. 1969 Cladoceran faunas associated with aquatic macrophytes in some lakes in northwestern Minn. Ecology, 50, 170-179.

Quade, H.W. 1971 Niche specificity of littoral cladocera: habitat. Trans. Amer. Micros. Soc., 90, 104-105.

Roback, S.S. 1970 XII. The Chironomidae. Trans. Amer. Philosoph. Soc., 60, 150-162.

Roth, J.C. and Stewart, J.A. 1973 Nearshore zooplankton of southeastern Lake Michigan, 1972. Proc. 16th Conf. Great Lakes Res., 132-142.

Saether, O.A. 1975 Nearctic chironomids as indicators of lake typology. Verh. Internat. Verein. Limnol., 19, 3127-3133.

Sebestyén, Olga 1974 Present and past occurrence of members of the genus *Filinia* bory de St. Vincent, 1826 (Testudinellidae Rotatoria) in Lake Balaton. Annal. Biol. Tihany, 41, 289-315.

Stahl, J.B. 1959 The developmental history of the chironomid and *Chaoborus* fauna of Myers Lake. Invest. Indiana Lakes and Streams, 5, 47-102.

Stahl, J.B. 1969 The uses of chironomids and other midges in interpreting lake histories. Mitt. Inter. Verein. Limnol., 17, 111-125.

Stark, D.M. 1976 Paleolimnology of Elk Lake, Itasca State Park, Northwestern Minnesota. Arch. Hydrobiol./Suppl., 50, 208-274.

Sudzuki, M. 1964 New systematical approach to the Japanese planktonic Rotifera. Hydrobiologia, 23, 1-124.

Synerholm, C.C. 1974 The chydorid cladocera from surface lake sediments in Minnesota and North Dakota. M.S. Thesis, University of Minnesota, 37 p.

Thienemann, A. 1913 Der Zusammenhang zwichen dem Sauerstoffgehalt des Tiefenwassers und der Zusammensetzung der Tierfauna unsere Seen. Int. Rev. Hydrobiol., 6, 243-249.

Thienemann, A. 1922 Die beiden Chironomusarten der Tiefenfauna der norddeutschen Seen. Arch. Hydrobiol., 13, 609-646.

Tsukada, M. 1967 Succession of cladocera and benthic animals in Lake Nojiri. Jap. J. Limnol., 28, 107-123.

Tsukada, M. 1972 The history of Lake Nojiri, Japan. Trans. Conn. Acad. Arts and Sci., 44, 339-365.

Vitousek, P.M. and Reiners, W.A. 1975 Ecosystem succession and nutrient retention: a hypothesis. BioScience, 25, 376-381.

Warwick, W.F. 1975 The impact of man on the Bay of Quinte, Lake Ontario, as shown by the subfossil chironomid succession (Chironomidae, Diptera). Verh. Internat. Verein. Limnol., 19, 3134-3141.

Watts, W.A. and Bright, R.C. 1968 Pollen, seed and mollusc analysis of a sediment core from Pickerel Lake, northeastern South Dakota. Geol. Soc. Amer. Bull., 79, 855-876.

Wells, L. 1960 Effects of alewife predation on zooplankton populations in Lake Michigan. Limnol. Oceanogr., 15, 556-565.

Whitehead, D.R. and Crisman, T.L. 1977 Paleolimnological studies of small New England (U.S.A.) ponds. I: Late glacial and postglacial trophic oscillations. Pol. Arch. Hydrobiol., (in press).

Whitehead, D.R., Rochester, Jr., H., Rissing, S.W., Douglass, C.B. and Sheehan, M.C. 1973 Late glacial and postglacial productivity changes in a New England pond. Science, 181, 744-747.

Whiteside, M.C. 1970 Danish chydorid cladocera: modern ecology and core studies. Ecol. Monogr., 40, 79-118.

Whiteside, M.C. and Harmsworth, R.V. 1967 Species diversity in chydorid (cladocera) communities. Ecology, 48, 664-667.

5. FOSSIL MAMMALS IN THE RECONSTRUCTION OF QUATERNARY ENVIRONMENTS WITH EXAMPLES FROM THE AUSTRALIAN FAUNA

by R.T. Wells

SUMMARY

The short duration of the Quaternary period makes the use of small evolutionary changes in terrestrial vertebrates to date fossil sequences extremely difficult. Problems also arise in the use of radiometric techniques. Palaeontologists must rely on geological evidence for a relative time scale. The most useful chronologies are derived from regional sedimentary sequences related to world climatic change. Potentially useful markers are discussed with examples provided from typical Australian fossil sequences. I stress the value of using the post-cranial skeleton in addition to the taxonomically important elements such as teeth, jaws and skulls for inferring the ecological requirements of fossil mammals. Examples are given to support a suggestion that community structure can be inferred only from an analysis of the faunal assemblage and its mode of accumulation. The determination, significance and reliability of negative evidence (i.e. absence of taxa from a fauna) is discussed. Finally, the limitations of modern mammalogy in providing a basis for comparison of fossil faunas is outlined with Australian examples.

INTRODUCTION

Mass extinction is suggested as the cause of the great discontinuities in the fossil record and indeed is the basis upon which palaeontologists divide geological time into eras, periods and epochs. That these discontinuities may be interpreted as extinctions is more a reflection of the inevitable decrease in chronological resolution of increasingly older sediments, rather than evidence of ancient cataclysmic events. With an increase in the variety of absolute dating techniques and the number of dated sites, many of these apparent discontinuities can be resolved in a sequence of smaller extinction events covering broad periods of time.

The Quaternary is that period of earth history with both the greatest chronological resolution and the shortest duration; it poses special problems for the palaeontologist. When examined at high resolution, the differences between faunas are often subtle and differences between taxa rarely exceed those used to distinguish Recent species. Such small differences make it extremely difficult to use the morphological variety of fossils to date strata. This difficulty is further accentuated by variations in the geographic distribution of species within this relatively short period of geologic time. Thus in Quaternary studies the palaeontologist must rely heavily upon the geological evidence for a relative time scale. For these reasons the study of non-marine Quaternary fossils takes on a different aspect from the more traditional palaeontological studies. In the Quaternary we have the greatest opportunity to study, at the species level, changes which have determined the balance of present-day faunas.

It is in some ways a paradox that arising from the extensive chronology of late Pleistocene extinction in North America, we observe a return to theories of catastrophic "overkill" (Martin 1967, 1973) and communicable disease (Edwards 1967) as explanations for faunal change. That such hypotheses, no matter how accurate their explanation of North American events, fail adequately to account for events on other continents relegates them to a lower level in the hierarchy of generalities.

In any study of the Quaternary the first assumption to be challenged is that there are sufficient criteria to distinguish the events of this period from those that preceded it. The base of the Quaternary, as in other systems, is defined by a substantial fauna change in a marine succession of foraminifera and molluscs (Blanc et al. 1953); it is perhpas fortuitous that the discontinuity in the nearby non-marine mammal fauna at Villafranca d'Asti should be coincident in time. Undoubtedly the change in ocean temperature indicated by the marine fossils bears some relationship to changes in global climate but evidence of earlier glacial development in Iceland (McDougall and Wensink 1966) and North America (Dalrymple 1972) seems to indicate an earlier beginning to the decline in world temperature. Indeed oxygen isotope ratios from benthic and pelagic foraminifera in deep sea cores (Emiliani 1961, 1966; Shackleton and Opdyke 1973) indicate a decline in world temperature commencing in the early Tertiary. Thus perhaps, in broad terms, it is more fruitful to consider Quaternary events as a portion of yet another world-wide glaciation, akin to those of the Permo-Carboniferous, Ordovician, Precambrian, etc., a glaciation with its beginnings in the early Tertiary. The adoption of this broader hypothesis lends perspective to the fine detail of the Quaternary. This approach is not novel; it is a theme used extensively by Kurten (1968, 1972).

If the Quaternary represents a continuation of events which began in the Tertiary some explanation must be provided for the apparent increase in extinction rate of large mammals. From detailed studies of past distributions of plant and animal species still extant, or their near relatives, it is possible to draw inferences about past climate. Similarly inferences can be drawn from a study of non-marine sediments and palaeosols, e.g. in the Northern Hemisphere fluctuations in climate have been related to sedimentary sequences which contain moraines, varves and loess deposits all indicative of periods of glacial advance and retreat (Zeuner 1959; Wright and Frey 1965; West 1968; Wright and Moseley 1975). This has led to a relative chronology which has been dated in part using radiocarbon, K/Ar and Uranium series techniques in association with changes in the magnetic polarity of the earth. These Quaternary climatic fluctuations, which are superimposed on the general curve of falling world temperature, are shown also in fluctuations in the surface temperatures of the oceans as indicated by the oxygen isotope ratios of pelagic foraminifers from deep sea cores (Devereux 1968; Shackleton and Opdyke 1973). A mechanism for these oscillations in world temperature, based on changes in solar radiation due to perturbations in the earth's orbit, was originally proposed by Milankovitch (1930, 1938a, 1938b) and is now receiving renewed attention (Wigley 1977; Berger 1977; Hays et al. 1976; Veeh and Chappel 1970). It is against this background that further reconstructions of Quaternary environments must be tested and faunal evidence compared. But before this can be done many problems must be overcome in the interpretation of faunal evidence, particularly mammalian remains.

PROBLEMS IN THE INTERPRETATION OF MAMMALIAN REMAINS

The interpretation of mammalian remains may be pursued at many levels but to maximise its usefulness in the reconstruction of past environments it is essential to have the following data in association with the fossils:

(a) a detailed analysis of the environment of deposition,

(b) a knowledge of the geographic range of the species represented in the fauna, and

(c) a relative or absolute chronology for each fauna.

Stratigraphic and chronologic problems.

Although absolute dates for any fauna are desirable, it is impossible with current techniques to obtain these for many deposits. The establishment of a relative chronology is therefore of prime importance. It is dependent on detailed studies of the depositional environment. Thus careful selection of sites is critical to the early establishment of a chronology of continental faunas.

The best understood chronological sequences for Quaternary mammals are found in the Northern Hemisphere. Due to widespread glacial advances, much of the stratigraphic evidence is based on cold climate sediments including moraines, loess, solifluction deposits and gravel terraces. Glacial phenomena were restricted in the Southern Hemisphere and climatic fluctuations are evidenced more by the advance and retreat of low latitude deserts. On the Southern continents chronological sequences for Quaternary mammals are poorly understood; Quaternary stratigraphy is still in its infancy.

Fossil mammals were first discovered in Australia at the Wellington caves in New South Wales and, although this remains a productive site, the extensive collections have yet to be assigned to the known sequence of fissure-fills (Frank 1971). The lack of stratigraphic, or even of detailed geographic, information is exemplified by the wealth of material collected from southeast Queensland and labelled by De Vis simply as "Darling Downs". The major collection areas have been distinguished and evidence for the superposition of the faunas has been presented (Woods 1960) while elements in the collections have been assigned to these sites on the basis of their preservation (Bartholomai 1963). However no detailed stratigraphic studies are yet available of the intra-site detail in spite of the proximity of one site at Chinchilla to the steeply incised banks of the Condamine River.

Detailed stratigraphic studies of Australian mammalian faunas started with the work of Gill (1953) and Stirton (1955) but although the number of Tertiary sites investigated in this manner is steadily increasing (Stirton et al. 1961, 1967; Lundelius 1963; Tedford 1967; Woodburne 1967; Archer 1972; Archer and Baynes 1972; Wilkinson 1972; Gill 1973; Marshall 1973; Tedford 1973; Archer 1974a; Wells 1975; Goede and Murray 1975, 1977; Archer and Wade 1976; Tedford et al. 1977) much of the work is concentrated on Tertiary sediments. Regrettably there are few accurately dated Quaternary sites.

Environments of deposition. Most Quaternary fossil sites within Australia occur in four general sedimentary environments, namely:

(a) cave fills,

(b) swamp sediments,

(c) lacustrine and fluvial sediments, and

(d) aeolian sediments.

Cave fills. Cave fills and fissure fills remain the richest source of Quaternary fossil mammals in Australia. Interpretation of the depositional sequences within these fills poses special problems. The mode of accumulation of sediment at each site may be controlled by many factors and different factors may operate at different times. Caution must be exercised in any interpretation. Some general points for investigations of these sites can be made.

(a) Excavation should be carried out in a series of carefully controlled levels.

(b) Under no circumstances should all the fill be removed. Benches where the stratigraphy may be checked by future workers or further excavation undertaken should be left.

(c) The site of excavation should be selected with care; post-depositional changes are common, e.g. the winnowing of sedimentary cones into alluvial fans is common where water action is prevalent. This process redistributes fossils and charcoal (Archer 1974b) leading to a mixing of assemblages and underestimation of ages respectively. The use of hollow plastic pipe to retrieve cores, or hand augering on a grid pattern, allow wise selection of excavation sites and can provide valuable data for subsurface correlation of sedimentary horizons.

(d) All the sediments from each level of an excavation should be collected and screened since:

(i) the concentration of limestone clasts in a given horizon may represent a break in sedimentation with slow accumulation of fragments from spalling of the roof of the cave, and the concentration of clasts on a sedimentary cone may represent a lag and be indicative of water flow down the cone,

(ii) the organic content and nature of the sediments in the profile can be indicative of the soil development outside the cave at time of deposition and this record may well be preserved in the local soil mantle, and

(iii) the appearance of speleothems (flowstone, stalagmites) within the sediments may be important markers for both relative and absolute chronology (Uranium series dating, Harmon et al. 1975) as well as palaeotemperature (oxygen isotope ratios, Thompson et al. 1976). However, care should be exercised as carbonate deposition frequently occurs long after sedimentation and may be simultaneous at several different levels. Unfortunately the oxidised nature of many cave sediments precludes the use of palynology although new techniques (Juvigne 1975) have increased the yield of otherwise sparse pollen grains.

Swamp deposits. The organic sediments which accumulate around the margins of lakes and springs often yield large quantitites of well preserved bone, pollen and carbon. These sediments are ideally suited to palynological and radiocarbon dating techniques. Stratigraphic analysis of these organic muds is difficult due to their amorphous nature, but their colour and organic content may provide an indication of the conditions under which they accumulated, e.g. gray to gray-black muds rich in organisms such as diatoms are indicative of partly overgrown but well aerated ponds or lakes. Skeletal remains associated with these sediments are usually somewhat decomposed.

The accumulation of plant material around the margins of water bodies inhibits aeration by wave action, leads to increased accumulation of debris and ultimately to reducing conditions with the production of hydrogen and iron sulphides and the formation of black muds. Trapped animals are rapidly buried in these unconsolidated, accumulating sediments; the resulting fossils are usually well preserved and darkly stained. Skeletons from these deposits should be excavated with care as the sediments in the region of the body cavity of herbivores may contain pollen and plant cuticles from the alimentary canal and thus provide some insight into the dietary preference of the fossil species and the nature of the surrounding vegetation. Rich and Partridge (Rich 1976) used this approach to determine the age of fossil macropods from the Morwell coal fields in Victoria.

Lacustrine and fluvial sediments. Lacustrine and fluvial sediments often contain large numbers of fossil vertebrates, e.g. Lake Callabonna, South Australia; Bingara, New South Wales. Fluctuations in lake and stream levels recorded in such sediments are evidence of changing hydrologic regimes and the climates that controlled them, but it is often difficult to place such sediments in a rigid stratigraphic sequence if they do not have associated aeolian landforms or pedogenic structures. Chronological and palaeoecological information can often be obtained from studies of animal and plant remains contained in the sediments, e.g. molluscs (La Rocque 1960), cladocerans (Goulden 1969), conchostracans (Tasch and Zimmerman 1971), pollen (Dodson 1974, 1975; Kershaw 1970, 1974, 1975; Hope 1974; Tutin 1960) and diatoms (Hecky and Kilham 1973).

Aeolian sediments. The most powerful tool for the unravelling of climatic sequences on southern continents is developing from the study of aeolian landforms. The sediments and soils associated with the origin, expansion and stabilisation of dunes have been related to changes in hydrologic regimes in Australia (Bowler 1976), Africa (Butzer 1971) and South America (Tricart 1975). Fossil sites occurring in these sedimentary environments are potentially the most important in the reconstruction of faunal sequences due both to the regional nature of the sediments and their established chronology.

Such deposits should be excavated with care as the more obvious bone accumulations are often the result of dune deflation leading to the mixing of faunas. The use of dune depressions as camp sites by man can lead to incorrect association of artifacts with extinct faunas (Cooper 1959).

Problems in the analyses of the fauna.

Without a framework of detailed stratigraphic and chronologic studies, faunal studies can be of only local interest. However even given sound environmental, geographic and chronologic information, the interpretation of faunas is still fraught with difficulties. To make environmental interpretations from mammalian remains it is essential to:

(a) adopt a uniformitarian approach, i.e. to assume that similarity in form implies similarity in function such that niche and habitat can be inferred,

(b) to be confident that negative evidence, i.e. absence of a particular taxon from a fauna, is not an artifact of sampling but indicates a difference in habitat.

Although the piling of assumption upon assumption appears to lack the rigour of experimental science, hypotheses based on these assumptions can be tested in terms of their predictive ability. Thus, from an analysis of the fauna it should be possible to reconstruct an environment which should also be consistent with geological evidence.

Thus environmental interpretations based on fauna generally proceed in the following sequence:

(a) species identification, generally based on diagnostic features such as skulls, teeth and jaws,

(b) determination of the minimum number of individuals of each species present,

(c) determination of the proportion of each age class and sex, if possible, of each species,

(d) interpretation of the mode or modes of accumulation of the fossil assemblage,

(e) determination of the niche occupied by each species,

(f) reconstruction of the habitat based on faunal assemblage, and

(g) comparison with faunas of similar age.

Morphological similarlity and inferred habitats. The reconstruction of past environments from vertebrate fossils relies heavily on the assumption that morphological similarities between fossil and living forms are indicative of similar ecological requirements. A well used example is that of teeth of herbivores. High crowns and elaborate folding of the enamel is interpreted as an adaptation to wear and is normally found in extant grazing herbivores whose diet contains large quantities of silica, whereas low crowns and simple cusp patterns are found amongst browsing herbivores whose diet consists of soft plant tissues. Often these dental features correlate well with limb structure. Grazing animals, e.g. horses, are found in flat open grassland, and their defence from predators lies in their ability to travel at sustained high speeds. In these animals selection has favoured long, slender limbs with much lengthening of the distal elements thereby increasing stride length and hence speed. Their high centre of gravity is no disadvantage when travelling over hard ground in flat or gently undulating country. Even with extinct forms of unknown affinity such deductions could be made from a simple study of the skeleton and so the broad adaptive type could be defined. Indeed, during the last century, great comparative anatomists such as Owen were able to make remarkably accurate predictions of the structural adaptations of fossil mammals from scant evidence.

Thus, by comparisons of anatomy, both extinct and extant forms may be assigned to broad adaptive types. However, while the ecological requirements of the extant forms may be determined from neontological studies, those of extinct forms may only be deduced from detailed comparative studies of their functional anatomy.

In attempting to reconstruct palaeoenvironments it is therefore useful to divide the fauna into extinct and extant forms.

The extinct forms. In assigning an extinct mammal to a taxon the most important diagnostic characters are generally those associated with the teeth and skull. It is therefore understandable, but regrettable, that even today these diagnostic portions are often the only skeletal elements collected and described from fossil localities. In many cases due to durability of enamel they may be the only elements remaining but in the Quaternary deposits such instances are rare. While the teeth of a mammal may provide many clues to its diet and hence its general habitat preferences, detailed studies of the post-cranial morphology may provide information on its behaviour and hence its more precise ecological requirements. The following examples are drawn from the fossil record of Australian mammals.

The three species of wombat which now survive in Australia are burrowers and are but the remnants of a much broader radiation. As many as 14 fossil morphospecies have been described since 1866 (Mahoney and Ride 1975). Based on tooth and skull dimensions these animals range in size from forms half that of the modern *Vombatus ursinus* to giants like *Phascolonus gigas*, twice its size. Without an analysis of the post-cranial material little can be said of their habits, but even *P. gigas* exhibits characteristics of the modern burrowing forms i.e. short limbs, well developed olecranon process on the ulna. If we could determine which of these forms were burrowers we could infer much about the limitations to their range from physiological considerations of the extant forms. The latter are relatively sedentary, do not move long distances to water but rely on that which is available in their food (Wells 1973, 1978). Today wombats range from wet sclerophyll forest to the boundary between the semi-arid and arid zone. A large fossil wombat exhibiting the characteristics of extant burrowing forms would therefore not be expected to occur in an area of unpredictable rainfall.

The extinct sthenurine kangaroos are represented by two genera, *Procoptodon* and *Sthenurus*. Species of *Procoptodon* and of the subgenus *Simosthenurus* show adaptations to a browsing habit (Tedford 1966). They are brachycephalic and have bilophodont molars, short diastemata and a crescentic array of small incisor teeth ideally suited for plucking vegetation. From a study of the skulls alone one might conclude that these animals are likely to be restricted to a forest habitat yet, paradoxically, studies of the post-cranial skeleton reveal a functionally monodactyl pes, an adaptation exhibited by the open-grassland dwelling red kangaroo of today. Perhaps they were browsers of the "sclerophyll-semi-succulent" understory (Specht 1972) so common in the woodlands and shrublands of Australia. The extinct *Protemnodon* species of kangaroos exhibit many of the dental characteristics typical of the browsing forms, e.g. simple bilophodont cheek teeth, and an horizontal occlusal plane related to the simple orthal movement of the jaw. However in combination with these characteristics are robust forelimbs and a pes which, whilst retaining the distinctive syndactyly of the Diprotondonta, has relatively short and divergent metatarsals and phalanges. This pes is more typical of an animal which has forsaken speed for stability. It is not typical of a dweller of open plains; indeed its post-cranial morphology is not unlike that of the modern tree kangaroos.

The family Diprotodontidae includes a diverse group of large, extinct quadrupedal herbivores which are well represented in both Tertiary and Quaternary deposits in Australia. Three of the four subfamilies, the Diprotodontinae, Nototheriinae and Zygomaturinae have simple bilophodont molar teeth suggestive of browsing. Within this broad indication special adaptations have been inferred. Thus Scott (1915), in his reconstruction of *Zygomaturus trilobus (Nothotherium tasmanicum)* Scott, considered the general shape of the skull, the unusual structure of the nasals, the muscle scars on the frontal bones, the large pre-orbital foramen and the pre-alveolar extension of the incisor teeth as evidence for the possession of a short tapir-like trunk.

Pleistocene species of the subfamily Palorchestinae are unique among the Diprotodontidae in having a strong mid-link developed between the lophs of the molar teeth. This feature originally led to their inclusion by Owen in the Macropodidae. The form and configuration of the incisor teeth and the nature of the premolars led Woods (1953) to include them in the Diprotodontidae yet they differ from all other members of the family in the possession of obvious grazing adaptations. Little is known of the post-cranial skeleton of Palorchestines; Bartholomaii (1962) assigned some caudal vertebrae to the large Pleistocene species *Palorchestes azeal* Owen. In a similar way, in my study of the extensive Victoria Cave fossils from South Australia, by a process of elimination I have assigned to this species a relatively long humerus and some rather laterally compressed, scimitar-like claws and robust, fore-shortened phalanges. An almost complete but undescribed foot bearing similar claws was recovered from Wee Jasper Caves, New South Wales (A. Ritchie pers. comm.). A tentative reconstruction from this scant evidence suggests a large, quadrupedal grazing animal with longish limbs and plantigrade feet bearing large scimitar-like claws, features that are reminiscent of the extinct chalicotheres of the American Lower Miocene (R. Tedford pers. comm.). The adaptive significance of this unusual foot may only become apparent when more of the skeleton is known.

Although all these fossil forms are by virtue of their dentitions readily assignable to known families of marsupials, it is from a knowledge of the post-cranial morphology that we may refine our speculations about their habits and the environments in which they lived.

The extant forms. Fossil mammals are referred to extant species on the basis of their morphological similarities. When dealing with Late Pleistocene fossils of the small dasyurids, peramelids and phalangerids, Smith (1971, 1972) and Merrilees (1968) have found no significant difference in either size or morphology. However, among the larger animals, the Pleistocene forms, although morphologically similar, are often considerably larger than the extant species. In many lineages the size difference has been considered sufficient to name new morphospecies. Thus, based on tooth dimensions, the dasyurid *Sarcophilus laniarius* (Owen) was 15% larger than *S. harissii* (Boitard) (Gill 1953), the koala *Phascolarctos stirtoni* Bartholomai was about 30% larger than *P. cinereus* (Goldfuss) (Bartholomai 1968) and the wombat *Phascolomys magnus* Murie was 1.8 times the size of *Vombatus ursinus* (Shaw) (Stephenson 1966). Amongst the kangaroos, Marshall (1973) indicated that between the morphologically similar pairs, *Macropus titan* Owen was 30% larger than living species *M. giganteus* Shaw, *Osphranter cooperi* Owen was 25% larger than living *M. robustus* Gould, and *Halmaturus vishnu* De Vis was 10% larger than living *Wallabia bicolor* (Desmarest). It is interesting to note the relatively greater difference in size between the larger forms.

Morphoclines are not uncommon in modern populations. I have discerned clines in the adult skull size (basilar length) of koala skulls in Australian museum mammal collections. The largest forms, from the more mesic habitats in Victoria are 40% bigger than the smallest in the more xeric habitats of Queensland and South Australia. Before we can propose convincing explanations for the selective mechanisms involved in apparent dwarfing, we need to know considerably more about the variation in size of both modern and fossil populations of the species in question. Favoured hypotheses include maintenance of population density in the face of severe limitations of habitat and/or food supply (Kurten 1968; Guilday 1967), relative increase in surface area with decrease in body size leading to improved thermo-regulation under conditions of increasing temperature, Bergmann and Allen's rules (debated by Scholander 1955, 1956) and predation by man, large animals being easier targets and producing higher yields of protein (Edwards 1967).

Notwithstanding the size differences in these samples, it is generally assumed that food and habitat requirements of the species are similar to those of present day equivalents and thus differences in geographic distribution can be attributed to past changes in habitat distribution and hence climate.

Observations of modern faunas show that the requirements of species are rarely restricted to a single specialized environment. Calaby (1971) in his survey of the mammals of the Clarence River district in New South Wales has recorded 45 species of mammals occurring within a range of 3885 square km, with up to 10 species of macropodids occurring within areas of 809 hectares or less. Yet the range of individual species such as the grey kangaroo may extend from tropical woodland around Cooktown (Queensland) to arid woodland along water courses north of Broken Hill (N.S.W.) to the alpine regions of the Snowy Mountains (N.S.W.). Of the 45 species only 8 have a relatively restricted habitat requirement and even 6 of these are restricted more by the structural nature of the vegetation community rather than its composition while two, the platypus and water rat, are confined to permanent streams (Calaby 1971). Thus interpretation of changing environmental conditions may only be made from a study of the composition of a series of contemporaneous fossil assemblages at a number of broadly distributed geographic locations. In this way shifts in the composition of faunas towards forms favouring particular combinations of vegetation structure (e.g. grassland, shrub steppe, savanna, woodland, forest) can be detected.

The nature of the sample.

Any local accumulation of fossils may be the result of a single catastrophic event or may have occurred over a longer period. It is only through a detailed study of the stratigraphic relationships that the fauna can be sorted into contemporaneous assemblages.

In the broadest approximation, the faunal assemblage should comply with a general Eltonian pyramid. However the likelihood that a species from a given community will appear in the deposit will vary not only with its relative density and proximity to the site of deposition but also with the mode of accumulation of the fossils. In compiling a faunal list it is the absence of forms indicative of a particular habitat which greatly influences our environmental reconstructions. One has to decide whether the absence of particular forms is an artifact of the initial accumulation or the present sampling, or whether it is indicative of a real absence from the habitat in the past. Modern sampling techniques involving screen washing (McKenna 1962) and heavy liquid separation allow for collection of the remains of the smallest vertebrates and even invertebrates. These techniques, combined with detailed studies of the sedimentary sorting processes which have occurred in a deposit, allow collecting to be undertaken so as to minimise sampling errors. Absence of particular forms of animals may then be interpreted from an understanding of the mode of accumulation of the fossils.

Evidence may be gained either from the general geological and sedimentary studies or from the fauna itself, or both. The Quaternary deposits of Lake Callabonna (Tedford 1973) consist of blue green gypseous clays containing the skeletons of many large vertebrates. Most of these animals were bogged in the ancient lake; excavation reveals the well preserved feet and lower portions of the limbs either adjacent to or beneath the vertebral column. The clays surrounding the animals often contain contorted lenses of sand where the trapped animals have flailed at the surface of the swamp in attempts to escape. The same process can be observed today where animals have wandered too far across the thin veneer of sand and salt which covers these old lake beds. Here the sorting process is apparent; only those animals whose mass is too great to be supported by the crust are trapped whilst smaller forms are not.

Many cave deposits in Australia have accumulated from animals and sediments falling through roof openings (Merrilees 1968; Wells 1975). These pitfall traps may select a broad range of terrestrial animals. Often the drop is quite small, two to three metres, but this may be sufficient to trap the larger animals even though they are not injured in the fall.

Many of the smaller forms are able to scale the walls and ceilings of a cave and escape. Nonetheless cave deposits often contain vast accumulations of small bone as well as large. Smith (1971) analysing the small vertebrates from Victoria Cave in South Australia detected distinct size classes of small animals, the upper limits of which transgressed age classes of some species. She concluded that many of these small forms were brought in by predators, possibly the owls whose remains were also found in the cave. The upper limit of size of the small mammals represented the maximum size of prey and was independent of species. The total absence of the large ubiquitous and arboreal brushtail possum (*Trichosurus vulpecula* (Kerr)) could be explained by this hypothesis, although the ease with which these animals enter and leave the caves today, together with evidence of recent skeletons, suggests a real absence from the earlier fauna.

Some indication of the mode of accumulation of the large mammals in Victoria Cave was indicated by the disposition of their skeletons. The number of partially articulated skeletons increased with distance from the entrance shaft suggesting that some animals survived the fall, the skeletons of those which perished furthest from the entrance being least disturbed by later events. Our recent excavation of fossil material began in the area furthest from the original entrance and continued until we reached surrounding sediments representing the winnowed fine material at the far end of an alluvial fan which was of limited stratigraphic value. At this time our analysis of the large fauna revealed 33 individuals of 4 species of extinct browsing marsupials and 79 individuals of 4 species of large grazing marsupials. Associated with these were the fossils of many arboreal forms, koalas and possums, suggesting an environment comprising sclerophyll forest on the surrounding ridges and savanna woodland in the valleys, a pattern similar to that found in the area in historic times (Wells 1975). Current excavation at the base of the sedimentary cone adjacent to the entrance is producing an entirely different environmental interpretation. Here over 450 large specimens are associated with sedimentary layers reflecting changing pedogenic processes which have occurred outside the cave, an association which was not apparent in the winnowed sediments of the fan. Wells and Moriarty (in prep.) relate the pedogenic evidence to changing climatic regimes in this area of southern Australia.

Preliminary dating using radiocarbon and racemisation techniques yields dates ranging from 2610 ± 80 B.P. to ~50,000 B.P. (table 5.1).

Table 5.1 Radiocarbon (H. Polach pers. comm.) and racemisation dates (J. Bada pers. comm.) for bone and charcoal samples from the top metre of the Victoria cave fossil deposit.

	Depth of sample below surface			
	0-30 cm		30-105 cm	
^{14}C Charcoal	16,700 $^{+3000}_{-2180}$	ANU-1858	>25,100 (~35,500)	ANU-1852
Bone $CO_3^{=}$	2,610 ± 80	ANU-1855	6,650 ± 100	ANU-1859
Apatite	23,000 ± 600	ANU-1856	19,300 $^{+1690}_{-1390}$	ANU-1860
'Collagen' (acid insoluble)	>23,900 (~38,400)	ANU-1857	26,600 $^{+2560}_{-1940}$	ANU-1861
Racemisation D/L aspartic acid	~50,000			

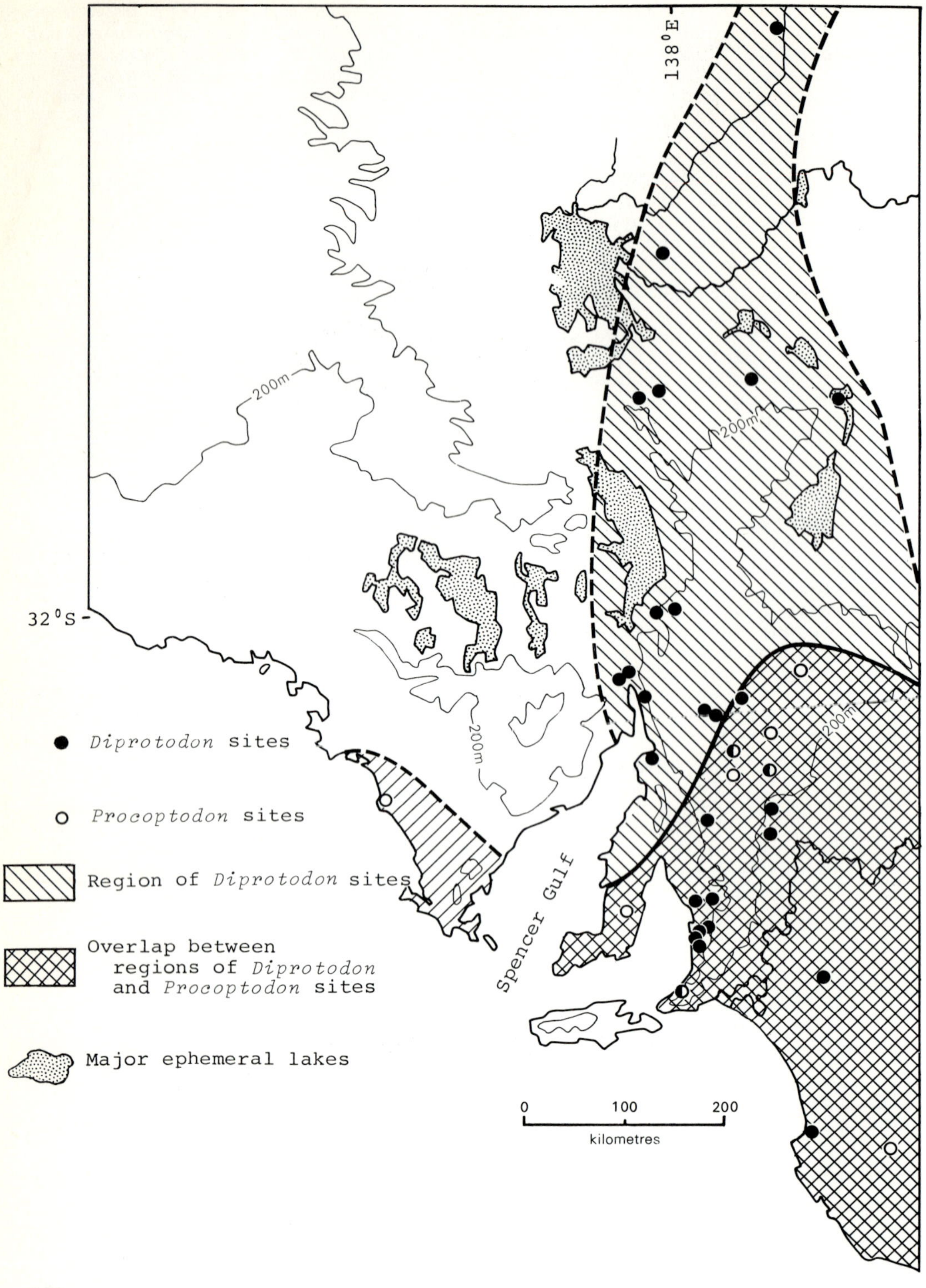

Fig. 5.1 Distribution of fossil sites yielding *Diprotodon* sp. and *Procoptodon* sp. in South Australia.

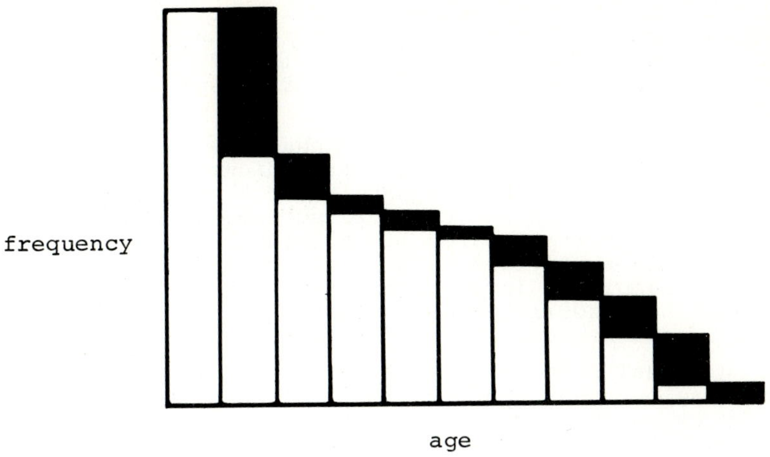

Fig. 5.2a Theoretical age distribution of a living population of ungulates with a potential longevity of 11-12 years.

The white bars are proportional to the numbers of individuals of each age while the dark bars represent the number which must die at each age to account for survivors in next age group (after Voorhies, 1969).

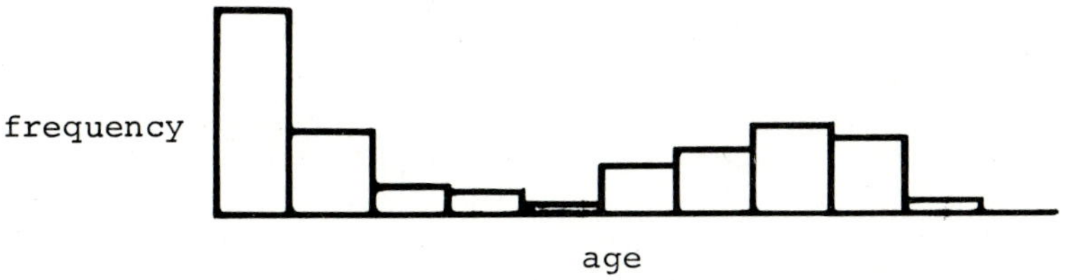

Fig. 5.2b The age distribution of a fossil population accumulated by attritional mortality from fig 2a.

Despite the small charcoal samples there is a gradation in age in the charcoal profile, the bone carbonate indicates exchange with soil carbonate down the profile, and the apatite results indicate some carbonate contamination possibly due to exchange with soil carbonate (H. Polach pers. comm.). 'Collagen' dates required kilogram samples of bone yet even these were insufficient to obtain accurate estimates of potentially the most important dates.

Although variation between these dates emphasises the caution that must be exercised in any chronological interpretations based on single radiocarbon dates, it is interesting to note that the top metre of the 4 metre deep deposit lies within the time of vegetation changes reflected in pollen analyses of cores from nearby lakes (Dodson 1975, 1977). As we can be reasonably certain that Victoria Cave acted as a pitfall trap accumulating a random sample of large herbivores, the fauna can now be analysed in terms of the association of browsing and grazing forms with particular pedogenic, and hence inferred climatic, regimes.

When comparing faunal samples from similar sedimentary sequences a different approach may be used. For example lunette deposits may be difficult to interpret as they often appear to be depauperate. Careful analysis of the associated lunette sediments may give an indication of the climatic regime which existed at times of accumulation of such fossils as may be found (Bowler 1976) but even these can be confused by the presence of burrowing forms such as wombats and bettongs. In times of drought large herbivores tend to congregate around drying water holes, stripping the surrounding countryside of its remaining ground cover and eventually dying and being buried rapidly by wind-blown sediments. The attendant carnivores ultimately suffer a similar fate (Horton 1978). The potential use of such sites to deduce the tolerance range of extinct forms is pointed out by Williams (1976). Here we have the highest probability of an association between extinct and extant forms at their limits of range. The following comparison is used to illustrate this point.

Williams (1976) has compared the fauna and sediments of a closed drainage basin, Dempsey's Lake South Australia, with those of Lake Menindee New South Wales, adjacent to the Darling River (Tedford 1967). Geologically the sites are similar; both lakes occur at approximately the same latitude 32° 20' S, both are bordered by dunes containing fossiliferous calcareous soil horizons. Minimum ages for the faunas range from 15,000 ± 900 BP (ANU-1771) (on charcoal from the fossiliferous calcareous sediments) for Dempsey's Lake to 26,300 ± 1500 BP (LJ-204) and 18,800 ± 800 BP (Gak 335) on *in situ* charcoal for Lake Menindee (Tedford 1967).

The 12 genera which occur at Dempsey's Lake are also found at Lake Menindee (table 5.2). The Menindee site contains extant forms suggestive of a riverine woodland (Tedford 1967) while the fauna of Dempsey's Lake is more typical of sclerophyll scrubland and open plains. Since the geological settings are so similar and the large taxa are relatively abundant, we can infer that the absence of other large forms such as sthenurine kangaroos from Dempsey's Lake is real and a reflection of a more arid environment.

Thus from a careful appraisal of these faunal associations it may eventually be possible to define the limits of range of extinct species in terms of our knowledge of extant forms such that older faunas will be of greater use in environmental reconstructions.

Many Quaternary sites are represented in museum collections by fragments of extinct large mammals. These specimens, often readily assignable to known genera, are the remains of isolated skeletons exposed in stream channels, road cuttings and lake beds. Figure 5.1 shows the distribution of *Diprotodon* sp. (n = 29) and *Procoptodon* (n = 9) sites in South Australia. 79% of *Diprotodon* sites are found in stream channels within or at the edges of the Mt Lofty and Flinders Ranges, while 21% are associated with lake or swamp deposits. It is impossible to tell whether this represents the range of the species or the distribution of fossilisation sites. However, if we compare the distribution of *Procoptodon* sp. (67% in channels, 33% in lakes

Table 5.2 Comparison of the faunas from Lake Menindee, N.S.W. (Tedford 1967) and Dempsey's Lake, S.A. (Williams 1976)

	Lake Menindee	Dempsey's Lake
DASYURIDAE		
Dasyurinae	*Antechinomys sp.*	
	Dasycercus sp.	
	Dasyurus sp.	*Dasyurus* cf. *viverrinus*
	†**Sarcophilus* cf. *harrisii*	†**Sarcophilus harrisii*
Thylacininae	†**Thylacinus* cf. *cynocephalus*	
Myrmecobiinae	*Myrmecobius sp.*	
PERAMELIDAE	*Perameles sp.*	
	Isoodon sp.	
	Chaeropus sp.	
	Macrotis sp.	
PHALANGERIDAE	*Trichosurus sp.*	
PHASCOLARCTIDAE	*Phascolarctos sp.*	
THYLACOLEONIDAE	†*Thylacoleo* cf. *carnifex*	
VOMBATIDAE	†*Phascolomis (Phascolonus) sp.*	†*Phascolonus sp.*
	Lasiorhinus sp.	*Lasiorhinus latifrons*
MACROPODIAE		
Hypsiprymnodontidae	†*Propleopus oscillans*	
Potoroinae	*Bettongia lesueur*	*Bettongia lesueur*
	Bettongia penicillata	*Bettongia penicillata*
	Caloprymnus cf. *campestris*	
Sthenurinae	†*Sthenurus atlas*	
	†*Sthenurus andersoni*	
	†*Sthenurus tindalei*	
	†*Sthenurus sp.*	
	†*Procoptodon goliah*	
Macropodinae	*Lagorchestes leporides*	
	Lagorchestes conspicillatus	
	Onychogalea lunata	
	Onychogalea fraenata	
	Wallabia sp.	
	†*Protemnodon brehus*	†*Protemnodon brehus*
	Macropus rufus	*Megaleia rufa*
	†*Macropus birdselli*	*Macropus eugenii*
	†*Macropus ferragus*	†*Macropus ferragus*
		Macropus robustus
DIPROTODONTIDAE	†*Diprotodon sp.*	†*Diprotodon* cf. *minor*

† extinct

†* extinct on the Australian mainland

or swamps) it is apparent that there is an area of overlap where both species occur together in some deposits while in other areas in which there is equal potential for fossilisation there is no overlap. We might then conclude that the northern boundary for *Procoptodon* in the area of overlap is more likely to be a reflection of real limit of its range rather than an artifact of sampling.

Community structure as inferred from an analysis of the faunal assemblage. In order to make a statement about the composition of the community represented by fossils, it is essential to estimate the minimum number of individuals of each species present. The minimum number is arrived at by examining the most abundant recognisable elements, usually teeth, skull, jaw or limb portions. The relative abundance of a particular species may be an indication of the proximity of its preferred habitat to the site of deposition.

Shotwell (1955) used the degree of completeness of preservation of a species to indicate the proximity of the habitat. Voorhies (1969) has demonstrated the difficulties with this approach as several different mechanisms may be responsible for the fragmented and incomplete nature of the fossil remains; they may be results of selective movement and sorting of portions of skeletons by water action, slumping, etc. or a result of predation. The former may be deduced from the sediments themselves, the latter from forensic studies of the remaining elements (e.g. Brain 1970). Nonetheless studies of a faunal assemblage can provide a more accurate indication of the vegetation structure existing at the time of deposition of the fossils than studies of single species, although even these determinations require careful interpretation of the nature of the faunal sample. If the fauna represents a catastrophic accumulation it is less likely to be representative of the total community e.g. grazing herbivores tend to move in herds and so have a greater probability of occurring in large numbers in any death assemblage than do forest or savanna forms which tend to be more solitary. If the fauna has accumulated by normal attritional mortality it is more likely to be representative of the total community provided the deposit accumulates over a long enough period to include the larger long-lived mammals.

Van Valen (1963, 1964) and Voorhies (1969) have used age frequency distributions to distinguish catastrophic from attritional mortality (figure 5.2). The determination of the age of individuals is critical to these analyses. Fossils of extant mammals can generally be aged using contemporary studies of tooth eruption and wear; this method can be applied in extinct species to determine discrete age classes only if they were seasonal breeders with a seasonal mortality. Therefore the determination of seasonal mortality in fossil faunas depends on either the presence of discontinuous seasonal growth factors or the knowledge that births occurred in a short season. Restricted seasonal breeding is known to occur in Bennett's wallaby *(Macropus rufogriseus)* on Tasmania and the tammar *(M. eugenii)* on Kangaroo Island (Tyndale-Biscoe 1973). Kirkpatrick (1964, 1965) described a method for estimating the age (in days) of many kangaroo species based on molar progression. Knowledge of a restricted breeding season based on daylength, in addition to an age estimate in days, would enable seasonal mortality to be determined in fossil populations of these species. Unfortunately mainland populations of *M. rufogriseus* do not exhibit the restricted breeding period of the island form (M. Smith pers. comm.).

The age frequency distribution of *M. rufogriseus* from Victoria Cave is plotted in figure 5.3. The data indicate catastrophic mortality consistent with the earlier hypothesis of a pitfall trap.

THE LIMITATIONS OF MODERN MAMMALOGY IN PROVIDING A BASIS FOR COMPARISON OF FOSSIL FAUNAS

In the preceding sections I have touched upon the problems encountered in the use of fossil mammal assemblages to reconstruct past environments. I have stressed the need for good stratigraphic information to determine both the environment of deposition and the relative age of the deposit and

I have discussed the limitations of defining past habitats from the inferred ecological requirements of single species within faunal assemblages.

Finally the importance of negative evidence (i.e. the absence of indicator species from contemporaneous faunas) in deducing past environmental changes has been stressed. The major difficulty facing vertebrate palaeontologists in Australia today lies in the poor understanding of the factors limiting the range of living species. Indeed we still know very little about the distribution of many small mammals. In recent years a number of species of mammals thought to be extinct have been discovered, e.g. *Gymnobelideus leadbeateri* McCoy, *Sminthopsis psammophilus* Spencer (Aitken 1971a); *Burramys parvus* Broom (Lane and Richards 1967); *Antechinus apicalis* (Gray) (Morcombe 1967); *Macropus parma* Waterhouse (Ride 1970); extensions of range have been reported for others e.g. *Planigale tenuirostris* Troughton (Aitken 1971b); *Wyulda squamicaudata* Alexander (Calaby 1957); *Cercartetus lepidus* (Thomas) (Aitken 1974, 1977). Moreover, new species and even a new genus of mammals have recently been described e.g. *Planigale gilesi* Aitken (Aitken 1972); *Ningaui timealeyi* Archer and *N. ridei* Archer (Archer 1975).

Ride (1968) pointed out that "almost a fifth of the species (44 of 224 of modern Australian mammals) have been collected less than a dozen times this century". In spite of the dearth of information on the range of modern species, attempts have been made to correlate faunal distributions with climate. As long ago as 1896, Baldwin Spencer (1896) assigned the mammalian genera to three major regions: Bassian, Eyrean and Torresian. The Bassian subgroup, confined to southeastern Australia, was thought to have its affinities with South America, the Torresian subgroup of north-eastern Australia showed affinities with the Papuan region while the large Eyrean subgroup represented the Australian continental fauna which evolved in the vast drainage system surrounding Lake Eyre. Although doubt has been cast on Spencer's grouping of species according to a hypothetical site of origin, Marlow (1958) and Keast (1962) saw these faunal assemblages to be influenced more by climate with the Torresian forms adapted to habitats experiencing summer rainfall, Bassian to those experiencing winter rainfall and Eyrean to areas of unreliable rainfall. Ride (1968) criticized the use of these simplistic correlations, citing many exceptions based on recent collections. Calaby (1971) notes that the mammal distributions can be correlated more easily with structure of the vegetation rather than its floristics and yet even this generalisation tends to be refuted. The pigmy possum *Cercartetus lepidus* once thought to be confined to the cool moist forests of Tasmania has in recent times been collected in heath scrub in Tasmania on Kangaroo Island and from the dry mallee country near Pinnaroo, South Australia (Aitken 1974, 1977), as well as western Victoria (Dixon 1978). Thus modern species distributions alone are of limited use and, with the continuing destruction of habitat by man, we may never know the full extent of the range of many mammals in historical time. The most useful guide to the limits of ranges of many species may well come from the combination of modern physiological and ecological studies together with studies of recent sub-fossils.

This final point brings me back to the beginning of the paper. Palaeoenvironmental reconstructions are based on uniformitarian principles and it is thus only from an understanding of the distribution and limits to ranges of present species in conjunction with careful stratigraphic studies of Quaternary fossil deposits that the ebb and flow of evolutionary events may be unravelled.

ACKNOWLEDGEMENTS

I am indebted to Dr Meredith Smith, Mr Dominic Williams and Alice Wells for their helpful discussion and criticism of the manuscript. During the symposium Drs Michael Archer, Jeanette Hope and Rhys Jones also provided constructive comments on the manuscript. Research funds for studies of fossil marsupials have come from the Australian Research Grants Committee, the South Australian Government and the Flinders University; this support

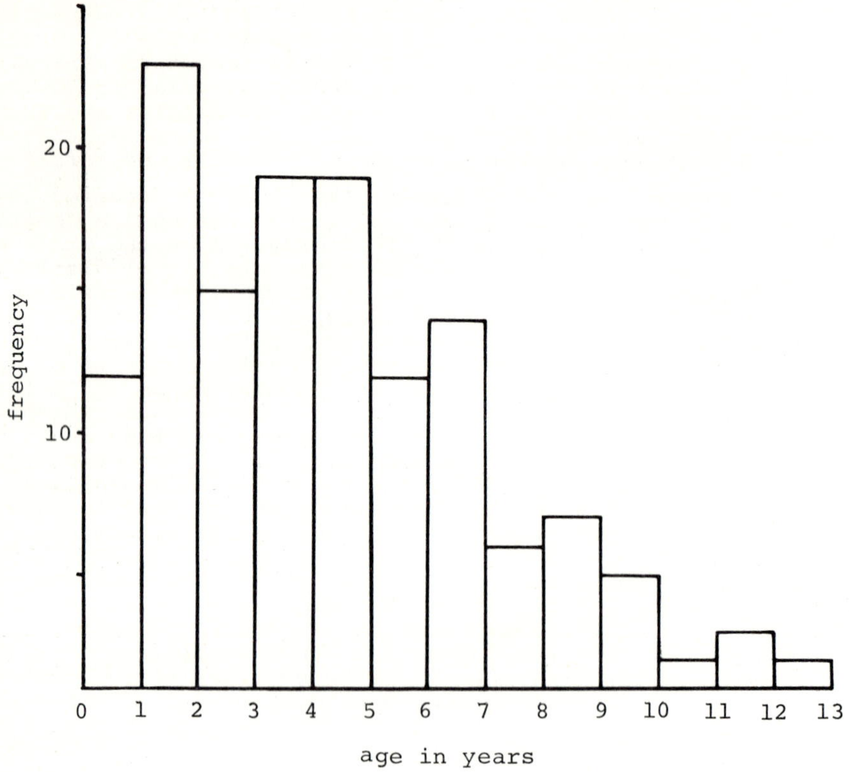

Fig. 5.3 The age/frequency distribution of a fossil assemblage of red-neck wallabies *Macropus rufogriseus* from Victoria Cave, Naracoorte, South Australia.

is gratefully acknowledged.

REFERENCES

Aitken, P.F. 1971a Rediscovery of the large desert Sminthopsis (*Sminthopsis psammophilus* Spencer) on Eyre Peninsula, South Australia. Vic. Nat., 88, 103-111.

Aitken, P.F. 1971b *Planigale tenuirostris* Troughton. The narrow-nosed planigale, an addition to the mammal fauna of South Australia. Sth. Aust. Nat., 46, 18.

Aitken, P.F. 1972 *Planigale gilesi* (Marsupialia, Dasyuridae); a new species from the interior of South eastern Australia. Rec. S. Aust. Mus., 16, 1-14.

Aitken, P.F. 1974 The little pigmy possum *(Cercartetus lepidus* (Thomas)*)* on Kangaroo Is. South Australia. Sth. Aust. Nat., 48, 36-43.

Aitken, P.F. 1977 The little pigmy possum*(Ceracartetus lepidus* (Thomas)*)* found living on the Australian mainland. Sth. Aust. Nat., 51, 63-66.

Archer, M. 1972 *Phascolarctos* (Marsupialia, Vombatoidea) and an associated fauna from Koala Cave, Yanchep, Western Australia. Helectite, 10, 49-59.

Archer, M. 1974a Excavations in the Orchestra Shell Cave, Wanneroo, Western Australia. Arch. and Phys. Anthrop. in Oceania, 9, 156-162.

Archer, M. 1974b Apparent association of bone and charcoal of different origin and age in cave deposits. Mem. Qld. Mus., 17, 37-48.

Archer, M. 1975 *Ningaui,* a new genus of tiny dasyurids (Marsupialia) and two new species, *N. timealeyi* and *N. ridei,* from arid Western Australia. Mem. Qld. Mus., 17, 237-249.

Archer, M. and Baynes, A. 1972 Prehistoric mammal faunas from two small caves in the extreme South-west of Western Australia. Roy. Soc. W.A., 55, 80-89.

Archer, M. and Wade, M. 1976 Results of the Ray E. Lemley Expeditions, Part 1. The Allingham Formation and a New Pliocene vertebrate fauna from Northern Queensland. Mem. Qld. Mus., 17, 379-397.

Bartholomai, A. 1962 A new species of *Thylacoleo* and notes on some caudal vertebrae of *Palorchestes azael.* Mem. Qld. Mus., 14, 33-40.

Bartholomai, A. 1963 Revision of the extinct macropodid genus *Sthenurus* Owen in Queensland. Mem. Qld. Mus., 14, 51-76.

Bartholomai, A. 1968 A new fossil koala from Queensland and a reassessment of the taxonomic position of the problematic species *Koalemus ingens* De Vis. Mem. Qld. Mus., 15, 65-71.

Berger, A. 1977 Power and limitation of an energy-balance climate model as applied to the astronomical theory of palaeoclimates. Palaeogeography, palaeoclimatology and palaeoecology, 21, 227-235.

Blanc, A.C., Tongiorgi, E.L. and Trevison L. 1953 Le Pliocene et la Quaternaire aux alentours de Rome. Livret-guide IV Congr. Int. INQUA, Rome.

Bowler, J.M. 1976 Aridity in Australia: age, origins and expression in aeolian landforms and sediments. Earth Science Reviews, 12, 279-310.

Brain, C.K. 1970 New finds at the Swartkrans Australopithecine site. Nature, 225, 1112-1119.

Butzer, K.W. 1971 Environment and Archaeology, an ecological approach to prehistory. Aldine Atherton, Chicago.

Calaby, J.H. 1957 A new record of the scaly-tailed possum (*Wyulda squamicaudata* Alexander). W.A. Natur., 5. 186-192.

Calaby, J.H. 1971 Man, Fauna, and Climate in Aboriginal Australia. pp 80-93 in Aboriginal Man and Environment in Australia (eds J. Mulvaney and J. Golson), Australian National University Press, Canberra.

Cooper, H.M. 1959 Large archaeological stone implements from Hallett Cove. Trans. Roy. Soc. S.A., 82, 55-60.

Dalrymple, G.B. 1972 Potassium-argon dating of geomagnetic reversals and North American glaciations. pp 107-134 in Calibration of Hominoid Evolution (eds W.W. Bishop and J.A. Miller), Scottish Acad. Press, Edinburgh.

Devereux, I. 1968 Oxygen isotope palaeotemperatures from the Tertiary of New Zealand. Tuatara, 16, 41-44.

Dixon, J.M. 1978 The Little Pigmy Possum in Victoria. Vic. Nat., 95, 4-7.

Dodson, J.R. 1974 Vegetation and climatic history near Lake Keilambete, western Victoria. Aust. J. Bot., 22, 709-717.

Dodson, J.R. 1975 Vegetation history and water fluctuations at Lake Leake, south-eastern South Australia. II. 50,000 BP to 10,000 BP. Aust. J. Bot., 23, 815-831.

Dodson, J.R. 1977 Late Quaternary Palaeoecology of Wyrie Swamp, south-eastern South Australia. Quat. Res., 8, 97-114.

Edwards, W.E. 1967 The late-Pleistocene extinction and dimunition in size of many mammalian species. pp 141-154 in Pleistocene extinctions: the search for a cause (eds P.S. Martin and H.E. Wright), Yale University Press, New Haven.

Emiliani, C. 1961 Cainozoic climatic changes as indicated by the stratigraphy and chronology of deep-sea cores of globigerina-ooze facies. Annals. N.Y. Acad. Sci., 95, 521-536.

Emiliani, C. 1966 Palaeotemperature analysis of Caribbean cores P6304-8 and P6304-9 and a generalised temperature curve for the past 425,000 yrs. J. Geol., 74, 109-126.

Frank, R. 1971 The clastic sediments of the Wellington Caves, N.S.W. Helectite, 9, 3-26.

Gill, E. 1953 Distribution of the Tasmanian devil, the Tasmanian wolf and the dingo in S.E. Australia in Quaternary time. Vic. Nat., 70, 86-90.

Gill, E. 1973 The Geology and Geomorphology of the Murray River Region between Mildura and Renmark, Australia. Mem. Nat. Mus. Vic., 34, 1-97.

Goede, A. and Murray, P. 1975 Vertebrate Remains from Pleistocene Cave near Montagu, N.W. Tasmania. Unpublished manuscript of paper presented at Int. Symp. of Cave Biology and Cave Palaeontology, Int. Union of Speleology, S. Africa.

Goulden, C.E. 1969 Interpretative studies of cladoceran microfossils in lake sediments. Mitt. Int. Verein Limnol., 17, 43-55.

Guilday, J.E. 1967 Differential extinction during Late Pleistocene and recent times. pp 121-140 in Pleistocene extinctions: the search for a cause (eds P.S. Martin and H.E. Wright), Yale University Press, New Haven.

Harmon, R.S., Thompson, P., Schwarez, H.P. and Ford, D.C. 1975 Uranium-Series Dating of Speleothems. Nat. Speleological Soc. Bull., 37, 21-33.

Hays, J.D., Imbrie, J. and Shackleton, N.J. 1976 Variation in the earth's orbit and pacemaker of the Ice Ages. Science, 194, 1121-1132.

Hecky, R.E. and Kilham, P. 1973 Diatoms in alkaline, saline lakes: ecology and geochemical implications. Limnol. Oceanogr., 18, 53-71.

Hope, G.S. 1974 The vegetation history from 6000 B.P. to present of Wilson's Promontory, Victoria, Australia. New Phytol., 73, 1035-1053.

Horton, D.R. 1978 Extinction of the Australian megafauna. Aust. Inst. Aboriginal Studies Newsletter, New Series 9.

Juvigne, E. 1975 Note on Pollen Extraction from Coarse Sediments. Quaternary Research, 5, 121-123.

Keast, A. 1962 Vertebrate speciation in Australia: some comparisons between birds, marsupials and reptiles. pp 380-407 in The Evolution of Living Organisms (ed G.W. Leeper).

Kershaw, A.P. 1970 A pollen diagram from Lake Euramoo, north-east Queensland, Australia. New Phytol., 69, 785-805.

Kershaw, A.P. 1974 A long continuous pollen sequence from north-eastern Australia. Nature, Lond., 251, 222-223.

Kershaw, A.P. 1975 Stratigraphy and pollen analysis of Bromfield Swamp, north eastern Queensland, Australia. New Phytol., 75, 173-191.

Kirkpatrick, T.H. 1964 Molar progression and macropod age. Qld. J. Agric. Anim. Sci., 21, 163-165.

Kirkpatrick, T.H. 1965 Studies of Macropodidae in Queensland 2. Age determination in the grey kangaroo, red kangaroo, the eastern wallaroo and the red-necked wallaby, with notes on dental abnormalities. Qld. J. Agric. Anim. Sci., 22, 301-317.

Kurten, B. 1972 The Ice Age. Hart-Davis, London.

Lane, E.A. and Richards, Aola M. 1967 *Burramys parvus* Broom, a living fossil. Helictite, 5, 30-34.

La Rocque, A. 1960 Molluscan faunas of the Flagstaff Formation of Central Utah. Mem. Geol. Soc. Amer., 78, 1-100.

Lundelius, E.L. Jr. 1963 Vertebrate remains from the Nullarbor Caves Western Australia. J. Roy. Soc. W.A., 46, 75-80.

McDougall, I. and Wensink, H. 1966 Palaeomagnetism and geochronology of the Pliocene-Pleistocene lavas in Iceland. Earth Planet. Sci. Lett., 10, 179-185.

McKenna, M.C. 1962 Collecting small fossils by washing and screening. Curator, 5, 221-235.

Mahoney, J.A. and Ride, W.D.L. 1975 Index to the Genera and Species of Fossil Mammalia Described from Australia and New Guinea between 1838 and 1968. Spec. Publs. West Aust. Mus., 6.

Marlow, B. 1958 A survey of the marsupials of N.S.W. C.S.I.R.O. Widl. Res., 3, 71-114.

Marshall, L.G. 1973 Fossil vertebrate faunas from the Lake Victoria region, S.W. New South Wales, Australia. Mem. Nat. Mus. Vic., 34, 151-172.

Martin, P.S. 1967 Prehistoric Overkill. pp 75-120 in Pleistocene Extinctions: the search for a cause (eds P.S. Martin and H.E. Wright), Yale University Press, New Haven.

Martin, P.S. 1973 The Discovery of America. Science, 179, 969-974.

Merrilees, D. 1968 Man the destroyer: Late Quaternary changes in the Australian marsupial fauna. Proc. R. Soc. West. Aust., 51, 1-24.

Milankovitch, M. 1930 Mathematische Klimalehre und astronomische Theorie der Klimaschwankungen. Handb. Klimatol., 1, 176.

Milankovitch, M. 1938a Neue Ergebnisse der astronomischen Theorie der Klimaschwankungen. Bull. Acad. R. Serbe Sci. math. nat., Belgrade (A), 4, 41.

Milankovitch, M. 1938b Astronomische Mittel zur Erforschung der erdgeschichtlichen Klimate. Handb. Geophys., 9, 593-698.

Morcombe, M.K. 1967 The re-discovery after 83 years of the Dibbler *Antechinus apicalis* (Marsupialia, Dasyuridae). West. Aust. Nat., 10, 102-111.

Murray, P. and Goede, A. 1977 Pleistocene Vertebrate Remains from a cave near Montagu, N.W. Tasmania. Rec. Q. Vic. Mus., 60, 1-30.

Rich, T.H. 1976 Recent Fossil Discoveries in Victoria. Vic. Nat., 93, 198-206.

Ride, W.D.L. 1968 On the past, present and future of Australian mammals. Aust. J. Sci., 31, 1-11.

Ride, W.D.L. 1970 A guide to the native mammals of Australia. Oxford University Press, London.

Scholander, P.F. 1955 Evolution of climatic adaptation in homeotherms. Evolution, 9, 15-26.

Scholander, P.F. 1956 Climatic rules. Evolution, 10, 339-340.

Scott, H.H. 1915 A monograph of *Nototherium tasmanicum*. Rec. Geol. Surv. Tasm., 4, 1-47.

Shackleton, N.J. and Opdyke, N.D. 1973 Oxygen isotope and palaeomagnetic stratigraphy of equatorial Pacific core V28-238: Oxygen isotope temperatures and ice volumes on a 10^5 and 10^6 year scale. Quaternary Research, 3, 39-55.

Shotwell, J.A. 1955 An approach to the palaeoecology of mammals. Ecology, 36, 327-337.

Smith, M.J. 1971 Small Fossil Vertebrates from Victoria Cave, Naracoorte, South Australia. I. Potoroinae (Macropodidae), Petauridae and Burramyidae (Marsupialia). Trans. R. Soc. S. Aust., 95, 185-198.

Smith, M.J. 1972 Small Fossil Vertebrates from Victoria Cave, Naracoorte, South Australia. II. Peramelidae, Thylacinidae and Dasyuridae (Marsupialia). Trans. R. Soc. S. Aust., 96, 125-137.

Specht, R.L. 1972 The Vegetation of South Australia (2nd Edn). Govt. Printer, Adelaide.

Spencer, W.B. 1896 Summary of the zoological, botanical and geological results of the Expedition. pp 139 and 196-199 in Part I of Report on the work of the Horn Scientific Expedition to Central Australia (ed W.B. Spencer), Dulau, London.

Stephenson, N.G. 1967 Phylogenetic trend and speciation among wombats. Aust. J. Zool., 15, 873-880.

Stirton, R.A. 1955 Late Tertiary marsupials from South Australia. Rec. S. Aust. Mus., 11, 247-268.

Stirton, R.A., Tedford, R.H. and Miller, A.H. 1961 Cenozoic stratigraphy and vertebrate palaeontology of the Tirari Desert, South Australia. Rec. S. Aust. Mus., 14, 19-61.

Stirton, R.A., Tedford, R.H. and Woodburne, M.O. 1967 A new Tertiary formation and fauna from the Tirari Desert, South Australia. Rec. S. Aust. Mus., 15, 427-462.

Tasch, P. and Zimmerman, J.R. 1961 Fossil and living conchostracan distribution in Kansas-Oklahoma across a 200-million year time gap. Science, 133, 584-586.

Tedford, R.H. 1966 A review of the macropodid genus *Sthenurus*. Univ. Calif. Publ. Geol. Sci., 57, 1-72.

Tedford, R.H. 1967 The fossil Macropodidae from Lake Menindee, N.S.W. Univ. Calif. Publ. Geol. Sci., 64, 1-156.

Tedford, R.H. 1973 The Diprotodons of Lake Callabonna. Aust. Nat. Hist., 17, 349-354.

Tedford, R.H., Archer, M., Bartholomai, A., Plane, M., Pledge, N.S., Rich, T., Rich, P. and Wells, R.T. 1977 The discovery of Miocene vertebrates, Lake Frome area, South Australia. J. Aust. Geol. Geophys., 2, 53-57.

Thompson, P., Schwarcz, H.P. and Ford, D.C. 1976 Stable isotope geochemistry, geothermometry, and geochronology of speleothems from West Virginia. Geol. Soc. Am. Bull., 87, 1730-1738.

Tricart, J. 1975 Existence de periode seche au Quaternaire en Amazonie et dans les regions voisins. Rev. Geomorphol. Dyn., 4, 145-158.

Tutin, W. 1960 The usefulness of pollen analysis in interpretation of stratigraphic horizons, both Late-glacial and Post-glacial. Mitt. Int. Ver. Limnol., 17, 154-164.

Tyndale-Biscoe, H. 1973 Life of Marsupials. Edward Arnold, London.

Van Valen, L. 1963 Selection in natural populations: *Merychippus primus*, a fossil horse. Nature, 197, 1181-1183.

Van Valen, L. 1964 Age in two fossil horse populations. Acta Zoologica, 45, 94-106.

Veeh, H.H. and Chappell, J. 1970 Astronomical theory of climatic change: support from New Guinea. Science, 167, 862-865.

Voorhies, M.R. 1969 Taphonomy and population dynamics of an Early Pliocene vertebrate fauna, Knox County, Nebraska. Univ. of Wyoming Contributions to Geol. Special Paper No. 1.

Wells, R.T. 1973 Physiological and behavioural adaptations of the hairy-nosed wombat (*Lasiorhinus latifrons* (Owen)) to its arid environment. Ph.D. thesis, University of Adelaide.

Wells, R.T. 1975 Reconstructing the Past. Excavations in Fossil Caves. Aust. Nat. Hist., 18, 208-211.

Wells, R.T. 1978 Thermoregulation and activity rhythms in the hairy-nosed wombat *Lasiorhinus latifrons* (Owen) (Vombatidae). Aust. J. Zool., (in press).

West, R.G. 1968 Pleistocene Geology and Biology. Longmans, London.

Wigley, T.M.L. 1977 Spectral analysis and the astronomical theory of climatic change. Nature, 264, 629-631.

Wilkinson, E. 1972 The Duck Ponds fossil marsupial fauna, Hovell's Creek, Lara, Victoria, Australia. Mem. Nat. Mus. Vic. 33, 41-46.

Williams, D.L.G. 1976 The vertebrate palaeontology and stratigraphy of Dempsey's Lake (Pt. Augusta) and its relationship to Quaternary faunal succession in southern Australia. B.Sc. thesis, Flinders University, South Australia.

Woodburne, M.O. 1967 The Alcoota fauna, central Australia: An integrated palaeontological and geological study. Aust. Bur. Miner. Resour, Geol. Geophys., Bull., 128, 1-187.

Woods, J.T. 1958 The extinct marsupial genus *Palorchestes* Owen. Mem. Qld. Mus., 13, 177-193.

Woods, J.T. 1960 Fossiliferous fluviatile and cave deposits, in Chapter XIII, Upper Cainozoic, Geology of Queensland. Bull. Geol. Soc. Aust. 7, 393-403.

Wright, H.E. and Frey, D.G. (eds) 1965 International Studies on the Quaternary. Special Papers, 84, Geological Society of America, New York.

Wright, A.E. and Moseley, F. (eds) 1975 Ice Ages: Ancient and Modern. Geol. J. Special Issue No. 6.

Zeuner, F.E. 1959 The Pleistocene Period, its climate, chronology and faunal successions. Hutchinson, London.

6. SPECIAL PROBLEMS IN THE INTERPRETATION OF THE HUMAN REMAINS

by A.G. Thorne

To many Quaternarists, and certainly the majority of non-specialists, man and his ancestors are basic or essential ingredients in reconstructions of past environments. It is common, for example, to see the glacial phases of Europe reconstructed as a landscape in terms of a sombre Neanderthal staring into icy space while a cave bear lurks menacingly in the background. The Middle Pleistocene Asian tropics are similarly endowed with groups of hairy Pithecanthropines tarzaning through the jungle past incredulous stegodonts. Given the widespread interest in palaeoanthropology it is not surprising how many books, museum dioramas, models and "imaginative reconstructions" have man in the foreground or at least somewhere in the scene. What is surprising, however, is how little of the reconstruction stems from hominid data directly and how many assumptions or misunderstandings go into the ideas that are conveyed, intentionally or otherwise, to specialist and layman alike.

Quaternary hominid remains, I believe, tell us precious little, if anything at all, about the environments of which they were parts, or at least very little that can be understood at present. The reasons are numerous but include difficulties with available samples and their interpretation, population densities and movements, phenotypic responses to particular conditions and the interplay or culture and biology. While *Homo sapiens* and his forebears are but one limited source for biological information about the past, it is likely that the general and special problems of the human remains are mirrored to some extent by other animal and plant data used in environmental reconstruction.

The most fundamental problem lies in the size and nature of the fossil sample. Compared with many other areas of palaeontology, hominid research is based on extremely small and unevenly distributed samples. The Neanderthals again illustrate the sample size difficulty, but they are by no means an extreme case. It is the practice for this well-known group of hominids to be described in quite general biological terms, as though there was a wealth of skeletal and other data permitting rounded statements. The actual remains however, consist of not more than 20 reasonably complete crania, some of which have associated mandibular parts. Of those preserving part of the dentition, some have relatively unworn teeth. There are fewer postcranial skeletons than crania and in all Neaderthal Man is based on the anatomically diverse remains of about 100 individuals. These materials stem from sites as widely separated as Gibraltar and the U.S.S.R., north Germany and Israel. Some of the remains may be less than 25,000 years old while others date to more than 100,000 years ago. While there are infants and juveniles represented in the available study material, there are also the remains of aged individuals, including the old man from La Chappelle-aux-Saints, around whom so many of the earlier misconceptions about Neanderthals developed.

This then is the sample of a so-called population, which only a generation ago was regarded as a major phase in human evolution. It would no doubt come as a surprise to many to realise that in the last decade more Pleistocene hominids have been recovered in Australia than all the Neanderthals put together. While there are some Australian individuals represented by fragments there are numerous near-complete skeletons, few are subadult and the total time distribution for the series is only 20,000 years. They are drawn from an area in southeastern Australia that is roughly the size of France and, if the Keilor and Talgai individuals are excluded, more than 100 individuals have now been recovered from an area that is smaller than Switzerland.

Lest the comparison of the Australian and late Pleistocene European hominids be taken to imply that the understanding of the former skeletal sequence is clear, it should be remembered that within the Swiss-sized chunk of Australia there were substantial environmental changes between, say, 30,000 and 10,000 years before present and a series of morphological changes in the human population of that period that were as dramatic as those involving the Neanderthal and Cro-Magnon populations. Yet the reasons behind the changes in skeletal form among early Australians are, to put it mildly, far from clear. There is no agreement by osteologists as to the source of the changes; whether they stem from an inflow of new people from outside Australia who joined or replaced long-established inhabitants, whether the changes were the result of important environmental alterations taking place within Australia that are reflected in human morphology, or some combination of these processes.

The group of fossils known as *Homo erectus* presents similar problems. It also illustrates the difficulty of assessing the biological or environmental nature of particular sites from the data they produce. The number of *erectus* sites has expanded significantly in recent years and there are now some 20 localities in Europe, Africa and Asia at which skeletal discoveries have been made. Yet how typical are these sites? Very few are on or close to the sea coast (of that period) yet much of the cultural evidence drawn from *erectus* sites suggest lifestyles similar to that seen in near-contemporary hunting and gathering societies, where coastal or riverine groups are the most representative, at least in terms of demography. Lacking the data, it is not possible to demonstrate that Pleistocene man was essentially a coastal species, although his earliest movements from Africa to southeast Asia suggest that for much of the Pleistocene his highest population densities were in or close to coastal regions. If this can be shown to be the case, much of our current skeletal data may prove atypical in some respects. At present we risk a view of Pleistocene man that resembles contemporary views of the Australian Aborigines where the people of the central desert are regarded as good examples of Australian biology and society simply because so much information has been obtained from the arid interior relative to coastal areas.

If the spread of *erectus* during middle Pleistocene times was achieved because he liked to be beside the sea, then we lack the crucial sites. Without them, how can we say whether the north Chinese population represented by the Choukoutien sample is typical or atypical of China at the time, assuming we knew what the time period was, which we do not? How can we hope accurately to interpret the possible environmental implications of particular Pleistocene morphologies when we remain so ignorant of the role that recent environments have played in forming human populations? There are many clues to the reasons for the physical differences between major human groups today, but as environmental evidence they remain circumstantial.

Perhaps the clearest demonstration of the difficulties facing those attempting to relate hominid and environmental data is the present status of hominid taxonomy for the Quaternary period. There are a number of taxonomic views of the hominids that need to be rationalised before morphological variability can be assessed. At one extreme are specialists who regard modern man as a development from ancestral forms owing nothing to *Homo erectus*. Among those who recognise *erectus-sapiens* as a subgeneric or subspecific sequence there is considerable disagreement about rates of morphological and general evolutionary change, when and where the significant changes took place and the status of particular specimens. There is the somewhat localised African problem of the robust and gracile early Pleistocene hominids (assuming that certain Javan material is ignored) with no general agreement on the taxonomy or the implications of specific, subspecific or general distinctions that might be made. With unresolved anatomical problems of this magnitude the notion that facial shape, skin colour or other distinguishing feature of contemporary populations reveals something of their development and of its environmental component, is simple indeed.

If the purely biological problems outlined above were not enough, students of Quaternary hominids face an additional set of questions, namely the bearing of social organisation and cultural development on morphology. While "society and culture" are undoubtedly far more important to palaeo-anthropologists than to other palaeontologists, it is likely that even here social flexibility and its effects on morphological adaptability is too often the subject of assumption rather than investigation. Man is the most flexible species behaviourally but not the only one capable of social and cultural insulation. Nevertheless, the capacity of contemporary man, hunter-gatherer or no, to speed up, slow down or in some way influence his biology, makes analysis and comparison of historic data difficult, let alone that of prehistoric or lower Pleistocene material. The relevance of tools, clothing, weapons, transport, food preparation and storage, are too obvious to require examples of the influences they may play in shielding individuals or populations from stable or changing conditions that men may want to enter, remain in or leave. The basic issue here is one of slowing down or minimising environmental effects on morphology. The capacity of human societies to do the reverse, to speed up morphological change or narrow its trends, is another element in this complex of problems. To take this difficulty into account anthropologists should not only be able to recognise situations of this kind but be able to measure its effects. They appear unable to do either at present.

Much of what we know or are learning about Quaternary environments is the result of information gathered during the investigation of archaeological sites or areas that are being examined for traces of man. Attempts to describe the environments in which past human populations moved depend in many instances on the debris produced by man. It is ironic that man himself is almost silent on the subject.

7. POPULATION GENETICS AND RATES OF EVOLUTION

by I.R. Franklin

SUMMARY

Evolutionary change can be measured from the fossil record, or inferred from the genetic relationships between extant species. Rates of evolution have been estimated from differences in protein or DNA sequence (the molecular level), from changes in the phenotype (the morphological level), or from first and last appearances in the fossil record (the taxonomic level). Change at each of these levels is discussed.

INTRODUCTION

It is often thought that evolutionary processes are too slow to be observed directly. While this is true of many of them, there are numerous examples of rapid evolution, many of which are direct responses to man's impact on world ecosystems. Eutrophication of fresh water lakes, for example, leads to marked changes in the diversity and abundance of plant and animal communities. Disturbance of the habitat often results in local and sometimes global extinction. During the last 2000 years over one hundred mammalian species or subspecies have become extinct, and over two thirds of these extinctions have occurred since the mid-nineteenth century. These, of course, are dramatic changes but subtler evolutionary processes are occurring continually, a number of which are well documented and well understood. Resistance to insecticides is now widespread in many insect species (Brown 1967). Marked changes in the susceptibility of rabbits, and the virulence of myxoma, occurred rapidly following the introduction of myxomatosis to Australia (Fenner 1965). Populations of the grass *Agrostis tenuis* which are tolerant of lead, zinc, nickel, or copper have been found on the tailings of abandoned mine workings (McNeilly and Bradshaw 1968). A classic example is that of industrial melanism in Lepidoptera. Between 1800 and 1900 a dark form of *Biston betularia*, known to carry a dominant gene (*carbonaria*), spread through the industrial areas of England. The original light form is inconspicuous to predatory birds on a background of pale lichens, but in industrial areas, where the lichens have been killed and tree trunks blackened by soot, the light form has become conspicuous. In these areas the dark form is at an advantage. Kettlewell (1956) calculated the predation rates for light and dark coloured moths, and Haldane (1956), using a model developed in the 1920's, showed that the observed selection intensities could account for the increase in the carbonaria gene.

Industrial melanism has served as a paradigm for the evolutionary process. From the geneticist's point of view evolutionary change is any alteration in gene or genotype frequency in a population of individuals. A mathematical theory has been developed to explain single gene substitutions, and it is presumed that the same theory can account for changes in continuously varying (quantitative) traits, where the inheritance is less well understood. Almost all measurements derived from the fossil record fall into this latter category.

RATES OF EVOLUTION CALCULATED FROM MORPHOLOGICAL DIFFERENCES IN THE FOSSIL RECORD

The fossil record provides us with two kinds of information on evolutionary rate, both of which are covered by Simpson in his classic work 'Tempo and Mode in Evolution' (1944). The first is the measurement of traits in a phylogeny which are thought to be involved in the adaptation and evolutionary success of the lineage. Simpson, for example, chose to study the teeth of horses, which are readily identified in fossils and are important in grazing efficiency. Measurements of the length of the outer ridge (ectoloph) and height of the main cusp (paracone) characterise the grinding surfaces, and the two measures have been shown to vary at different rates. The second kind of information is the rate of taxonomic extinction and origination, and this will be discussed later.

Simpson showed that any trait or combination of traits could change markedly in the course of phyletic evolution, and proposed that within a larger taxonomic group (such as order or class), there is generally a modal rate typical of the group. The distribution of rates within such a group is typically negatively skew, i.e. the mode is greater than the mean, and often close to the maximum for the group. Simpson called this distribution of standard rates the horotelic rate. In some lineages Simpson observed very slow rates, which he considered to be statistically discontinuous from the standard (horotelic) rates, and these he called bradytelic rates. Very rapid rates, on the other hand, were called tachytelic.

An example of bradytelic rates can be found in the fossil record of the lungfish. This group showed rapid rates about 300 My ago, but has changed very little in the last 200 My (Simpson 1953). Perhaps the most extreme conservatism is found in the brachiopod *Lingula*, which has persisted with little change for 500 My. The evolution of the clubmosses (*Lycopodium* spp.) is an example of bradytelic rates in the plant kingdom. Tachytely is observed more rarely due to sparseness of the fossil record, and Simpson suggested that such rates occur when populations shift from one adaptive zone to another.

Measurement and comparison of evolutionary rates.

These measures of morphological change mean very little to the population geneticist since they tell practically nothing about the selective forces involved. The first problem is one of scale. We know from selection theory that response to selection for any trait is a function of its genetic variability, and the generation time. We should therefore standardise the rates by dividing by these two quantities, neither of which can be calculated from the fossil record. If the rates are measured on character complexes, we need even more information (e.g. correlations).

Haldane (1949) made a number of suggestions about the measurement of rates, none of which are completely satisfactory. One possibility is to measure morphological change using percentage rather than absolute values. Statistically this is equivalent to measuring changes in logarithms. For a trait that has changed from x_1 to x_2 in t years, the appropriate measure is:

$$\frac{\log_e(x_2) - \log_e(x_1)}{t}$$

This procedure will correct for the dependence of variability (and hence opportunity for selection) on the mean, which is a common observation in metric traits. Where such dependence does not exist the above formula would be inappropriate.

Alternatively we could take as the unit of change the movement through one standard deviation. This suggestion of Haldane's was followed up by Lerman (1965). The major difficulty here is that standard deviations are less accurately measured than means.

Even if we could obtain reliable estimates of rates, appropriately corrected for genetic variability and generation time (through measurement and analogy with extant species) we are still far from a meaningful genetic interpretation. How do we compare, for example, rates of change in unworn third molars during the evolution of the horse with the changes in cranial capacity in the genus *Homo*? What are the selective forces involved, and do we know if we are looking at a primary adaptive trait, or a correlated change due to selection for some other characteristic? We shall see later that information at the protein sequence level allows more meaningful comparisons, since these can be directly related to genetic change. The question of the selective forces involved still remains however.

Notwithstanding all the above objections, natural selection operates at the level of the phenotype, not the genotype, and morphological changes provide a necessary insight into, and a description of, the course of evolution. For the present the best we can do is examine the trends, and the rates, and ask if these are adequately explained by current genetic theory.

In order to simplify comparisons, we shall use yet another suggestion of Haldane's, namely that a change by the factor e (base for natural logarithms) in one million years (1/1000 in 1000 years) should be called a <u>darwin</u>. Using this measure, the average rate for changes in paracone height and ectoloph length in Simpson's data is about 0.04 darwins. Kurten (1959), in a review of evolutionary rates in Tertiary mammals, calculated the average change in size over four groups to be 0.023 darwins.

<u>Evolutionary rates in the Quaternary</u>.

A commonly measured quantity in the Quaternary fossil record is brain volume in the genus *Homo*. *Homo habilis*, which is now estimated to have existed two million years ago, had an estimated brain volume (from three fossils) of 640 ml. *Homo sapiens* currently has an average value of 1360 ml. (*Homo erectus*, dating about 500,000 years ago, had mean capacity of 930 ml, with a standard deviation about 100 ml). The ratio of capacity of *H. sapiens* to *H. habilis* is 2.13, hence the rate of change in darwins is 2.13/2e = 0.39. This rate is an order of magnitude greater than the values for Tertiary mammals.

Kurten (1959) in an examination of fossils from the Pleistocene, has found rates of change even greater than this. Figure 7.1 shows the changes in size of brown bears (*Ursus etruscus* to *Ursus arctos*) during their Pleistocene history. There are marked fluctuations in size, with increases during glaciations and decreases during the interglacials. Rates of change vary from 0.3 to 3.2 darwins over this period, and there is a rapid decrease in size during the postglacial at a rate of 13.8 darwins. Kurten showed that other European mammals also changed dramatically in size during the postglacial.

Aside from the dramatic changes in the postglacial, it appears that evolutionary rates during the Quaternary (both in morphological change, and in extinction rates) have been much greater than in the Tertiary. Kurten's data suggest that cyclic changes in mammalian size have followed the climatic variations in the Pleistocene. Only a good fossil record with close spacing between samples could have detected these episodic changes.

Finally it should be mentioned that the rates calculated by Kurten, while fast by comparison with previously estimated rates, are many orders of magnitude slower than can be achieved in animal breeding programs. As Haldane pointed out, artificial selection produces responses that must be measured in kilodarwins.

FITNESS AND THEOREMS OF NATURAL SELECTION

A lot of confusion surrounds the concept of fitness, and this has resulted in serious problems of communication amongst population biologists. It is unfortunate that this concept, which is central to evolutionary theory, should not have a commonly agreed definition. Most usually fitness is used by geneticists to describe the expected contribution of an individual to the next generation. This definition has several difficulties, the most serious of which is that an individual (diploid genotype) does not reproduce itself, but produces offspring which have a different array of genes. Theoretical ecologists have used concepts such as rate of population increase (r), or carrying capacity (K) as measures of fitness (see MacArthur and Wilson 1967).

In this paper, I have chosen to define fitness in terms of the logarithmic rate of increase in number (of genes, or representatives of a species), a definition first used by Fisher (1930) and called the 'Malthusian parameter of population increase' (m). At the species level m is often known as the intrinsic rate of increase. Note that if a population is unchecked, it will tend to increase (or decrease) exponentially, which means that

$$\frac{1}{N}\frac{dN}{dt} = \frac{d\log_e N}{dt} = \text{constant (m)}$$

Hence after time t, the population size in relation to the size at time zero (N_o) is:

$$N_t = N_o e^{mt}$$

Time (t) is most meaningfully measured in generations.

The fundamental theory of natural selection.

In 1930, Fisher showed that, under certain idealised conditions, the rate of change in the mean fitness of a population is:

$$\frac{d\bar{m}}{dt} = V_A(m)$$

where $V_A(m)$ is the additive genetic variance in fitness (sometimes called the genic variance in fitness). V_A is the variance associated with the average effects of genes on a trait, in this case fitness. Since a variance is necessarily non-negative, it follows that $d\bar{m}/dt$ is also, hence fitness defined in these terms will always tend to increase. This relationship, the 'Fundamental theorem of natural selection', has been much debated and much misunderstood. A proof and extensive discussion can be found in Kimura (1958).

Suppose now that there is a trait (x), and that the relationship between the trait and fitness is a simple linear one, i.e.,

$$m = a + bx$$

where a and b are constants. Any change in x, Δx, will be associated with a change in fitness

$$\Delta m = b\Delta x$$

Now it can be shown that the rate of change of the mean (in generations) is:

$$\frac{d\bar{x}}{dt} = bV_A(x)$$

where $V_A(x)$ is the additive genetic variance in x. Since $d\bar{x}/dt$, and $V_A(x)$ are in principle measurable quantities, we can calculate b, the regression of fitness on the trait.

Consider for example the increase in human brain capacity at a rate which we have already indicated has been quite rapid as evolutionary changes go. Suppose that there has been a linear increase in capacity

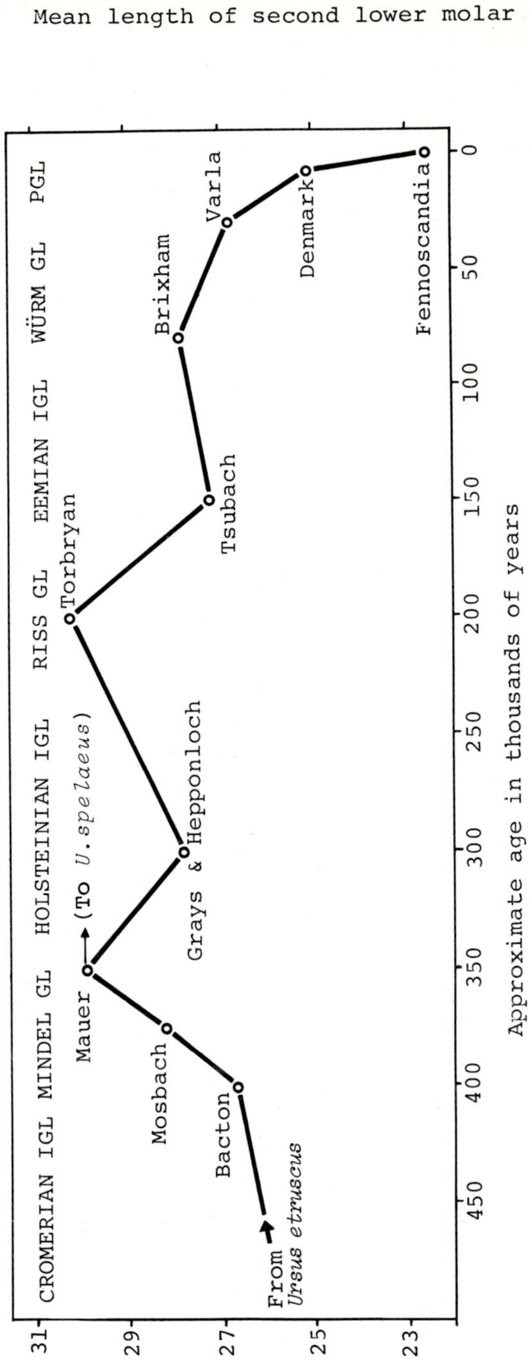

Fig. 7.1 Size trends in the Pleistocene history of the brown bears (*Ursus etruscus* to *Ursus arctos*) in Europe, based on length of second lower molar. Redrawn from Kurten(1959).

of 720 ml over the period 2 x 10^6 years. If the generation interval is 20 years, we have:

$$\Delta\bar{x} = 720 \text{ ml}/10^5 \text{ generations}$$

therefore

$$\frac{d\bar{x}}{dt} = .0072 \text{ (ml/generation)}.$$

Let the phenotypic variance, $V_P = 10^4$ (corresponding to an average standard deviation in capacity of 100 ml), and let the heritability be 0.25 (a typical figure for many quantitative traits), then:

$$V_a = .025V_P$$

$$= 2.5 \times 10^3$$

hence

$$b = 7.2 \times 10^{-3}/2.5 \times 10^3$$

$$= 2.9 \times 10^{-6}$$

which is a very small figure indeed. In fact, if we were to accept the above analysis, the total increase in fitness associated with a change in brain capacity of 720 ml is 0.002. Unfortunately, we have no data, for any organism, which would allow us to estimate the possible rates of change in mean fitness. Spuhler (1963), using an index derived by Crow (1958), estimated that the phenotypic variance in fitness in tribal populations ranges from 0.6 to 3.7. If only one percent of this variance could be ascribed to genic differences, this would allow rates of change in mean fitness of the order of 0.01 per generation (one thousand in 10^5 generations). Speculations of this kind lead to all kinds of unanswered questions, and it is pointless to push this argument much further. The point of these calculations, however, is to illustrate that the apparent selective forces, calculated from the fossil record, are surprisingly small. Another way of looking at the observed rate of change is to compare it with what is possible under artificial selection. For example, if we culled 1% of the population on brain capacity (a modest culling rate for an animal breeder), we could change the mean by 0.7 ml per generation.

Why then are evolutionary rates so slow? One possibility is that natural selection rapidly exhausts additive genetic variance so that it is generally close to zero. Rarely, however, do we find any trait which lacks genetic variability - the problem arises when we try to consider any quantitative trait in isolation. Any change in a quantitative trait will necessitate changes in many other characters. Increase in brain capacity for example will require changes in neck vertebrae and musculature, and an increase in the size of the pelvic girdle. Also it is possible that some of the increase in brain size is a correlated response to an increase in body size. Hence a model which assumes a linear relationship between the trait and fitness will seldom apply, and in general we might suppose a quadratic relationship of the kind

$$m = a - b(\phi - x)^2$$

where ϕ is an optimum value. This optimum may be continually changing in response to changes in the environment (external or internal), but if $\bar{x}$ is close to ϕ, evolutionary rates will be slow despite considerable differential survival. In domesticated species many of these developmental constraints have been removed, and breeders are able to concentrate on a single characteristic. Quadratic relationships between survival and birth weight, and stature and reproductive performance, have been demonstrated in human populations (Haldane 1954; Cavalli-Sforza and Bodmer 1971).

The cost of natural selection.

Haldane (1957) approached the problem of the slow evolutionary rates from a completely different point of view. It is a common experience in animal breeding that selection intensity, and hence progress, is limited by the reproductive potential of the species. Haldane argued that a similar constraint applies under natural selection, since each gene substitution requires some differential survival or reproduction in the evolving population. The total selective death over the course of a gene substitution was defined by Haldane to be the cost of natural selection. Perhaps even more than Fisher's theorem of natural selection, this theorem has been the subject of intense debate and controversy. A good account of the theory, aside from Haldane's original paper, can be found in Crow (1969) and Felsenstein (1971).

In a population of constant size N, and with small but constant selection coefficients, the total cost in substituting one allele for another from an initial gene frequency of p_o is at least $-2N.\log_e p_o$ which is independent of the selection coefficient. For example, if the initial gene frequency is 10^{-4} (the sort of frequency we might expect from a newly arisen allele, or a gene maintained by a balance between mutation and selection) the cost is >19N. Haldane suggested that 30N is a reasonable figure, and argued that a species could support an excess in fertility of 10 percent in the process of gene substitution. Therefore, a species would be limited to a gene substitution on the average every 300 generations.

This argument has had a great influence on the interpretation of evolutionary rates inferred from molecular data. Sved (1968) and Maynard Smith (1968) showed that if natural selection is largely competitive, and that if competitive ability is controlled by a large number of loci, then rates of substitution may be much greater than predicted by Haldane. In competitive models the proportion of selective deaths is determined by extrinsic factors, such as the carrying capacity of the environment, and is largely independent of the array of genotypes in the population. Haldane's model, which assigns absolute selective values to genotypes, is inappropriate under such conditions.

Recent discussions of the cost of natural selection include O'Donald (1969), Wallace (1970), Grant and Flake (1974) and Darlington (1977).

MOLECULAR EVOLUTION

Techniques in immunology and in protein and DNA biochemistry have, in the last ten years, contributed enormously to our understanding of evolutionary rates. The most important information has come from protein sequence data, since it has allowed precise reconstruction of evolutionary phylogenies and, perhaps more importantly, direct information on rates of gene substitution. DNA sequencing will eventually be even more valuable, because very rapid sequencing techniques are available and direct comparisons can be made at the genic level. Ultimately we can expect information not only for the genes which code for proteins (structural loci) but also on regulatory sequences which are not transcribed but may be more important in morphological change.

Evidence from comparison of protein sequences.

Zuckerkandl and Pauling (1965) and Margoliash and Smith (1965) were the first to estimate evolutionary rates by comparison of polypeptide sequences. The estimated rates (in hemoglobins and cytochrome C), and the apparent constancy of these rates over different phylogenetic trees, led Kimura (1968) and King and Jukes (1969) to propose that much of the observed differences were due to neutral amino acid substitutions. I will present the method of analysis, and the argument in some detail. Much of the following is drawn from the papers of Kimura (1969) and King and Jukes (1969).

Consider two homologous polypeptides, each with n amino acid sites. At a given site we suppose that the probability of substitutions since the two lines diverged is a Poisson variable with mean k. Then the expected number of identical amino acid sites is e^{-k}. If n_i of the n amino acids are identical, we can estimate e^{-k} by n_i/n.

Hence $$k = -\log (n_i/n)$$

The rate of substitution is therefore

$$\lambda = k/2t$$

where t is the estimated time since the two proteins (or species) diverged. A commonly presented example is the comparison of carp and human α hemoglobin. Of the 140 sites, 72 are identical.

Hence $$k = -\log_e(72/140)$$
$$= .665$$

It is estimated that these two lines diverged in the Devonian, approximately 350-400 My ago, i.e. $2t \simeq 750 \times 10^6$

and $$\lambda = 8.9 \times 10^{-10}\text{/amino acid site/year}$$

A comparison of human and horse hemoglobin yields an almost identical estimate, namely $\lambda = 8.8 \times 10^{-10}$. We can also compare α and β hemoglobins, and these give the same values for λ. Rates of substitution for cytochrome C can be calculated, and these give values of λ of the order of 3×10^{-10}.

King and Jukes surveyed seven proteins from different mammalian orders, and assuming that these proteins diverged 75 My ago, they obtained an average rate of 16×10^{-10}/amino acid site/year. Since these studies, rates have been calculated for a number of other proteins (see McLaughlin and Dayhoff 1972). The slowest rates are observed for histone IV (6×10^{-12}), and the highest for fibrinopeptide A (9×10^{-9}). Excluding histone IV, which is outside the normal range, the average rate from McLaughlin and Dayhoff's data is 13.2×10^{-10} ($\pm 1.2 \times 10^{-10}$). This is not very different from King and Juke's original estimate.

The hypothesis that amino acid substitutions are selectively neutral hinges on two observations from these data.

(a) For each protein, the rate of evolution in terms of amino acid substitutions is approximately constant per year per site for various lines, as long as the function and tertiary structure of the molecule remain essentially unaltered (Kimura and Ohta 1974).

This view has recently been disputed by a number of investigators (e.g. Langley and Fitch 1974; Goodman et al. 1975; Homquist et al. 1976). These authors have used statistical procedures to estimate the relative rates of substitution in each of the branches of a phylogenetic tree. Langley and Fitch chose an *a priori* phylogenetic relationship and then used this tree to estimate the relative rate of substitution for each of four proteins, and the relative rate for each branch of the tree. They were then able to compare the observed number of substitutions in each branch with that expected on the assumption of constancy. A χ^2

test showed a significant departure from expectation. Recently Moore et al. (1976) and Homquist et al. (1976) have used two methods, called the 'Stochastic' and 'Maximum Parsimony' procedures, to construct remarkably detailed phylogenetic trees. These authors unequivocably reject the hypothesis of constant rates. In particular, they conclude that cytochrome C has evolved three to four times faster in protozoa, fungi and plants than in birds and mammals, and that the globins changed more rapidly in cold blooded vertebrates than in placental mammals and birds.

(b) The observed rates of evolution are so high that if all substitutions were adaptive there would be an unbearable genetic load in the population.

I have already discussed the argument of cost of evolution. The observed average rate of 13 x 10^{-10} substitutions/amino acid/year corresponds to about 6 x 10^{-10}/ nucleotides/year. The mammalian haploid genome contains about 3 x 10^{9} nucleotides, so the substitution rate per genome is 2 nucleotides per year. DNA hybridisation data suggest that the rate is even higher than this (Kohne et al. 1972). Such rates are incompatible with Haldane's limit of one in three hundred generations. However, since we have reason to believe that much of this DNA does not code for protein, it may not be reasonable to calculate rates on the basis of total nucleotides. There may be as many as 4 x 10^{4} loci coding for protein, with an average size of 300 amino acids. The rate then, for structural loci, would be 13 x 10^{-10} x 4 x 10^{4} x 300 = 1.6 x 10^{2}, i.e. about one substitution in 60 years. This is still one to two orders of magnitude higher than Haldane considered possible.

The hypothesis that most substitutions are neutral solves the problem of substitutional load, and predicts constancy in rates of protein evolution. The theory of neutrality has an attractive simplicity. It is quite simple to show that the substitution rate is equal to the mutation rate to neutral genes, and the observed rates are compatible with known mutation rates. This is particularly true of rapidly evolving polypeptides such as fibrinopeptide A which has no known enzymatic function. Since the expected time to fixation of a neutral gene is $4N_e$ generations, where N_e is the effective population size, the theory also predicts that many 'neutral genes' will be undergoing substitution in a population, and the distribution of such genes can be predicted. The observed non-constancy of rates throws doubt on the view that almost all substitutions are neutral, and I am convinced by the arguments of Sved (1968) and Maynard Smith (1968) that rates faster than one in three hundred generations are possible under selection. We cannot discard the neutral theory entirely however. The 'evolutionary clock' principle still gives remarkable agreement with paleontological phylogenies, and it is possible that many substitutions are neutral or nearly so. Extensive discussions of neutral theory can be found in Kumura and Ohta (1971), Nei (1975) and a recent review by Ewens (1977).

DNA HYBRIDISATION, COMPLEMENT FIXATION, ELECTROPHORESIS AND BLOOD GROUPING

Each of these techniques has provided information on the relative similarity of species, subspecies or races, and hence can be used to construct phylogenetic trees and estimate rates of divergence. DNA hybridisation techniques can be used to detect mismatches in DNA sequence, and hence provide estimates of nucleotide differences. Man and chimpanzee, for example, differ at approximately 2.5% of their base pairs, whereas man and gibbon are estimated to differ at 5.1% (Kohne et al. 1972). The immunological technique, complement fixation, has been used to quantify antigenic differences between pairs of proteins. A measure called the immunological distance has been shown to be linearly related to the proportion of amino acid differences between two proteins (Prager and Wilson 1971) and has also been shown to be approximately linearly related to the time since evolutionary divergence (Sarich and Wilson 1967). Sarich and Wilson used this technique to construct a phylogeny of hominid evolution. These data and protein sequence information appear to conflict with the fossil record, and in a comprehensive review of fossil and molecular data Uzzell and Pilbeam (1971) suggest that molecular evolution may have slowed down in primate evolution. This may be related to the increase in generation time. Prager et al. (1976) have recently extended these techniques to compare evolutionary rates in conifers.

Gel electrophoresis, which is used extensively to study genetic polymorphism in natural populations, can also be used to estimate the genetic distance between subspecies, species, or genera. It is of limited use in constructing evolutionary trees, but is very useful in comparing closely related species or different populations of the same species. The same is true of blood groups. The first phylogenetic tree, using blood groups, was constructed by Cavalli-Sforza and Edwards (1964) to derive an evolutionary history of the races of man.

All of these methods can be compared with each other, and with measurements of morphological differences, to search for consistencies or anomalies. Recently King and Wilson (1975) drew together all of the above information to compare *Pan troglodytes* and *Homo sapiens*. They concluded that these two species are very similar at the protein level (greater than 99% identity between homologous proteins), and that this difference was not sufficient to account for the observed dissimilarities in morphology. Their conclusion is that evolutionary divergence has occurred at the level of regulation of gene action rather than differentiation at structural loci.

CHROMOSOME REARRANGEMENT AND CHANGES IN NUMBER

No discussion of the evolutionary process is complete without considering karyotypic evolution. Rearrangements of the chromosome complement are so ubiquitous - in flowering plants, insects, and in mammalian evolution - that their importance in the evolutionary process cannot be questioned. The mechanisms through which these changes arise is understood, but often the adaptive significance can only be guessed at. In plants, interchanges (as in some species of *Oenothera*) are believed to be important in maintaining structural heterozygosity, and polyploidy, particularly following hybridisation of different species, is important in the generation of new types. Inversions are widespread in insect species, and it is thought that these rearrangements serve to preserve favourable gene combinations. Chromosomal rearrangements appear also to have played a significant role in mammalian evolution.

Since substitution of one arrangement for another must be preceded by a polymorphism, these observations pose (at least) two problems for evolutionists:

(1) Why should a particular gene arrangement be at an advantage to the pre-existing order?

(2) How does the organism cope with polymorphisms in arrangement?

Special mechanisms (regular alternate disjunction in plants, no crossing over in males of some insect species, and loss of fragments in the polar bodies of females) have apparently evolved to minimise fertility loss in chromosomal heterozygotes. Nevertheless some gametic loss is inevitable. Even in man, and certainly in primate evolution, chromosomal inversions are frequent. The significance of changes in chromosome number, common in the evolutionary history of all metazoans, is a greater puzzle.

Wilson et al. (1975) compared rates of chromosomal evolution in a variety of lineages. They concluded that rates of change in chromosome number in placental mammals have been much higher than in other vertebrates. For example, the rate of change in chromosome number has been about 7 per 10^8 years for primates and rodents, compared with 0.6 per 10^8 years in non-mammalian vertebrates and marsupials. Changes in chromosome number in conifers have occurred at rates which are an order of magnitude slower again (0.04 per 10^8 years) (Prager et al. 1976). Rates of karyotype evolution seem to be more closely related to morphological change than to protein evolution, and Wilson et al. (1974) proposed that chromosomal rearrangements are associated with changes in gene regulation.

SPECIATION, EXTINCTION AND COMMUNITY STABILITY

To many biologists, speciation and species extinction are the essence of the evolutionary process, but in many population genetics texts the subject is barely mentioned. The species concept is a difficult one for geneticists to come to grips with. At best the quality which separates one species from any other is the presence of a sexual isolating mechanism; at worst there is no barrier to crossing other than geographical isolation. Higher categorisation, into genus or family, means even less; the study of evolution at the molecular level proceeds with little attention to taxonomic relationship.

The mechanisms of speciation are quite well understood (Mayr 1963), but the selective forces involved are not. The most likely reason for the development of sexual isolation is to preserve adapted gene combinations which have arisen in response to unfilled niches. Mathematical theory in this infant field of ecological genetics is poorly developed.

Extinction is even more an ecological concept, since in general the chance of extinction is not a property of the species alone. In a completely uniform environment, biotic and physical, an unfortunate assumption of much genetic theory, the probability that a species will become extinct must be very remote. Extinction will occur when

(1) a species cannot evolve fast enough to keep up with a changing physical environment, large scale extinction being usually associated with environmental upheaval, or

(2) a species cannot evolve fast enough to compete with a new species, or to keep pace with evolutionary changes in other species in the community.

Rates of extinction and origination.

The fossil record shows, as we might expect, a balance between the origination rate and extinction rate of taxonomic groups. Groups differ, however, in their turnover rate. Simpson (1953) showed that the average survival time of extinct genera of bivalve molluscs (*Pelecypoda*) was 78 My, but genera of the order Carnivora survived on the average only 8 My. Van Valen (1973), in an extensive survey of the fossil record, showed a remarkable linearity of survivorship curves when plotted on the log scale. The inference here is that extinction probability at any time depends only on the proportion still surviving at that time, analogous to the process of radioactive decay.

Here we run into the problem of interpreting survivorship measured in taxonomically convenient units (genera or families) in terms of units which are more biologically meaningful (species). Linearity at the genus level probably implies linearity at the species level, since a genus will become extinct when all the species in that genus disappear. The relationship between origin of genera, and origin of species, is unclear. We also have to distinguish between phyletic extinction and origination, in which through gradual evolution a taxon becomes reclassified into a new one (anagenesis), and discontinuities in the lineage (cladogenesis). Van Valen (1973) believes that phyletic extinction in survivorship curves is of minor importance. Stanley (1975) discusses survivorship curves at the species level, and suggests that species duration for mammals is about 2 My in the pre-Pleistocene and about half this in the Pleistocene. The species duration for marine invertebrates has been about 6 My. Origin of species by phyletic change cannot account for the 'explosive' radiation of species well documented in the fossil record (Simpson 1953; Rensch 1959; Stanley 1975). Very rapid speciation is well known, especially in habitats such as fresh-water lakes. Lake Lanao, in the Philippines, was formed about 10,000 year ago, and in this time a single species of carp has evolved into 18 species, five of which have been placed in four endemic genera (Myers 1960).

Community stability and extinction.

The theory of community stability is crucial to any understanding of species extinction but at this stage the conditions for stability of an ecosystem are poorly understood. Most theoretical models are simple, involving two or three species at one or two trophic levels, but communities involve large numbers of species, often with many predators and prey amongst them. It is generally believed that increased trophic complexity leads to increased community stability, but this is not proven (May 1973). In a complex ecosystem the loss of one species may lead to the loss of many others. Thus Paine (1966) removed a predator from an intertidal community of 15 species, and the number of species dropped to eight in two years.

MacArthur and Wilson (1967) developed an elegant theory to describe the relationship between number of species, colonisation rates and extinction in island populations as a function of area, migration rates, etc., and this theory may provide a basis for understanding speciation and extinction on an evolutionary time scale. However, even if we knew the conditions for stability of a complex community, we need a theory of co-evolution, incorporating evolutionary changes in the constituent species and the effect on the community. Species occupying similar niches will tend to evolve at the same rate, but stochastic changes will occur, and extinction rates will depend on the stability of the system.

In general each species in a community will, through natural selection, be highly adapted to its environment, so almost any change in its physical or biotic environment will be deleterious. Such changes continually occur, either through accumulation of harmful genes by mutation, by geological or climatological changes, or as a result of increases in fitness of other species. Hence a species' environment is continually deteriorating, (Fisher 1930). The last cause is probably the most important. For example, if one prey species evolves a strategy to avoid predation, there will be a loss in fitness among the predators and an increased predation of other prey species. Each species needs to evolve continuously in order to keep up with its enemies and competitors. Van Valen (1973) (see also Maynard Smith 1976) suggested that the observed constant extinction rates can be explained in terms of constant deterioration of the biotic environment. The argument is essentially that used by Fisher, with the additional assumption of constancy, and the view that each increase in fitness in one species is distributed over the rest of the community as a loss in fitness. Van Valen called this 'The Red Queen's Hypothesis', from Through the Looking Glass: "Now here, you see, it takes all the running you can do, to keep in the same place". This view suggests that competition is much more important that mutualism in species interactions.

Certainly the most dramatic cause of extinction that we know is competition. Many marsupial species have disappeared through competition with placental mammals. All large mammals (terrestrial and marine) die out in competition with man, and at a rapid rate. Even in prehistoric times, the massive extinction of large terrestrial herbivores in the Late Pleistocene may have resulted from competition and predation by man (Martin 1967).

CONCLUSIONS

There are, as we have seen, at least three kinds of evolutionary rates, all interrelated, but each may show very different behaviour in a given phylogeny.

Molecular rates.

These are revealed only at the genic level, or inferred from the immediate products of gene action, the proteins. Even in this category it is evident that we need to distinguish between two kinds of genes: those which code for protein (structural genes) and those which regulate gene action through unknown developmental processes. Structural genes seem to evolve at a roughly constant rate in a variety of lineages, but the variation in rate seems to be about twice as great as would be expected if this were truly a random process. Regardless of the mechanism (neutral substitution or adaptive substituion under relatively uniform selective pressures), changes at this level provide a very useful yardstick against which other rates can be measured. Less is known about changes in regulatory genes, although it is apparent that greater differences exist in the DNA of related species that can be accounted for by differences in transcribed proteins. There may even be a third level, the highly repeated DNA sequences, which show very different patterns in closely related species.

Morphological rates.

Here we are concerned with phenotypic differences. Some of these differences are accessible in the fossil record, but many important traits, such as behaviour and reproductive performance, are not. When the genetics of the trait is known, as in industrial melanism, the changes can be directly related to gene substitution. More often than not the formal genetics is unknown. We do know, however, from some of the examples cited in the introduction and from artificial breeding experiments, that quantitative traits can be changed very rapidly. A geneticist would not be surprised by findings in the fossil record which indicated rates of evolution which were an order of magnitude greater than anything previously recorded.

In this category our observations are on the same level as natural selection, since the success of a new gene or gene combination can only be tested in the whole organism. Molecular rates certainly underlie morphological traits, but regulatory genes are probably more important than structural genes.

Chromosomal rates fall somewhere between these two categories. The karyotype is in some sense part of the phenotype, but it also reflects changes in the physical arrangement of the genetic material.

Taxonomic rates.

In the preceding discussions taxonomic rates have been considered separately from molecular and morphological rates. Clearly, however, these processes are not independent. Phyletic extinction occurs when a population changes to such a degree that it is classified into a different taxon. Here the taxonomic rate is directly related to the degree of morphological change. On the other hand, cladogenesis, which most evolutionists believe is more important in taxonomic origination, is not necessarily associated with morphological change. As we have already indicated, speciation requires only the presence of a sexual isolating mechanism, i.e. a genetically determined barrier to mating. There are numerous examples of insects which are almost identical morphologically (sibling species), although it can be shown that there are numerous differences at the protein and DNA level. Compared to mammals, frog species differ very little morphologically.

However, it has recently been suggested that morphological evolution is closely related with speciation events (rectangular evolution), rather than a slow and even process throughout the lineage (phyletic gradualism). This is implicit in Mayr's (1963) view of the importance of speciation; - "it is in the process of creating so many species which leads to evolutionary progress" a hypothesis extended by Eldredge and Gould (1972) and Stanley (1975, 1976). While there appears to be considerable paleontological support for this view, Avise (1977) examined electrophoretic variation in species-rich and species-poor phylads of living fishes, and concluded that there was little evidence for rectangular evolution. Note, however, King and Wilson's (1975) suggestion that considerable morphological change can occur despite little change at the protein level.

Finally, there is a question which is fundamental to the theme of this conference, namely the assumption that plants and animals with identical morphological traits have the same physiological tolerances today as in the Quaternary past. There is every reason to doubt the validity of this hypothesis. In fact the rapid development of resistance to insecticides and antibiotics in previously susceptible species, and adaptation by grasses to mine tailings, are observations which directly contradict this assumption. Constancy in morphology gives us no assurance that there have not been changes in colouration, behaviour, or physiological tolerance, since any of these traits can change independently. It is therefore quite likely that there will be physiological adaptation to even rapid changes in the physical environment, especially if the competitors and enemies of the species are similarly affected.

There may, however, be some constancy at a higher level, namely the preservation of relationships between one organism and another in a community. If we accept the Red Queen's hypothesis, competition between species will act to preserve the structure of the ecosystem, since the further a species lags behind the others, the greater will be the selection pressure to restore the species to its previous status. This principle, which needs validation, will tend to react against change in morphology, and perhaps even physiological change, relative to the other species in an undisturbed community. Today, of course, the whole picture has changed.

ACKNOWLEDGEMENT

I am very grateful to Dr J.M. Rendel for discussion and criticism of the manuscript.

REFERENCES

Avise, J.C. 1977 Is evolution gradual or rectangular? Evidence from living fishes. Proc. Nat. Acad. Sci., 74, 5083-5087.

Brown, A.W.A. 1967 Genetics of insecticide resistance in insect vectors. pp 505-552 in Genetics of Insect Vectors of Disease (eds J.W. Wright and R. Pal), Elsevier Publishing Co., Amsterdam.

Cavalli-Sforza, L.L. and Bodmer, W.F. 1971 The Genetics of Human Populations, W.H. Freeman and Company, San Francisco.

Cavalli-Sforza, L.L. and Edwards, A.W.F. 1964 Analysis of human evolution. pp 923-933 Genetics Today. Proc. 11th Int. Cong. Genet. The Hague, Pergamon Press, Oxford.

Crow, J.F. 1958 Some possibilities for measuring selection intensities in man. Human Biol., 30, 1-13.

Crow, J.F. 1969 Genetic loads and the cost of selection. pp 128-177 in Mathematical Topics in Population Genetics (ed Ken-ichi Kojima), Springer, Berlin.

Darlington, P.J. Jr. 1977 The cost of evolution and the imprecision of adaptation. Proc. Nat. Acad. Sci., 74, 1647-1651.

Eldredge, N. and Gould, S.J. 1972 Punctuated equilibria: an alternative to phyletic gradualism. pp 82-115 in Models of Paleobiology (ed T.J.M. Schopf), Freeman-Cooper, San Francisco.

Ewens, W.J. 1977 Population genetics theory in relation to the neutralist-selection controversy. pp 67-134 in Advances in Human Genetics, 8 (eds H. Harris and K. Hirschhorn), Plenum Press, New York.

Felsenstein, J. 1971 On the biological significance of the cost of gene substitution. The American Naturalist, 105, 1-11.

Fenner, F. 1965 Myxoma virus and *Oryctolagus cuniculus*: Two colonising species. pp 485-499 in The Genetics of Colonising Species (eds H.G. Baker and G.L. Stebbins), Academic Press, New York.

Fisher, R.A. 1930 The Genetical Theory of Natural Selection. The Clarendon Press, Oxford.

Goodman, M., Moore, G.W. and Matsuda, G. 1975 Darwinian evolution in the genealogy of haemoglobin. Nature, 253, 603-608.

Grant, V. and Flake, R.H. 1974 Solutions to the cost-of-selection dilemma. Proc. Nat. Acad. Sci., 71, 3863-3865.

Haldane, J.B.S. 1949 Suggestions as to quantitative measurement of rates of evolution. Evolution, 3, 51-56.

Haldane, J.B.S. 1954 The measurement of natural selection. Proc. 9th Inter. Cong. Genet. Bellagio (Como) Pt. 1, 480-487.

Haldane, J.B.S. 1956 The theory of selection for melanism in Lepidoptera. Proc. Roy. Soc., B 145, 303-304.

Haldane, J.B.S. 1957 The cost of natural selection. J. Genet., 55, 511-524.

Holmquist, R., Jukes, T.H., Moise, H., Goodman, M. and Moore, G.W. 1976 The evolution of the globin family genes: Concordance of stochastic and augmented maximum parsimony genetic distances for α hemoglobin, β hemoglobin and myoglobin phylogenies. J. Mol. Biol., 105, 39-74.

Kettlewell, H.B.D. 1956 A resume of investigations on the evolution of melanism in the Lepidoptera. Proc. Roy. Soc., B, 145, 297-303.

Kimura, M. 1958 On the change of population fitness by natural selection. Heredity, 12, 145-167.

Kimura, M. 1968 Evolutionary rate at the molecular level. Nature, 217, 624-626.

Kimura, M. 1969 The rate of molecular evolution considered from the standpoint of population genetics. Proc. Nat. Acad. Sci., 63, 1181-1188.

Kimura, M. and Ohta, T. 1971 Theoretical Aspects of Population Genetics. Princeton University Press, New Jersey.

Kimura, M. and Ohta, T. 1974 On some principles governing molecular evolution. Proc. Nat. Acad. Sci., 71, 2848-2852.

King, J.L. and Jukes, T.H. 1969 Non-Darwinian evolution. Science, 164, 788-798.

King, M.C. and Wilson A.C. 1975 Evolution at two levels in humans and chimpanzees. Science, 188, 107-116.

Köhne, D.E., Chiscon, J.A. and Hoyer, B.H. 1972 Evolution of mammalian DNA. Proc. 6th Berkeley Symp. Math. Statist. Probab. V, 193-209.

Kurten, B. 1959 Rates of evolution of fossil mammals. Cold Spring Harbor Symp. Quant. Biol., 24, 205-215.

Langley, G.H. and Fitch, W.M. 1974 An examination of the constancy of the rate of molecular evolution. J. Mol. Evol., 3, 161-177.

Lerman, A. 1965 On rates of evolution of unit characters and character complexes. Evolution, 19, 16-25.

MacArthur, R.H. and Wilson, E.O. 1967 The Theory of Island Biogeography. Princeton University Press, Princeton, New Jersey.

Margoliash, E. and Smith, E.L. 1965 Structural and functional aspects of cytochrome C in relation to evolution. pp 221-242 in Evolving Genes and Proteins (eds V. Bryson and H.J. Vogel), Academic Press, New York.

Martin, P. 1967 Prehistoric overkill. pp 75-120 in Pleistocene extinctions: the search for a cause (eds P.S. Martin and H.E. Wright), Yale University Press, New Haven.

May, R.M. 1973 Stability and Complexity in Model Ecosystems. Princeton University Press, Princeton, New Jersey.

Maynard Smith, J. 1968 'Haldane's Dilemma' and the rate of evolution. Nature, 219, 1114-1116.

Maynard Smith, J. 1976 A comment on the Red Queen. The American Naturalist, 110 325-330.

Mayr, E. 1963 Animal species and evolution. The Belknop Press of Harvard University Press, Cambridge, Mass.

McLaughlin, P.J. and Dayhoff, M.O. 1972 Evolution of species and proteins: a time scale. pp 47-66 in Atlas of Protein Sequence and Structure. Vol. 5 (ed M.O. Dayhoff), Natl. Biomed. Res. Foundn., Washington, D.C.

McNeilly, T. and Bradshaw, A.D. 1968 Evolutionary processes in populations of copper tolerant *Agrostis tenuis* Sibth. Evolution, 22, 108-118.

Moore, G.W., Goodman, M., Callahan, C., Holmquist, R. and Moise H. 1976 Stochastic versus agumented maximum parsimony method for estimating superimposed mutations in the divergent evolution of protein sequences: Methods tested on cytochrome C amino acid sequences. J. Mol. Biol., 105, 15-37.

Myers, G.S. 1960 The endemic fish fauna of Lake Lanao and the evolution of higher taxonomic categories. Evolution, 14, 323-333.

Nei, M. 1975. Molecular Population Genetics and Evolution. North-Holland Publishing Co., Amsterdam.

O'Donald, P. 1969. Haldane's dilemma and the rate of natural selection. Nature, 221, 815-816.

Prager, E.M. and Wilson, A.C. 1971 The dependence of immunological cross-reactivity upon sequence resemblance among lysozymes. J. Biol. Chem., 246, 5978-5989.

Prager, E.M., Fowler, D.D. and Wilson, A.C. 1976 Rates of evolution in conifers (*Pinaceae*). Evolution, 30, 637-649.

Paine, R.T. 1966 Food web complexity and species diversity. The American Naturalist, 100, 65-75.

Rensch, B. 1959 Evolution above the Species Level. Methuen and Co. Ltd., London.

Sarich, V.M. and Wilson, A.C. 1967 Immunological time scale for hominid evolution. Science, 158, 1200-1203.

Simpson, G.G. 1944 Tempo and Mode in Evolution. Columbia University Press, New York.

Simpson, G.G. 1953 The Major Features of Evolution. Columbia University Press, New York.

Spuhler, J.N. 1963 The scope for natural selection in man. in Genetic Selection in Man. 3rd Macy Conference on Genetics 1961 (ed W.J. Schull), University of Michigan Press, Ann Arbor.

Stanley, S.M. 1975 A theory of evolution above the species level. Proc. Nat. Acad. Sci., 72, 646-650.

Stanley, S.M. 1976 Stability of species in geologic time. Science, 192, 267-268.

Sved, J.A. 1968. Possible rates of gene substitution in evolution. Amer. Nat., 102, 283-293.

Uzzell, T. and Pilbeam, D. 1971 Phyletic divergence dates of Hominoid primates: a comparison of fossils and molecular data. Evolution, 25, 615-635.

Van Valen, L. 1973 A new evolutionary law. Evol. Theory, 1, 1-30.

Wallace, B. 1970 The Genetic Load. Prentice Hall, Englewood Cliffs, N.J.

Wilson, A.C., Sarich, V.M. and Maxson, L.R. 1974 The importance of gene rearrangement in evolution: evidence from studies on rates of chromosomal, protein, and anatomical evolution. Proc. Nat. Acad.Sci. 71, 3028-3030.

Wilson, A.C., Bush, G.L., Case, S.M. and King, M.C. 1975 Social structuring of mammalian populations and rate of chromosomal evolution. Proc. Nat. Acad. Sci., 72, 5061-5065.

Zuckerkandl, E. and Pauling L. 1965 Evolutionary divergence and convergence in proteins. pp 97-166 in Evolving Genes and Proteins (eds V. Bryson and H.J. Vogel), Academic Press, New York.

8. DETERMINANTS OF TOLERANCE LIMITS IN ANIMALS AND MAN

by W.V. Macfarlane

SUMMARY

The basic setting of rates of water, energy and protein turnover of animals seems to be stable over long terms in an evolutionary sense. Camels have functioned for the desert for more than 3 million years.

But adjustment of animals to Quaternary heat, cold, aridity or pluvials probably arose mainly from habituation, enzyme induction, adaptive behaviour or endocrine responses. In any population there is a wide range of inbuilt functional competences (polyfunctionalisms) and under duress some of the rarer physiological types would survive.

The ecophysiological hierarchy of high rates of turnover for wet tropical ecotypes and low rates for desert forms does not necessarily show in the bony structure of fossils. Burrowing is an effective escape behaviour in the dry areas but it is not easy to deduce morphologically.

Migration, and reinvasion as climate changes, are tactics for specific survival.

Known extinctions of Tasmanian Aborigines, *Thylacinus* or the aurochs all seem to have involved the failure of residual males and females to sustain adequate breeding performance. Attrition of giant mammal populations by fire could have led to small-group reproductive failure.

Overall, jungle and desert ecotypes would be eliminated by major climatic reversals. But animals evolved for mesic environments with cold-hot, wet-dry components of climate, would adapt biochemically, physiologically and behaviourally to most Pleistocene changes.

Structurally, physiologically and biochemically man is a rather generalised animal, with a basic equatorial physiology which has in the past 3 million years been adjusted to high latitudes, hot dry, cold wet and frozen environments. The adaptations have not come through any fundamental change in function or structure, which are very conservative, but rather by changes in behaviour. Facile hands, tools to produce foods and teeth able to cope with meat, seeds or fruit remain a basis for behavioural adaptation and survival. Omnivorous food habits had led to an extraordinary range of foods being acceptable.

The considerable fluxes of temperature in the Pleistocene, forced man to change behaviourally towards making protective shelter and clothing, probably also to the storing of foods. Heat or cold habituation and active redistribution of blood between skin and core of the body would alter insulation.

Dependence on water is intrinsic to the human condition. Interestingly, this dependence is becoming greater rather than less, since the exponential rate of water use has a doubling time of 10 years (nearly 4 times the rate of population increment) to meet behavioural, not physiological, demands. In a neutral environment man might starve to death in 2 months but he would certainly be dead from water lack in 2 weeks. In a hot environment when water supplies fail, man in the sun is dead within the day.

Starvation in the cold is more rapid than in warm or neutral environments. In the extinctions of human species over the last 3 million years, infant deaths from starvation, virus epidemics or inbreeding, with genetic disease, are likely causes. There may have been reduction of group size to a level where inbreeding became a decimating influence and it is clear that Australian Aborigines, who survived some rigorous conditions, made some of the most rigid social conventions for preventing inbreeding that have been devised. The almost universal intolerance of inbreeding must have deep social roots in the early phases of small group behaviour through the Pleistocene.

The success of tactics and strategems for survival in changing environments, together with a tropical round-the-year breeding potential, has led to the present 4 thousand million humans being generated. This is a behaviour-based aspect of survival. Whether there is sufficient skill in the group aspects of human function to achieve an ecological balance within the next 100 years or not, is probably the major Holocene test being applied to this species.

INTRODUCTION

The last 3 million years have been unusual in several respects. There has been a fairly regular series of ice ages interspaced with warmer, wet or arid, periods. Land and ocean temperatures have become cooler (Kurten 1968). Mammals and birds have moved between the continents in surprising numbers; many species of large mammals and birds became extinct and the definitive emergence of fire- and tool-using, behaviourally adaptive and pervasive primates took place. Movement of ice fronts over more than 1000 kilometres of latitude took place every 20-40,000 years, and there were concomitant fluxes in air temperature, rainfall, cloud cover, ultra-violet input and supplies of both water and vegetation.

The oscillations of ecosystems (indicated by pollen studies, Turekian 1971) led to natural selection of animals and either extinction, or adaptation by a variety of physiological, biochemical and behavioural adjustments. Avoidances included migrations and reinvasion, torpor, changes of food habits or shelter-seeking.

Fossil oxygen ratios, pollen, fossil plants or bones are secondary signs of changing climates, and of the resultant alteration of ratios of desert to scrub land, or savanna to forest. Those in turn relate to plants and animals that functioned in past environments.

This paper is concerned with the functions of mammals now and with speculation on the development and roles of their physiological processes over the last 3 million years. Extrapolation from tooth structure to food habits is traditional in palaeontology. But some back projection of the rates of metabolic, water and electrolyte handling, or of the behaviour and tolerances of Pleistocene mammals is also possible, within limits. Many functions, however, are indicated by the bones. And since mammals are polymorphous and polyfunctional (genetically diverse within a species) as well as biochemically plastic and quite clever, these aspects of function need to be kept in mind in reconstruction exercises.

The rate of phenotypic evolutionary change has on the average been slow; but sequence changes in DNA or peptides are remarkably fast and uniform (Wilson et al. 1977). Some animal forms were apparently successful over many millions of years without change (brachipods, ammonites). In general, however, the evolution of distinctive generic characteristics required between 5 and 10 million years (Simpson 1953). There have been some more rapid evolutions. The human brain is apparently one of them. Between the Australopithecines of 3 million years ago, and *Homo sapiens*, less than 2 million years seem to have elapsed, while the brain more than doubled in size. The degree of evolutionary change taking place in the Quaternary is likely, however, to have reached only species or sub-specific imensions. So the major extant orders, families, genera and species (with their functional settings) are essentially those evolved during the late Pliocene or Villafrancian. Since it takes 1 to 3 million years for a species to evolve by natural selection, much of the impact of successive ice ages and pluvials during the last million years must have been taken up by the elastic adaptive processes of animals. There would be few structural differentiations. Adjustments of body temperature, cardio-vascular, renal and alimentary functions would be limited, but possible. More likely is the selection of isoenzymes and the induction of adaptive enzymes to meet changing physical and chemical conditions (Prosser 1970). The endocrine system also is essentially adaptive in function (Macfarlane and Good 1976). Quite rapid short- and long-term resettings of hormone levels are possible as the environment changes. And peptide sequence mutations proceed at a steady rate to alter hormone functions (Wilson et al. 1977). Similarly, behaviour is adjustable, in spite of the genetic control of more than half the components of behavioural patterns.

MAMMALS OTHER THAN MAN

Ecophysiological hierarchies.

Parallel moulding of animal functions to ecosystems by environmental pressures occurred in many orders and families. One recurrent evolutionary principle is the matching to environment of the rates at which water, energy and protein are turned over by organisms (Macfarlane 1976a). In wet tropics there is only minor seasonality of food and water and energy sources can be used at high rates with little or no need for storage (Macfarlane et al. 1971). Salt is leached from soils and in plants reaches only one tenth of the temperate zone levels (Howard et al. 1962) so biological salt tolerance is unnecessary. Reproduction takes place throughout the year, since at any season it is likely that there will be adequate food for the lactating mother to feed the new born mammal. Similarly a short coat is retained without seasonal shedding, and replacement of the reflecting, low-insulation pelage occurs gradually throughout the year. Since there is always plenty of water, renal concentration mechanisms are poorly developed, the concentrating loops of Henle are short, and the kidney is relatively unresponsive to vasopressin (Macfarlane 1976a).

This complex of linked functions occurs in herbivorous animals evolved in the tropics, and is exemplified by water buffalo, cattle, deer, pigs, hippopotami and elephants. Amongst antelopes, the water buck and the eland function similarly to cattle, with high rates of water, energy and probably protein turnover.

The tropical evolutionary hearth has remained the habitat for many of these animals. On the other hand, some moved to neighbouring territories presumably under pressures of successive ice age or desert climates and they have made short-term non-genetic adaptations to different food water, and day-length components of the environments. Much of this adaptation has been behavioural, as in the sub-tropical or temperate forms of cattle, pigs and deer which have sought shelter from the sun under the trees of the savanna, and have remained near water sources in drier country. In addition, however, there appears to have been some fairly labile evolutionary organisation of fat stores (humps and fat tails) against the seasonal flux of food supply. Seasonal responses of endocrine and pineal mechanisms also occurred so that breeding behaviour was induced at times when the young would be able to obtain reasonable sources of food. The shedding of hair also became seasonal, for summer and winter coat adjustment (Yeates 1957) in temperate latitudes.

Carnivores are automatically provided with water from their food, which yields about 70% of its weight as water. Metabolic water from burning hydrogen in fats, sugars or aminoacids adds to the total water available. The carnivore can live, accordingly, with little drinking water. Dogs, cats and bears have modest water and energy demands, and they concentrate urine to the same extent as sheep, around 3000 mosmol/l. This helps account for the remarkable resilience of the dingo, a tropical dog from south India, that thrives in the arid zone of Australia on food of nearly the same composition as its own body. By eating small mammals it obtains concentrated vitamin supplies. Yet it is also resiliant in intensely cold environments.

The great deserts of the world lie between 15 and 30° north and south from the equator. They are characterised by intense solar radiation highly intermittent water and energy supplies, wind and dust, lack of shelter and low carrying capacity. Animals that evolved in the desert use water, food and protein at half to one third the rates of turnover of tropical animals (Macfarlane et al. 1971; Macfarlane and Howard 1974).

Behaviourally, small animals (up to 30 kg) may escape underground. Larger forms have coats of sand-grey camouflage, and reflecting surfaces to reject solar energy (Macfarlane 1968). Reproduction is opportunistic in the sense that it is likely to be turned on by an adequate supply of food occurring intermittently (as in camels) rather than by a regular switching procedure triggered from day-length (sheep, deer). In the higher desert latitudes also, there is accumulation of salt, usually in the zone of 26-32° latitude. The xerophytic plants in these regions tend also to be halophytes. Mammalian tolerance to salt is well developed in these regions. Behaviourally salt plants are accepted as food, the cellular enzymatic resistances of animals and NaK ATPase pumping mechanisms for salt tolerance have developed in the kidney as well as gut and liver to protect from sodium intoxication (Macfarlane 1971).

Since food supplies are intermittent, evolutionary pressures have led to establishment of fat reserves and protein recycling. These are highly developed in desert animals, where camels can carry more than 6 months food in the hump, and where fat-tail and fat-rump sheep may have a local reserve (up to one fifth of the body weight) which adds little to body insulation because of its highly localised distribution (Siebert and Macfarlane 1975). Goats, on the other hand, are excellent desert animals but their fat reserves are negligible; they evolved in the mountains of Asia. The dependence of herbivores on specific vegetation (Griffiths and Barker 1966) which changes with drought, fire or climate is illustrated by the euro and red kangaroo interactions with sheep in the Pilbara (W.A.) over the last century (Ealey 1967). Euros survive on *Triodia* grass; red kangaroos prefer green shoots, while sheep will eat dry grasses as well as some *Triodia* species but not others. Overstocking with sheep reduced the amount of grass and eliminated some *Triodia* (but all were damaged by burning at inappropriate times, before seeds were shed). So the red kangaroo declined in numbers with the sheep. By the 1950s euros were numerous because they ate *Triodia*, the spines and lignin of which were unacceptable to the sheep or kangaroo. Human intervention in grassland and forests over the past half to one million years may have been more damaging to animals by imperceptive or ritual burning than by hunting. Fire could have incinerated giant animals which rarely would be fast enough to move at 30 km per hour. But perhaps more important was their large demand on vegetation for food which would be destroyed by forest fires, and long periods of food deprivation would follow burns. This would increase death rates and reduce fertility leading downwards towards extinction.

On either side of the desert, towards the jungle and towards the wet temperate latitudes, there are mesic animals with intermediate rates of water, energy and protein use and of salt tolerance (Rogerson 1968; Richmond et al. 1962; Macfarlane and Howard 1972). Such animals are sheep, goats, wildebeest, kongoni, horses, donkeys, kangaroos, wallabies, possums and bandicoots. Metabolic and water turnover rates, renal concentrating functions and salt tolerance, fall between the jungle and desert types. Some members of each group have moved either towards the deserts or towards the wetter country and adapted reasonably well. If survival is a measure of successufl adaptation, then exposure of animals of different ecophysiological types to 40°C in the sun without water provides a useful criterion. Cattle die within 4-6 days, sheep within 8-10 days, and camels between 15 and 20 days. These rates of mortality are inversely proportional to the turnover rates of energy and water in each genus (Macfarlane 1964, 1968). In severe drought, elimination would be rapid in cattle and slow in camels; the latter in fact lived through the central Australian droughts (1958-1968) when cattle were out by 1962.

In the temperate zone, there is again intermittency of food supply, but nutrients are scarce in winter rather than in summer. Starvation is three or four times faster in a cold than in a hot environment (Morris et al. 1962). Cold temperatures do not put a strain on water supplies. In the past, cold temperate forms of tropical animals (the woolly rhinoceros, the woolly mammoth) have been eliminated (Kurten 1971). But tropical animals such as cattle, man and pig, have moved during the

Pleistocene into temperate zones, where fat was used for insulation. Subcutaneous fat deposition is wide-spread in cool environments and is readily increased by cold (Mount 1968). At the same time the pelage is adaptively increased in density and length for insulation in winter. Rats living in a beef cold-store become very fat, and their coats can reach 10 cm in length, by mobilising adaptive potential mechanisms for insulation. Yet breeds of sheep differ in fat deposition when living in the same environment, where the growing Merino puts on twice as much fat as the Dorset (Macfarlane and Howard 1972). Seasonal breeding by photoperiod control does not occur in cattle, pig or man because of their jungle origin, where day and night are the same length through the year. But breeding seasons are emphasised in the mesic types, so that the further from the equator they live, the shorter the breeding season and later in the year oestrus occurs. Subarctic sheep and deer mate only in November to produce young in spring.

The Australian fauna has an interestingly high proportion of mesic and desert mammals. *Macrotis* (bandicoot) and *Lasiorhinus* (wombat) are more desert-hardy and have lower metabolic rates and water turnovers than the camel. Among macropods the euro is a desert type while most wallabies and kangaroos are mesic, except for the wet-land agile wallaby. Possums, bandicoots and koalas are mesic. The dasyurid insect-eating carnivores evolved both desert forms *(Dasyuroides, Dasycercus)* and those with high water and oxygen turnover for wetter country *(Antechinus* or *Sminthopsis)*.

Sminthopsis crassicaudata exploits desert territories when there is food and water, but it dies in 5 days without water. The skeletally similar *Dasyuroides bernei* lives, however, in the desert without the need to drink water. Yet it would be difficult to obtain any clues on this from fossil remains. Similarly with burrowing. There is nothing morphologically to differentiate *Dasycercus, Macrotis, Notomys* or *Bettongia lesseur* which exploit the desert and burrow from *Sminthopsis* or *Bettongia pencillata* which do not burrow, but are skeletally similar to the other forms. Behaviours do not necessarily show in the skeleton.

The inappropriately high rate of consumption of water and energy for the desert, such as chat shown by *Sminthopsis crassicaudata,* allows exploitation in good years, but probably extinction of such ecotypes would follow recurrence of aridity in both the evolutionary hearth and the extended zone of exploitation.

In the Arctic deserts, animals evolved with functions suitable to the climate. The musk ox fattens in the summer on the taiga and it carries a coarse summer coat. As the days shorten towards winter, a very fine pelage of quivit develops for winter insulation and reproductive activity is induced so that there is a winter pregnancy followed by a spring birth, when food again becomes available to assist lactation. The musk ox has a low rate of water turnover and energy use like the hot-desert camels. Reindeer and moose, on the other hand, are deer derived from the temperate zone where water and energy supplies are relatively good and pass through at high rates. They are vulnerable to shortage of food or water. Although these deer have good winter insulation they have a light-induced shedding process which allows summer comfort by removal of the pelage.

Polymorphism and polyfunctionalism.

In any population there is diversity of genotype and phenotype. Rate functions on a genetic basis range ±30 to ±50% about the mean. This has implications for selection or survival or ecotypes. In a series of dry years, for instance, it is likely that animals with low water demands would breed and survive more than those with high requirements (Macfarlane et al. 1971). In a breeding group of animals the rates of use and synthesis of protein, of adrenal hormone production during stress, of ability to adapt to unfamiliar food or of endurance of dehydration may differ among individuals by 50% or more. Such wayout individuals could form the nucleus of a revised gene pool after climatic crises have passed.

Tolerance mechanisms.

Surviving mammals must in one way or another have tolerated large changes of food supply, water, salinity, temperature and predation or parasites during the Cenozoic. Some of the components of survival tolerance may be considered.

General adaptability. Adrenal steroids help cellular survival during cold, trauma or infection. But cortisol and the output of related steroids is reduced in the heat (Robinson et al. 1955; Macfarlane and Robinson 1957; Collins and Weiner 1968). Adequate tissue and circulating immunoglobulins protect against foreign proteins (or lead to allergic responses). Enzyme induction to inactivate poisonous plants or to pump sodium from cells (as in desert marsupials and camels, Macfarlane 1976a) is part of the resistance mechanism. Isoenzymes may help cope with changes of body temperature. And growth of insulation or reduction of pelage are fat and protein adjustments made automatically as environments change.

Genetic differences within a species are considerable. Among sheep breeds, Merino and Ogaden can survive daily temperatures up to 45°C in the sun. But Romney or Blackface are incapacitated by the heat in latitudes lower than 28°C. Rats and mice, with no cooling mechanisms, die at 35°C. They also are flaccid and reproductively inadequate at 32°C. Yet there are exceptional performers among mice. By selection Pennycuik (1969) has developed a strain of *Mus musculus* which is lively, metabolically normal and reproductively unimparied at 32°C. This required 12 generations of selection. Metabolic rate, water turnover and protein synthesis are the same as in mice at 22°C but the hot strain produces eggs, young and milk like cool mice. The adapted mice derive from a small part of the gene pool and are thus effective survivors in an environment that otherwise would have extinguished the species. Similar functional moieties of the human population were found in Japan by Yoshimura and Kobayashi (1975), some individuals having much greater survival potential than others.

DNA bases and protein aminoacid sequences are constantly changing (Wilson et al. 1977) but few of the changes may be effective for survival if there are sudden food or climatic upheavals. It is possible to start a new colony with only one breeding pair. Yet during extinction many breeding pairs seem not to reproduce.

Size. All small mammals have high metabolic rates, high protein and water turnovers (though desert species have one third the requirements of rainforest forms). This means that small forms have few reserves and starve or dehydrate rapidly so they are vulnerable to short-term failures of supply. Large animals have much slower rate functions (Kleiber 1961) but the total eaten is greater and more territory is needed for their survival. Density should affect their populations considerably. The young of all species are small and metabolically vulnerable because of their high rate functions (Macfarlane and Howard 1972).

Cold. Insulation by vasoconstriction, fat deposition and fine fur, is the primary defense against cold, rather than a change of metabolic rate. Cold induces long fur, though the mechanism is not clear. Shivering and cyclic AMP-adrenalin thermogenesis are only for emergencies. Escape from poor food supply by torpor and hibernation are of nervous system origin and they reduce the metabolic load. Cave animals appear sometimes to have hibernated but failed to reverse the process (Kurten 1968). Hibernation in echidna, voles and hamsters or torpor in dasyurids and mice are mechanisms for coping with both cold (sometimes heat) and poor food supplies by reducing metabolic demand (Kayser 1961; Godfrey 1968). Cell enzyme systems are required to function at both low and high temperatures for hibernation to be reversible. Migration of birds has allowed them to deal with cold periods, but few mammals could move significantly through other animals' territories in migratory escape from cold. When shivering fails to maintain temperature there is a quiet hypothermia and the heart ceases to beat at about 20°C in large mammals, about 5°C in small animals. Rats can be frozen, however, and thawed back to life (Smith 1961).

Heat. Reduced insulation and seasonally shed pelage help in heat tolerance. Small mammals (up to about 10 kg) have no evaporative cooling. Respiratory evaporation by panting has developed among wallabies and kangaroos, possums, wombats, dogs, pigs, antelope, deer and bovids of medium size. They have few sweat glands. Larger mammals, over 100 kg, rely more on sweating. Thus horses, cattle, camels and elephants all have sweat glands activated by adrenalin. The highest rates of sweating, however, were developed in the primates from sweat glands driven by acetyl-choline (Kuno 1956). **Man can sweat more than any other mammal; and without water, under the desert sun, this sweating is frequently lethal within a day.**

Improved tolerances of heat are induced by repeated exposure. Partly these arise from neuronal habituation, partly from adjustments of behaviour. Failure in the heat is either from dehydration and inadequate plasma circulation, or from loss of cooling power and thus hyperthermia with neuronal damage to cerebellum and the brain stem.

Food, water and salt. Lack of water for desert animals is compensated by small urine and faecal water losses, low metabolic rate so that respiratory water loss is minimal and relatively impermeable skin (a low transpiration rate due to the protein structure).

Fat reserves help with the intermittent food supply (as in *Dasycercus* and *Camelus*). As deserts came and went, the ability of herbivores to eat dry vegetation or, like the red kangaroo (Newsome 1965), to choose succulent components, probably helped survival.

On the fringe of deserts there are salt waters where halophytic plants accumulate. Camels have the highest tolerance to salt water, being unaffected by 5% NaCl in drinking water (Maloiy 1972). *Setonix*, the quokka, can use 2.5% NaCl; goats nearly 2% and sheep 1.3% in water (Peirce 1968; Macfarlane 1971). Desert dasyurids seem better able to induce enzymes to pump sodium from cells than non-desert forms. NaK ATPase in gut, liver and kidney cell membranes is probably protective against excess cellular sodium, but there may be other mechanisms as well.

Dehydration is a greater threat in dry country than starvation. In temperate grasslands about 200 tonnes of water are involved in growing 1 kg dry matter, while 1000 tonnes are required to produce 1 kg of wool (McMillan 1965). Apart from this ecological water flow, however, the individual animal is highly dependent on water supplies. Both man and beast centre on water sources in the desert, and the piosphere, the area of grazing encircling a water hole, is determined by the walking distance from water that an animal can achieve (Lange 1969; Squires and Wilson 1971; Wilson 1974). For sheep this is about 5 to 6 km and for *Bos taurus* cattle 7 to 8 km in summer. *Bos indicus*, zebu cattle, can walk up to 15 km from water. Some cattle and goats camp by the water sources or in nearby shade in hot weather. In the cool season kangaroos, camels and sheep do not need to drink but, once evaporative cooling begins, water must be taken. The time endured without water is a reciprocal function of the genetically determined rate of water turnover (Macfarlane and Howard 1970). Dehydration kills by reducing plasma volume to a level where circulation fails. Sheep survive a 45% reduction, but camels retain water in plasma by protein osmosis and so do not endanger circulation (Macfarlane and Howard 1972). Man does neither, and a 10 to 15% fall in plasma volume is lethal.

Starvation is rapid in small mammals with high metabolic rates (shrews, voles, mice) and in the cold. It is slow in warm climates among large well insulated animals with fat reserves. Slow starvation reduces all endocrine functions including reproduction and lactation, so the young are few and they quickly die. The adults become weaker and cannot forage. Finally in starvation, circulation by the heart ceases. Those animals with the drive to search further for diverse foods and the skills to conserve energy and keep warm are likely to survive. Up to 50% of initial body weight may be lost before death, if water is available.

This complex of ecophysiological characteristics runs parallel in marsupials, rodents, ungulates and birds. The rates of protein, water and energy turnover were evolved to suit each type of environment. And they are reciprocal to salt tolerance and renal concentration potential.

Migration and reinvasion.

During several phases of the Quaternary, there were large inter-continental movements of birds and mammals. Presumably, the movements en masse were the product of climatic changes and population pressures. A period of good rain and sun would induce vegetation levels at which breeding could be successful. But with build-up of population, over-grazing would occur in ungulate groups. Although the numbers of prey are useful for carnivores, the herbivores would either die or have to migrate. A prolonged period of drought would lead to differential death rates in terms of individual reserves of fat, rates of energy and water use, and in terms of behaviour which would allow migratory escape from unfavourable regions, or adjustment of reproductive performance. Since cycles of ice, cloud and rain must have been responsible for long crisis periods during the Pleistocene, stimuli encouraging migration would have been recurrent.

The migrations of Asiatic animals to America in the Pliocene was considerable. Deer, bison, saiga, bears, goats and sheep, moved from west to east; others moved from America to Asia, as well as the camel and horses. Some like caribou maintained a seasonal migrational pattern, moving thousands of kilometres between summer and winter grazing areas. Similarly in the mesic zones of Africa seasonal migrations induced by changing food supplies continue to drive antelope such as the springbok, kob, wildebeest, zebra and kongoni on mass movements.

One opportunistic tactic which seems to be used by the Australian *Sminthopsis crassicaudata* is that of exploitation of the desert river beds (Finck or Darling) during wetter periods. In drought years they die out but reinvade from their more mesic heartlands. *Sminthopsis* has high water and energy demands (Kennedy and Macfarlane 1971; Haines et al. 1974) about three times greater than desert dasyurids such as *Dasyuroides byrnei* which, like *Dasycercus*, does not need to drink at all. Local extinction, and later reinvasion probably has been common in the past, as climates changed. This is occurring with the koala. It takes place among cattle in the Australian Centre with each long drought (like that of 1958-1966) when all cattle there were moved out or died. With human intervention the water-dependent, marsh-derived Shorthorns or Herefords returned to the arid ecosystem. Australian cattle raising on the desert fringe is based on opportunistic exploitation and migration.

Camels and horses left the region of the Rocky Mountains in the late Pliocene (3 million years ago) but continued migrations through the Quaternary, before becoming extinct in America within the last 20,000 years. Camels and horses migrated to Asia and then to Africa. Earlier, llamas had moved into South America from the heartland. It is interesting that *Camelus* took to Afro-Asian desert ecosystems like those in which camel evolution had taken place, but the llama and guanaco groups in South America came to live in the cold Andes, or moved on to the savannas where food was reasonably well assured through the year and water quite plentiful. Yet desert physiological traits (such as low rates of water and energy turnover, high salt tolerance, with high levels of the red blood cell Na pump enzyme, NaK ATPase) have been maintained unchanged in camels, llamas and guanacos, in spite of 3 million years separation into the different environments (Macfarlane 1976a). This persistence of ecophysiological characteristics is useful evidence indicating that rate functions, once set, are not readily changed. From the present day measurements it is possible thus to approximate the physiology of the Pliocene ancestors of the camelids palaeophysiology. Unless there is strong evolutionary pressure against a mechanism it appears not to be modified although it is not necessarily ideal for the new environment in which the animals come to live. Cattle, for instance, survive as water-dependent animals in central Australia then die out in droughts, then reinvade. Adaptation in the physiological-biochemical sense takes place. Although it is not necessary for a camel living on temperate zone green pasture to have a low rate of water and energy use or to have a very high salt tolerance, it has not changed significantly in those rates of function since it left America. There is no threat to its survival by persistence with desert-type functions. The camelids have thus provided an interesting piece of palaeophysiological evidence for the persistence of rate functions in changed ecosystems.

Australia.

Australia's continental isolation was well-established by the Quaternary period. Migrations had provided a pan-Antarctic fauna and flora (*Nothofagus, Peripatus,* ratite birds and marsupials) from Gondwanaland. It is not clear whether monotremes evolved in Australia from therapsids or whether they also moved in during the Cretaceous (Keast et al. 1972).

Marsupials radiated into all ecological niches and showed a parallel type of functional hierarchy to that occurring in eutheria. There are fossorial desert forms such as *Lasiorhinus, Macrotis, Dasycercus* and *Dasyuroides* which have very low rates of energy, water and protein turnover. There are more mesic sheep-like animals of the wallaby and kangaroo group, koalas, bandicoots and possums, with intermediate rates of water and energy use and there are wet forest forms of carnivores such as *Sminthopsis, Antechinus* and *Satanellus* or of herbivores such as the agile wallaby. These evolutionary settings took place probably 50 million years ago. But it is not possible to estimate rate functions or burrowing habit from the skeleton as such. A given species or genus of known function may be presumed to have changed little in a million years.

Some eutheria (rodents and bats) appear to have reached Australia in the Miocene. They probably evolved into the desert rats and mice. In the Pliocene and Pleistocene the more water-dependent rats came across from the north, presumably by land bridges between New Guinea, Australia and south Asia. Ice age sea level fluctuations of 200 m would greatly aid successive invasions. Some of the rodents which came in retained tropical rates of water and energy use, the others adapted to the desert conditions of much of the continent, radiating to a total of 124 species (62 superspecies, Keast et al. 1972). It looks as though the very high concentrating power of the kidneys of *Pseudomys* and *Notomys* evolved in Australia since the renal mechanisms can reach 9 mosmol/l. in these forms, but the kidneys of their relatives in Africa or America did not reach more than about half that urinary concentration (Schmidt-Nielsen 1964; MacMillen and Lee 1970).

Function and survival.

It is not easy to generalize the survival potential of a species. Part of this comes from the fact that, in a specific breeding group, there may be more than 50% difference between individuals in any parameter measured, be it metabolic rate, water turnover, hormone concentration or resistance to dehydration or infection. Another factor is differences in adaptive potential. Enzyme induction, drive to migrate or to try new foodstuffs, or ability to breed in adverse circumstances, each has a wide range within a breed or species. Some mammals live not so much in a habitat or ecological niche as in a zone of exploitation. Man is the prime example, able to live anywhere. The echidna and the grey kangaroo live in a range of altitudes and climates, from cold savanna to tropical forest, so that their presence is no good index of the type of ecosystem. On the other hand, the platypus is confined to cool streams and the koala to *Eucalyptus,* especially *E. melliodora* and *E. punctata.*

Yet all of these species survived Pleistocene climatic changes effectively while a number of large mammals became extinct. For the extinctions the causes could be:

(a) unusual agents (fire, cold), predation (immigrant killers like dingo or man) or new infective agents (virus recombinations),

(b) limiting specialization of food supply, size or breeding requirements, and

(c) flaws in immunogenesis, fertility, behaviour or gene-derived biochemical adaptation.

It is not certain how long the gaint kangaroos and *Diprotodon* persisted. In the dry cold more than 40,000 years ago the Lake Calabonna bones of those animals became fossilised; but the dates of their later fossils (perhaps at 7-11,000 BP) are not at all certain. *Diprotodon* in peat on Kangaroo Island persisted to the height of the last glaciation. Whether they lived on to the warm dry Holocene is not clear. The Lake Colongulac date of 13,725 BP for the giant forms of marsupial (*Diprotodon* (wombat), *Thylacoleo* (lion) and *Procoptodon* (kangaroo) is the latest apparent date for those survivors. Associations with the dates are still under dispute. However, that would bring them to the end of the glacial period, but not later (Keast et al. 1972). The cause of their extinction is still not obvious. There is doubt about the possibility of fire and Aboriginal hunting as sources of extinction of *Diprotodon, Sthenurus* and *Phascolonus*. It is unlikely, however, that *Thylacoleo*, a carnivore, was eliminated in the Holocene by man, since it probably was not hunted and it would have as good a potential for escaping fire as the dingo.

Evolution in terms of function requires analysis not only of extinctions and their mechanisms, but also of the adaptations that allow survival. Extinctions have occurred at a fairly regular rate since the Cambrian. And it appears that the rate of protein sequence mutation has also been regular (Wilson et al. 1977) even though rates of phenotypic change have varied considerably (as in the different rates of change of ferns and flowering plants, fish and mammals, brachipods and insects). Functional changes probably arise from mutations in regulatory genes; from quantitative changes in enzyme production or from gene duplication. All these could allow adaptive biochemical and physiological action. Transfer of DNA by virus could produce major changes. On the other hand, chromosomal mutations are more likely to lead to sterility and set a course for extinction. Extinction may follow climatic change and ecological incompatibility (monkeys in Europe), intense predation *(Thylacinus, Ovibos)*, or disease and pressure of population on food supplies.

But there are other extinctions that are not readily explained. *H. neanderthalensis* had fire and tools, and a large brain. The species was widely dispersed over Europe and Asia. Yet it became extinct during the last glaciation while *H. sapiens* survived. Possible behavioural inadequacies developed. If building of shelters or refuge in caves did not become fashionable, or if incest taboos did not arise in small groups, there could have been steady attrition. There could have been aminoacid sequence changes that led to inadequate immunoglobulins for viruses or failure of luteinising hormone to release ova. Something of that sort seems to have occurred in *Bos primigenius*, the aurochs. It became extinct in Spain around 900 AD, in France about 1200 and in Poland in 1627. In Poland groups of 30 or more males and females were protected in the forest for half a century but they failed to breed and so became extinct. Failure of small groups to reproduce is an interesting problem. This seems to have occurred with the residual groups of Tasmanian men and women, who could not sustain a breeding population.

The *Thylacinus* in Tasmania was numerous last century (Guiler 1961; Tyndale-Biscoe 1973) and was hunted as vermin until 1908. Each year up to 1902 more scalps were taken, followed by a plateau till 1905, at 100 a year. Then there suddenly was depletion and extinction (apparently) by 1909. Probably disruption of breeding groups and too great a separation of individuals led to reproductive failure, under human predatory pressures. Certainly no juveniles were taken for 4 years before extinction.

Local extinctions of marsupials have occurred widely in Australia. By 1920 the koala was eliminated from South Australia, as was the woilie wallaby, coursed to extinction. Members of the species survived, however, in the other States, and are being redistributed. Some species of bandicoot, dasyurid and macropod have become extinct in the last century in each State, about one in four species overall (Serventy 1966). Direct killing, predation on the young and changes of habitat produced by agriculture, deforestation or pine forest replacement, all contributed.

In any extinction process the young are likely to be important, partly because their food and water demands are relatively greater than those of adults, their temperature susceptibilities are high and their immunity to viruses, bacteria or worms still undeveloped. In the failure of a species it may be the impact of environmental change (density, food supply) on the young that brings about the primary attrition. Then non-adaptive genetic protein changes, reduced fecundity and behavioural inadequacies could complete the extinction. Breeding may be adversely affected by long droughts or fire. Oestrus, for instance, occurs throughout the year in well-fed Merino sheep but it fails in spring if nutrition is poor for climatic reasons (Entwistle 1974). A hot environment as such reduces oestrus, implantation and foetal survival in sheep as well as in rats; and the effects of heat are additive to lack of food, of proteins or of vitamins. Cessation of adequate food supply in cold conditions can have the same overall effect on reproduction (Macfarlane et al. 1959). With starvation there is infertility, foetal loss, poor lactation and neonatal death.

The last glaciation appears to have been the most severe of the Pleistocene cold phases (Kurten 1968) giving an average drop of 10°C in temperature. If cold were the cause of mortality it would be expected that extinctions would be fewer in the 14,000 to 11,000 BP era as climates became warmer. But there was in that period a series of extinctions, presumably as mammals faced dry conditions and increased human neolithic skills. Perhaps fires were more damaging at that time. This late wave of extinctions of large mammals on all continents is difficult to explain. As Martin (1970) points out, if fossil rodents, reptiles, plants and pollen are effective indicators, no major ecological disturbance can be seen. But of the large mammals (more than 100 kg) two thirds became extinct. In Africa, large mammal extinctions began 100,000 to 50,000 years ago. Eurasian and American extinctions covered 14,000 to 8000 BP and the last megafauna losses occured in Madagascar and New Zealand in the last 2000 years (Martin and Wright 1967). These patterns certainly seem to link with the movements and cultural developments of *H. sapiens*.

HUMAN TOLERANCE LIMITS

Although tool-using Australopithecines evolved in the Pliocene, the major evolutionary product of the Pleistocene was man. Human palaeontology and ecophysiology both indicate that man evolved as a tropical animal. The tropical stigmata are lack of significant insulating fur, high rate of sweating and water turnover, low renal concentration maximum (1300 mosmol/l.), and reproduction throughout the year. These characteristics are likely to have persisted over 3 million years or more. Tropical hominids spread widely so that *H. erectus* and *H. sapiens* both reached 50°N latitudes. They were also subjected to the sequence of Pleistocene glaciations, pluvials and desert conditions.

The thermal neutral point for humans at which heat input equals output, is about 29°C for males and 30°C for females (the temperature of a comfortable bed, Macfarlane 1958). Without protective structures, human adaptations to cold and drought are essentially behavioural. The human speciality of large brain size with neo-cortical development at the rate of about 1 gm increment each 3000 years, is not only fast evolution, but also effective for coping with environmental diversity.

Early Pleistocene mammals in Eurasia and Africa appear to have evolved in warm conditions which were not too dry. Man would, therefore, be emerging from the tropics together with elephants, antelope, boars, giraffes, horses and rhinoceroses. These moved to the holarctic region, but with Pleistocene glaciation, beginning about 2 million years ago, many of them became extinct. Man-like creatures, together with horses and camels, lived through the pluvials of the Villafranchian and through the long sequence of subsequent glaciations. Peking man of half a million years ago in a cold period on latitude 42°N, may have used fire (Kowalski 1971; Turekian 1971). At least two major species of man became extinct in the last glacial period in spite of tools and fire. These were *H. neanderthalensis* and *H. erectus*. They had a wide geographical range into cool temperate zones but they failed to survive in any latitude. The residual form, *H. sapiens*, was presumably sufficiently biologically rugged, as well as cunning, to survive glaciations, pluvials and deserts. The time at which men began to use shelters, clothing and fire is not clear, but fire was certainly available for the last glaciation between 60,000 and 12,000 years ago.

The tolerance limits of the early *H. sapiens* can only be inferred from present functions and measurements, but there probably has been little essential change. Males, through the Pleistocene, probably weighed 50 kg and females 40 kg as adults. They would require on average at least 2 litres of water and 6500 KJ of food energy per day. Adults, at least, survived daily maximum temperatures over 50°C dry bulb in desert areas and -50°C in the glacial zone. On the edges of desert, where saline waters occur, drinking of about 1% NaCl solution would be tolerated. Infants would be less tolerant of heat, cold or salinity and the aged also.

Extrapolating the demographic process backwards, it would be likely that expectation of life was 30-35 years, and the number of children born per reproductive female would be in the order of 10. The greatest age specific death rates would be under 5 years. The tolerances of under 5-year old children are probably of less importance to the survival of the species than those of the adults, who can live to breed again in better circumstances. The problems of deciding on tests of heat or cold tolerance (short of mortality) are discussed by Yoshimura and Kobayashi (1975). They point out that populations differ in functions so that survival potential may be judged better by dividing large groups into inferior, average and superior components, for any parameter measured.

Infant ecophysiology and tolerance limits.

The metabolic rate and rate of water and protein turnover (either per unit of mass or of surface area) is about 3 times greater in the newborn than in the fully adult human (Macfarlane 1963; Cheek 1968; Young et al. 1975). Similarly the rate of sweating per unit area of new born is up to 6 times greater than in adults, so that water turnover is correspondingly higher in hot environments. Dehydration is rapid in heated infants, who have little response to vasopressin and do not readily concentrate urine. The newborn human and the infant have a relatively large water content in the plasma and interstitial spaces, up to 40% greater than in adults. This is a fluid reserve, and if nutrition is good there may also be a store of fat.

In conditions of duress (lack of food, lack of water, fluid loss by diarrhoea, heat or cold) the responses of infants take place at greater rates than those of adults. In hot environments, without water, there is very rapid loss of blood volume in spite of reserve fluids in the extracellular space. An air temperature of 40°C dry bulb for less than a day will dehydrate a child. Similarly dysentery and cholera produce catastrophic deaths from the same dehydrational cause. Tolerance to salt is low, so that the 40 mequiv/litre Na in cows milk may be lethal if used as a replacement fluid relative to the 12 mequiv/litre in human milk. Infants die when blood Na reaches 175 mequiv/litre (Danks et al. 1962).

The high metabolic rate of infants means that they starve quicker than adults. Starvation in the cold is, of course, faster than in hot environments because of the metabolic cost of shivering. In addition to pure energy starvation during which fat is used, there is depletion of muscle, liver and other proteins. The demand of infants for growth protein is relatively 2 to 3 times greater per unit of mass than in adults. If the milk supply fails before a child has access to legumes, meat or nuts, growth ceases and protein deficiency death is likely.

In cold environments on the other hand, there is a demand for vitamin D to increase the absorption of Ca for bone formation. With reduced solar UV light in cloudy pluvial or glacial environments, rickets would be common. This could lead later to difficult childbirth and infant deaths. The other likely vitamin deficiency of the ice ages would be vitamin C. Ascorbic acid is not plentiful in the diet of hunters. If they were besieged by snow for long periods in caves without access to fruit and herbs, scurvy deaths would be likely.

The effects of pure cold on infants is ambivalent. Because of their high metabolic rate, they can, if there is sun, play naked in the snow and balance the energy input with the radiation received. When there is persistent cloud and also cold air, however, they are vulnerable to death by cold if not insulated by clothing. Infants that kick off a covering during sleep and are exposed to air temperatures around 0°C lose body heat and die of hypothermia without protest within a few hours. At the other end of the thermal scale,infants are amongst the more vulnerable to heat in the human population. They readily suffer heat stroke (hyperthermia) if fluid replacement is insufficient, or if there is a slight infection with cessation of sweating. At air temperatures over 40°C they may cease sweating, body temperature rises, and they die of brain cell damage after 3 to 5 hours of body temperature around 44 to 46°C.

Human survivors of the Pleistocene must have been able to provide insulation and food for infants, shelter from both sun and cold and good sources of non-saline water during drought periods. Small animals as food provide all vitamins and minerals needed, but in large (500 kg) beasts the vitamins are in low concentration. Presumably relatively small groups of people provided some protection against enteric infections of the dysentery or cholera type. Zoonoses of the plague, Q fever or encephalitis groups could infect hunters, but they would also be less likely to devastate all of a region when the human groups were limited to 10-30 people. On the other hand, immunity to viruses like measles might be low in a particular group, and a few casual meetings between tribal elements could lead to high mortalities when non-immune populations became infected. Panum (1847) described such events in the Faröe, when measles spread from island to island often by one person acting as carrier.

Presumably through the Pleistocene a certain amount of predation by cave bears, cave lions, wolves and tigers would have taken place. But intraspecific hominid killing was probably an invention of the Holocene, after agricultural territories became established, and it may have reached its peak in the last 1000 years.

Adaptation to drought, pluvials and glaciations.

As a fundamentally wet tropical animal, the human species survived Pleistocene climate changes by using behavioural skills. In the warm dry times of the late Pliocene or Villafranchian, hominids presumably lived rather like desert Aboriginal groups. That is, they would have brushwood shelters against the sun and sleep through most of the heat of the day. Hunting would take place in the morning and evening when game would, in any case, be more likely to emerge. Women would search for insects, lizards and small mammals, as well as grass seeds, fruit and nuts, during the cooler parts of the day. In the hot season the groups would settle near water sources and could easily hunt out the surroundings of a water hole if no other water were available within less than one day's walk. Man has a limited piosphere. Summer migrations would be associated with death both of infants and of aged members of the group. The fixation on water during dry periods is an essential behavioural aspect of survival. The Aboriginal Australoids presumably spent most of their evolutionary life in sourthern Asia, where good water supplies were available. When they moved to the drought-prone continent in the south, they brought with them relatively high rates of sweating and water consumption which still persist (Macfarlane 1973). Desert tribes are still water dependent, and use water at twice the European rate (Macfarlane 1976b).

Processes.

There are three major adaptive processes which must have played a part in physiological survival over the last 3 million years. These are selection of genetic material suitable for hot or cold climates and for starvation or plenty. The second is habituation, the process of changing the levels of response of the nervous system to pain, heat or cold after birth (Glaser 1966). The third is biochemical and physiological elastic adaptations, allowing greater ranges of cellular tolerance to develop, largely through induced enzymes, isoenzymes and hormones. For instance, tolerance to salt or toxic substances arises from the generation of enzymes in liver cells so that the unwanted materials are inactivated or removed and the organism survives.

There is some evidence that large human groups differ in gene material relating to functions. The clearest case is that of metabolic rates (Roberts 1952; Macfarlane 1963). The Mongoloid group of Chinese and Japanese has the lowest metabolic rate of all human races, while the Amerinds have the highest. The difference is about 40%. Between these two come the Caucasians, Negroids and the Australoids. It seems likely that the Amerinds derived from north Asian cold country (Siberian) groups where the metabolic rates were and still are high. Relatively small incursions into the Americas across the Bering Strait led to the generation of tribes with a rather uniform high rate of metabolism. This small-group genetic drift set the basic rate. As the Amerind population moved from the cold north to the south they left a gradient from high metabolism in the cold areas to low in the equatorial zone. This has shown up particularly well in the Americas, but it occurs in all racial distributions, including Mongoloids and Caucasians. Eskimos and Athabascans have considerably higher metabolisms than Mayas and Nicaraguans or Columbian Indians. Similarly, Patagonians in the cold south are rather like Athabascans in their rate of metabolism. There have been two forces at work. One relates to remote gene pools setting a pattern which persists. There was a high rate of energy use among Siberians and low rate in the Chinese each from a separate gene pool. The second force is a gradient of temperature which brings about current adaptations so that polar peoples use energy at rest at a higher rate than equatorial people.

There is a greater rate of sweating in tropical people for a given heat stimulus than in those from cool temperate areas (Kuno 1956) yet this is not entirely intrinsic, since Caucasians moving to the equator increase their sweat rates about a third in the first one or two months of exposure to high temperatures (Macpherson 1960). Aboriginals form a different gene pool coming from the Asian tropics, however, and in the desert or in Arnhem Land they sweat more in the same environment than Europeans. This occurs when the two groups walk in the sun together (but it is not dependent on differences in skin temperature, since it occurs whether subjects are clothed or not clothed). There appears to be a racially determined difference in sweating rate (Macfarlane 1969, 1973, 1976b). Wyndham et al. (1964) and Fox et al. (1964) measured sweat rates in saturated air and their findings include a great deal of the effects of sweat suppression. It is clear that when the skin is thus saturated with water, sweating is inhibited (Brebner and Kerslake 1968; Itoh et al. 1972). What is interesting, however, is that the rate of suppression of sweating in Aboriginals is about 5 times faster than amongst Caucasians in the same environment (Macfarlane 1973). Presumably Aboriginal skin absorbs water more in the superficial layers possibly swelling the sweat ducts to prevent outflow of sweat. Aboriginal whole-body sweating measured by weight change when their skins are dry in the desert, is considerably greater, by a factor of two, than that of Caucasians (Macfarlane 1969, 1973). This presumably is a heritage of wet tropical function. Sweating estimated by weight would make Aborigines well acclimatised, but in a saturated environment they appear to be unacclimatised by Wyndham's standards.

Angolan Bantu coming to work in the hot mines (30°C wet bulb) of Johannesburg (Macfarlane et al. 1966; Wyndham 1970) became overheated during acclimatization. Those whose oral temperature rose least, and who were able to adapt to mine work without risk of heat stroke, were those with the highest rates of water turnover. For hot regions there is survival value in drinking water above the normal thirst demands. Man's thirst mechanism can lag 2 litres behind the actual loss of fluid (Macfarlane 1964). The highest daily intake of water by Caucasians in hot desert conditions recorded was measured by Hunt (1912) in India. With temperatures reaching 50°C his subjects drank up to 12 litres daily. Most fluid intakes in the desert range from 5 to 8 litres/24 hours. But lack of water replacement for sweat leads to circulatory failure within the day, when exercise is taken in the sun. Solar radiation is a greater source of heat than air temperature as such. So the Aboriginal habit of sleeping in the shade through the time of maximum insolation, is survival behaviour.

Metabolic and sweat patterns have implications for Pleistocene climates. It is clear that the Amerind Eskimos have been able to thrive with high metabolic rates in arctic Alaska and Canada, while Amerinds in humid jungle of central America and in the desert areas of Nevada and New Mexico have also lived effectively. Similarly, low metabolic rate Chinese thrive in the cold of northern China, as well as the heat of South China or Singapore. And a high rate of sweating may not seem appropriate for Aboriginals when water is scarce and yet they have survived in the desert for many thousands of years by remaining near water through the summer. Compensations for these rate functions have been made by behavioural adjustment. Food and water intakes have been regulated to the metabolic settings. The effects, also, of extremes of heat and cold have been avoided by use of shelter, fire or insulation.

There is, however, another protection that comes from habituation (Glaser 1966). All neurones can be changed in the amount of their response to repeated stimulation. This occurs in the spinal cord (Kozak et al. 1962) but it can also take place at the higher levels of the brain stem or limbic area. Each generation of children growing up in hot or cold environments, is exposed to considerable cold or heat at intervals. It is not necessary that this be for more than 5-10 minutes daily for habituation to occur. The input, however, has to be near the bearable limit in order to get quick and adequate habituation. The process was clearly illustrated in the case of Aboriginals who used, until the 1950s, to sleep without clothes near the fire with dogs during winter. They were habituated to sleep with foot temperatures down to 17° or 18°C. But body temperature did not fall, nor did they shiver (Hicks 1964; Scholander et al. 1958). This occurred in the Central Desert Aboriginals; but habituation did not take place in Arnhem Land Aboriginals where cold is rarely experienced (Hammel et al. 1959). Darwin observed that Patagonians were naked, but indifferent to snow falling on them and Hammel (1964) has discussed this adaptation. Davis and Joy (1962) habituated Americans to cold by 1-hour exposures daily over periods of months. This almost eliminated shivering, and the elevation of metabolic and pulse rates by cold was reduced to half. This type of habituation occurs in those who take to ritual ice-water bathing through the Baltic winter. And presumably it was acquired (at some cost of comfort) by primitive man during the glaciations.

Habituation to heat can also occur. Habituation is not inherited, but requires strong exposures to the stimuli in each generation. It arises from inhibitory processes amongst the interneurones of the spinal cord and brain stem. These cells prevent the onward transmission of sensory information so that pain and cold are not felt as strongly as in non-habituated subjects. Habituation also reduces the amount of shivering activity which passes outwards. At the same time reflex vasoconstriction of the skin or shell of the body takes place to increase insulation and conserve heat. This cold exterior is then not perceived as being cold, and it is possible to have not only comfort, but also to sleep, because of the habituation process. Interestingly, Eskimos are not very well habituated to cold except in the hands. Like fishermen who use their hands in cold water, there is a neuronal adjustment of a habituation type which ensures that, although the hands are cold, every few minutes there is a flood of warm blood from the arteries through arteriovenous anastomoses which warms the hand and allows activity to go on in spite of severely cold conditions (Burton and Edholm 1955). Apart from this the Eskimos are not ordinarily exposed to cold in a way which would provide very much habituation, since they use animal skins and shelters to provide local conditions of warm comfort in the Arctic. The time in human development when this level of sophisticated protection was invented, is not clear. Houses are known from 25,000 BP and caves certainly were inhabited well before that. Presumably some form of shelter from the snow was constructed by early man and presumably also during the first Quaternary glaciation there was considerable stimulus to use the furs of animals which had been killed for food. Experience of this could have been passed on through generations by example or by talk.

It appears that the Korean diving women who work in cold water for much of the day, and Australian Aborigines sleeping naked, both have little fat insulation, but can achieve levels of vasoconstriction which provide twice the insulation that is found in the European (Itoh et al. 1972). On their normal diet Eskimos are not fat. Japanese houses in winter are cold but people have some fat insulation. In early Pleistocene times, it is not likely that fat played a great part in protection against the cold as it depends upon excess food supply which may have been difficult to obtain (Itoh et al. 1972).

Another aspect of thermal adjustment to cold environments arises from non-shivering thermogenesis (NST). There is some brown fat available for this purpose in human infants but it is not well developed (Smith and Horwitz 1969). The release of nor-adrenalin mobilises free fatty acids from brown and white fat depots to provide rapid supplies of oxidisable carbon. Cyclic AMP and calcium are both involved in this release, as is serotonin. The human infant can get significant amounts of energy for immediate compensation against cold by this mechanism, but it is not clear whether it is better developed in people living like Aboriginals than in more domesticated forms of man. In spite of successive ice ages NST has not become a major human evolutionary protection against cold.

Biochemical adaptation

There are two main ways in which cellular enzymes are modified by environmental inputs. There is first a direct action of such substances as salt or plant toxins (alkaloids or glucosides) on cells, to induce enzymes that inactivate or reduce the cellular content of these damaging substances. Sodium acts through DNA to increase the amount of the enzyme NaK ATPase in gut, liver and kidney cells all of which are in immediate contact with high levels of absorbed salt. Isoenzymes may also be induced by the action of raised or lowered temperature on cells, as in fish, where different enzymes are formed in the same species reared at 5° and at 30°C. The range of temperature in which the enzyme works optimally is changed in accordance with the thermal conditions (Prosser 1970). Most changes amongst enzyme systems follow the action of hormones on cells. In this respect the steroid hormones (adrenal and gonadal) are most active, although thyroxine is also involved.

When humans are exposed to a cool environment, the amount of cortisol secreted from the adrenal is increased, but this decreases as air temperatures rise above 25°C (Macfarlane and Robinson 1957; Macfarlane 1964). The amount of cortical steroid secreted daily falls in summer to less than half that produced in winter (Collins and Weiner 1968). This has to do with the amount of protein and carbohydrate metabolism, as well as to some extent with electrolyte movements, all of which are involved in metabolic rate adjustment. Cortisol acts on DNA in the nucleus to induce RNA formation for the synthesis of enzymes in greater quantity for modulating metabolism. Without these steroids, resistance to trauma or cold stress is lowered and death is more likely.

Thyroxine acts in a similar fashion particularly on the enzyme chain concerned with oxidative processes (succinoxidase, succinic dehydrogenase). The amount of thyroxine used daily in most animals and people is greater during winter than in summer. This is because more food is eaten and therefore there is more energy to release, requiring additions to the enzymes of the oxidative chain governed by thyroxine. In any society there are very different rates of thyroxine synthesis and turnover amongst people of different economic status. In a temperate zone village in the New Guinea highlands we found that the amount of thyroxine in the blood and the rate of turnover of thyroxine were proportional to social status and work load. The low-status men had very low levels of both, and the high-status men about 3 times greater use of thyroxine. Women who did all the work and who had large food intakes because of this, had the highest rates of turnover and of thyroxine in the blood. Since the environment was the same thermally for all groups, the thyroxine functions were proportional to the amounts of food used. This has been shown clearly for sheep and rats also (Good et al. 1974; Heroux and Jobin 1971; Heroux and Brauer 1965; Yoshimura and Kobayashi 1975; Itoh et al. 1972). The larger amounts of food taken in the cold result in greater amounts of faecal excretion of thyroxine and there is a greater turnover of thyroid hormone which in turn leads to greater enzyme production. There is, however, a direct action of cold working through the nervous system on the pituitary to release TSH which gives an immediate output of thyroxine. This is not associated with enzyme induction, however, for some 36 hours. If thyroxine or T_3 were produced in cold environments regardless of the supply of food, its effect would be to burn out body tissues (as in hyperthyroidism) and to reduce the chances of survival except in reaction to a short cold exposure. It appears that the thyroid is not a major component of the adjustments of animals to cold. Non-shivering thermogenesis is probably of greater importance. Inadequate responses by hormones in cold eras could, however, reduce chances of survival and life in glaciated zones where iodine has been reduced by leaching is made less adequate by hypothyroidism.

Another aspect of the effects of heat and cold relates to reproduction. Since man is essentially an evolutionary product of the equatorial region, she and he reproduce throughout the year. There is no evidence of photoperiodic reproductive control in *H. sapiens*. There is, however, a seasonal change in reproductive potential and in yield of offspring. It is likely that continued hot environmental conditions reduce the amount of LH (luteinising hormone) secreted and thus the amount of ovulation. This is certainly clear in animals other than man (Macfarlane et al. 1959). Testicular function can also be inhibited by high temperatures. The main seasonality of human conception, however, appears to be related to temperature (Macfarlane 1970) but there could be some small input from the effects of food and vitamins. On the equator, there is about 20% greater conception in the hotter (27°C max. temperature) than in the cooler (24°C max.) seasons of the year. Away from the equator, however, this pattern is reversed so that in the 10-30° latitudes there is a summer depression of conception followed by a winter increment of up to 30%. This is very clear in India, northern Australia and the southern states of USA. In the 30-35° latitudes there may be little seasonality as in California and New South Wales. The cold temperate zones, however, are the reciprocal of the sub-tropics, with a summer maximum of 25-30% greater incidence of conception than the winter minimum. This occurs in Europe and Canada, as well as in New Zealand and Tasmania. There appears to be a mean temperature around 24°C at which conception is non-seasonal. A rise above or a fall below that mean leads to a reduction in conception. This may well be due to endocrine changes in the female resulting in fewer ovulations, less adequate implantation and greater abortion (Pennycuik 1964). Neither contraception nor air-conditioning has made any significant difference to these patterns of human conception in the last 50 years. It is likely, however, that through the Pleistocene there has been adequate reproduction in *H. sapiens* so that cold or heat have been of no great importance, although they modulate the number of young born. Post-natal deaths have been a much greater determinant of racial survival than the uterine phase of reproduction.

REFERENCES

Brebner, D.F. and Kerslake, D.M. 1968 The effect of soaking the skin in water at various temperatures on the subsequent ability to sweat. J. Physiol. Lond., 194, 1-12.

Burton, A.C. and Edholm, O.G. 1955 Man in a cold environment. Edward Arnold, London.

Cheek, D.B. 1968 Human Growth. Body composition, cell growth, energy and intelligence. Lea and Febiger, Philadelphia.

Collins, K.J. and Weiner, J.S. 1968 Endocrine aspects of exposure to high environmental temperature. Physiol. Rev., 48, 785-839.

Danks, D.M., Webb, D.W. and Allen, J. 1962 Heat illness in infants and young children. A study of 47 cases. Brit. Med. J., 2, 287-293.

Davis, J.R.A. and Joy, R.T.J. 1962 Natural and artificial cold acclimatization in man. pp 286-303 in Biometeorology (ed S.W. Tromp), Pergamon, Oxford.

Ealey, E.H.M. 1967 Ecology of the euro, *Macropus robustus* (Gould) in north-western Australia. CSIRO Wildlife Res., 12, 9-25.

Entwistle, K.W. 1974 Reproduction in sheep and cattle in the Australian arid zones. pp 85-97 in Studies of the Australian Arid Zone II Animal production (ed A.D. Wilson), CSIRO Melbourne.

Fox, R.H., Goldsmith, R., Hampton, I.F.G. and Lewis, H.E. 1964 The nature of the increase in sweating capacity produced by heat acclimatization. J. Physiol. Lond., 171, 368-376.

Glaser, E. 1966 The physiological basis of habituation. Oxford University Press, London.

Godfrey, G.K. 1968 Body temperatures and torpor in *Sminthopsis crassicaudata* and *S. larapinta* (Marsupialia-Dasyuridae). J. Zool. Lond., 156, 499-511.

Good, B.F., Howard, B. and Macfarlane, W.V. 1974 Body size and thyroxine secretion rate in grazing ungulates. Proc. Endocr. Soc. Aust., 17, 32.

Griffiths, M. and Barker, M. 1966 The plants eaten by sheep and by kangaroos grazing together in a paddock in south-western Queensland. CSIRO Wildlife Res., 11, 145-167.

Guiler, E.R. 1961 The former distribution and decline of the Thylacine. Aust. J. Sci., 23, 207-210.

Haines, H., Macfarlane, W.V., Setchell, C. and Howard, B. 1974 Water turnover and pulmocutaneous evaporation of Australian desert dasyurids and murids. Am. J. Physiol., 227, 958-963.

Hammel, H.T. 1964 Terrestrial animals in cold: recent studies of primitive man. pp 413-434 in Handbook of Physiology, 4 (ed D.B. Dill), American Physiological Society, Washington D.C.

Hammel, H.T., Elsner, R.W., LeMessurier, D.H., Andersen, H.T. and Milan, F.A. 1959 Thermal and metabolic responses of the Australian Aborigine to moderate cold in summer. J. appl. Physiol., 14, 605-615.

Heroux, O. and Brauer, R. 1965 Critical studies on determination of thyroid secretion rate in cold-adapted animals. J. Appl. Physiol., 20, 597-606.

Heroux, O. and Jobin, M. 1971 Seasonal dietary influences on thyroxine secretation rate in white rats kept outdoors. Proc. Inter. Union Physiol. Sciences, 9, 245.

Hicks, C.S. 1964 Terrestrial animals in cold: exploratory studies of primitive man. pp 405-412 in Handbook of Physiology, 4 (ed D.B. Dill), American Physiological Society, Washington D.C.

Howard, D.A., Burdin, H.L. and Lampkin, G.H. 1962 Variation in the mineral and crude protein content of pastures at Muguga in the Kenya highlands. J. Agric. Sci., 59, 251-256.

Hunt, E.H. 1912 The regulation of body temperature in extremes of dry heat. J. Hygiene, Cambridge, 12, 479-488.

Itoh, S., Ogata, K. and Yoshimura, H. 1972 Advances in climatic physiology. Igaku Shoin, Tokyo.

Kayser, C. 1961 The physiology of natural hibernation. Pergamon, Oxford.

Keast, A., Erk, F.C. and Glass, B. 1972 Evolution, mammals and southern continents. State University Press, N.Y., Albany.

Kennedy, P.M. and Macfarlane, W.V. 1971 Oxygen consumption and water turnover of the fat-tailed marsupials *Dasycercus cristicauda* and *Sminthopsis crassicaudata*. Comp. Biochem. Physiol., 40A, 723-732.

Kleiber, M. 1961 The fire of life. Wiley, New York.

Kowalski, K. 1971 The biostratigraphy and paleoecology of late Cenozoic mammals of Europe and Asia. pp 465-477 in Late Cenozoic glacial ages (ed K.K. Turekian), Yale University Press, New Haven.

Kozak, W., Macfarlane, W.V. and Westerman, R. 1962 Long-lasting and reversible changes in the reflex responses of chronic spinal cats to touch, heat and cold. Nature, Lond., 193, 171-173.

Kuno, Y. 1956 Human perspiration. Thomas, Springfield, USA.

Kurtén, B. 1968 Pleistocene mammals in Europe. Weidenfeld and Nicholson, London.

Kurtén, B. 1971 The age of mammals. Weidenfeld and Nicholson, London.

Lange, R.T. 1969 The piosphere: sheep track and dung patterns. J. Range Managt., 22, 396-400.

Macfarlane, W.V. 1958 Thermal comfort zones. Arch. Sci. Rev., 1, 1-14.

Macfarlane, W.V. 1963 Endocrine functions in hot environments. pp 153-222 in Physiology and Psychology in arid environments. UNESCO, Paris.

Macfarlane, W.V. 1964 Terrestrial animals in dry heat: ungulates. pp 509-539 in Handbook of Physiology, 4 (ed D.B. Dill), American Physiological Society, Washington D.C.

Macfarlane, W.V. 1968 Comparative studies in the adaptation of ruminants to hot environments. pp 164-182 in The adaptation of domestic animals (ed E.S.E. Hafez), Lea and Febiger, Philadelphia.

Macfarlane, W.V. 1969 The water economy of desert Aboriginals in summer. J. Physiol., 205, 13-14.

Macfarlane, W.V. 1970 Seasonality of conception in human populations. Biometeorol, 4, 167-182.

Macfarlane, W.V. 1971 Salinity and the whole animal. pp 161-178 in Salinity and Water Use (ed T. Talsma and J.R. Philip), Macmillan, London.

Macfarlane, W.V. 1973 Functions of Aboriginal nomads during summer. pp 49-68 in The human biology of Aborigines in Cape York (ed R.L. Kirk), Australian Institute of Aboriginal Studies, Canberra.

Macfarlane, W.V. 1976a Ecophysiological hierarchies. Israel. J. Med. Sci., 12, 723-731.

Macfarlane, W.V. 1976b Aboriginal palaeophysiology. pp 183-194 in The Origin of the Australians (eds R.L. Kirk and A.G. Thorne), Australian Institude of Aboriginal Studies, Canberra.

Macfarlane, W.V. and Good, B.V. 1976 Hormones and adaptation. pp 211-216 in Selected topics in Environmental Biology (eds B. Bhatia, G.S. Chhina and B. Singh), Interprint Publications, New Delhi.

Macfarlane, W.V. and Howard, B. 1970 The ecophysiology of water. pp 362-374 in Physiology of Digestion and Nutrition in Ruminants (ed A.T. Phillipson), Oriel Press, Newcastle.

Macfarlane, W.V. and Howard, B. 1972 Comparative water and energy metabolism of wild and domestic animals. Proc. Zoo. Soc. Lond., 31, 261-296.

Macfarlane, W.V. and Howard, B. 1974 Ruminant water metabolism in arid areas. pp 7-22 in Studies of the Australian Arid Zone II Animal production (ed A.D. Wilson), CSIRO, Melbourne.

Macfarlane, W.V., Howard, B., Haines, H., Kennedy, P.M. and Sharp, C.M. 1971 The hierarchy of water and energy turnover of mammals living in deserts. Nature, Lond., 234, 483-484.

Macfarlane, W.V., Howard, B., Morrison, J.F. and Wyndham, C.H. 1966 Content and turnover of water in Bantu miners acclimatising to humid heat. J. Appl. Physiol., 21, 978-984.

Macfarlane, W.V., Pennycuik, P.R., Yeates, N.T.M. and Thrift, E. 1959 Reproduction in hot environments. pp 81-96 in Recent progress in the endocrinology of reproduction (ed C.W. Lloyd), Academic Press, New York.

Macfarlane, W.V. and Robinson, K.W. 1957 Decrease in human excretion of 17-ketogenic steroids and 17-ketosteroids with increasing ambient temperature. J. Appl. Physiol., 11, 199-200.

MacMillen, R.E. and Lee, A.K. 1970 Energy metabolism and pneumocutaneous water loss of Australian hopping mice. Comp. Biochem. Physiol., 35, 355-369.

Macpherson, R.K. 1960 Physiological responses to hot environments. M.R.C. Report No. 298 H.M. Stationery Office, London.

Maloiy, G.M.O. 1972 Renal salt and water excretion in the camel *(Camelus dromedarius)*. Symp. Zool. Soc. Lond., 31, 243-260.

Martin, P.S. 1970 Pleistocene niches for alien animals. Bioscience, 20, 218-221.

Martin, P.S. and Wright, H.E. 1967 Pleistocene extinctions. Yale University Press, New Haven.

McMillan, J.R.A. 1965 Water, agricultural production and world population. Aust. J. Sci., 28, 135-141.

Morris, R.J.H., Howard, B. and Macfarlane, W.V. 1962 Interaction of nutrition and air temperature with water metabolism of Merino wethers shorn in winter. Aust. J. Agric. Res., 13, 320-334.

Mount, L.E. 1968 The climatic physiology of the pig. Arnold, London.

Newsome, A.E. 1965 The distribution of red kangaroos, *Megaleia rufa* (Desmarest) about sources of persistent food and water in central Australia. Aust. J. Zool., 13, 289-299.

Panum, P.L. 1847 Observations made during the epidemic of measles on the Faröe in the year 1846. Medical Classics, 3, 829-886.

Peirce, A.W. 1968 Studies on salt tolerance of sheep. VII. The tolerance of ewes and their lambs in pens for drinking waters of the types obtained from underground sources in Australia. Aust. J. Agric. Res., 19, 577-587.

Pennycuik, P.R. 1964 The effects on rats of chronic exposure to 34°C. IV. Reproduction. Aust. J. Biol. Sci., 17. 245-260.

Pennycuik, P.R. 1969 Reproductive performance and body weights of mice maintained for 12 generations at 34°C. Aust. J. Biol. Sci., 22. 667-675.

Prosser, C.L. 1970 Principles and general concepts of adaptation. pp 88-100 in Physiology, Environment and Man. (eds D.H.K. Lee and D. Minard), Academic Press, New York.

Richmond, C.R., Langham, W.H. and Trujillo, T.T. 1962 Comparative metabolism of tritiated water by mammals. J. Cell. Comp. Physiol., 59, 45-53.

Roberts, D.F. 1952 Basal metabolism, race and climate. J. Roy. Anthrop. Inst., 82, 169-183.

Robinson, K.W., Howard, B. and Macfarlane, W.V. 1955 Observations on seasonal fluctuations of 17-ketosteroid output in the subtropics and tropics. Med. J. Aust., 2, 756-760.

Rogerson, A. 1968 Energy utilization by the eland and wildebeest. pp 153-161 in Comparative Nutrition of Wild Animals. Symp. Zoo. Soc. Lond. No. 21 (ed M.A. Crawford), Academic Press, London.

Schmidt-Nielsen, K. 1964 Desert animals. Oxford University Press, London.

Scholander, P.F., Hammel, H.T., Hart, J.S., LeMessurier, D.H. and Steen J. 1958 Cold adaptation of Australian Aborigines. J. Appl. Physiol., 13, 211-218.

Serventy, V. 1966 Australia. A continent in danger. Andre Deutsch, London.

Siebert, B.D. and Macfarlane, W.V. 1975 Dehydration in desert cattle and camels. Physiol. Zool., 48, 36-48.

Simpson, G.G. 1953 The major features of evolution. Clarion, New York.

Smith, A. 1961 Biological effects of freezing and supercooling. Arnold, London.

Smith, R.E. and Horwitz, B.A. 1969 Brown fat. Physiol. Rev., 49, 330-425.

Squires, V.R. and Wilson, A.D. 1971 Distance between food and water supply and its effect on drinking frequency and food and water intake of Merino and Border Leicester sheep. Aust. J. Agric. Res., 22, 283-290.

Turekian, K.K. 1971 Late Cenozoic glacial ages. Yale University Press, New Haven.

Tyndale-Biscoe, H. 1973 Life of Marsupials. Arnold, London.

Wilson, A.C., Carlson, S.S. and White, T.J. 1977 Biochemical evolution. Ann. Rev. Biochem., 46, 573-639.

Wilson, A.D. 1974 Studies of the Australian arid zone. II. Animal Production. CSIRO, Melbourne.

Wyndham, C.H. 1970 Adaptation to heat and cold. pp 177-204 in Physiology, Environment and Man (eds D.J.K. Lee and D. Minard), Academic Press, New York.

Wyndham, C.H., McPherson, R.K. and Munro, A. 1964 Reactions to heat of Aborigines and Caucasians. J. Appl. Physiol., 19, 1055-1058.

Yeates, N.T.M. 1957 Photoperiodicity in cattle. II. The equatorial light environment and its effect on the coat of European cattle. Aust. J. Agric. Res., 8, 733-739.

Yoshimura, H. and Kobayashi, S. 1975 Physiological adaptability and nutritional status of the Japanese. Human Adaptability, 3. Japanese Committee for IBP, Science Council of Japan. University of Tokyo Press, Tokyo.

Young, V.R., Steffee, W.P., Pencharz, P.B., Winterer, J.C. and Scrimshaw, N.S. 1975 Total human body protein synthesis in relation to protein requirements at various ages. Nature, Lond., 253, 192-193.

9. ECOPHYSIOLOGY: TOWARDS AN UNDERSTANDING OF LATE PLEISTOCENE MARSUPIAL EXTINCTION

by A.R. Main

SUMMARY

Because ecophysiological studies contribute to understanding the fundamental Hutchinsonian niche rather than the realized niche, it is suggested that they do not contribute much to the understanding of late Pleistocene extinction of the large marsupials. Attention is drawn to the size reduction of some Pleistocene forms which persisted and the marked sexual dimorphism in the extant large kangaroos. It is suggested that these are adaptive responses in an oscillating environment when periods favourable for recruitment to the population are short. It is inferred that the large extinct fauna were unable to evolve smaller, earlier maturing, females and hence lacked the capacity to recruit to their populations and that this led to their extinction.

INTRODUCTION

Ever since the nature of the fossil record was recognised, workers have been faced with the problems of interpreting fossils in terms of biology and ecology, inferring the environmental conditions, especially climate, at the time of the deposition of the fossils, and explaining why so many have become extinct.

Wells (this volume) has made the point that while the Quaternary is only a short part of the geological record it offers our best chance of a complete fossil record. Moreover, since some of the fossil organisms are still extant, though often with altered ranges, there is a possibility that, provided uniformitarianism is assumed, ecological and physiological studies of extant species may account for both their present ranges and changes from the past distribution or even extinction of closely related forms.

This paper accepts the nature of the problem as outlined by Wells and sets out to evaluate the relevance of current ecological and physiological knowledge to our understanding of former ranges, past environments, extinctions and consequent changes in faunal composition. These matters cannot be discussed until the contribution of physiology and ecology to distributions has been explored.

In an ecological context the organism lives in a habitat in which it fulfils a metabolic role (e.g., crudely, may be herbivore, carnivore, etc.), has certain life history attributes, exhibits certain population parameters (e.g. total numbers, age and sex structure) and displays a geographic distribution with habitat preferences.

As a living organism an animal has certain physiological functions which will be performed simply in order to maintain its internal milieu. Additional to these attributes it will have certain adjustment responses which will maintain the internal conditions in the face of environmental changes which take place and so affect its chosen role and living place. These are eco-physiological responses. They differ from environmental physiology only in so far as analysis of the latter relates responses to extreme conditions, e.g. heat stress or water deprivation, which may be well outside what is experienced in nature and which may not be related to conditions ever experienced in the field in an ecological context.

The question arises as to whether the ability or inability to make physiological responses which lead to survival are sufficient to account for observed distributions in space and time. The problem first centres around the detail or scale of events that can be perceived and hence potentially interpreted by the palaeontologist, the neontologist and the organism. These aspects are discussed in turn.

THE SCALE OF EVENTS

The palaeontologist can discern: changes in fossil fauna, caused by extinctions or accessions; contrasts in faunal assemblages interpreted as reflecting habitat differences; and changes in the environment of deposition as reflected by the nature of the fossils or sediments. In complete sequences some or all of the foregoing can be placed in a relative order so that possible causes such as trends in climate which are readily discerned can be advanced to account for the observed record.

Other papers in this volume (e.g. Crisman on lacustrine environments and Davis on pollen data) indicate how such studies can support interpretation of climatic changes based on mammalian fossils, particularly if the climatic change be associated with trends in rainfall or temperature. Nevertheless, as Davis using pollen data and Watts macro-fossils show, there are many difficulties in interpreting ecological habitat and environmental conditions from either pollen or plant macro-fossils.

Chappell (this volume) points out that we are concerned with detail over 10^1 to 10^3 years which is the time interval for recurrence of extreme events in climate and landscape phenomena, events which by implication are related to the persistence or otherwise of animal populations. Nevertheless, apart from records of annual events, e.g. varves, the palaeontological record does not give us the ability to discern and date changes of the order of decades or centuries.

The neontologist is usually concerned with at most a few generations of a population. Many are concerned with an animal's response to extremes of heat or cold, or to water deprivation. Boundaries to distribution are frequently correlated with climatic factors, e.g. annual precipitation, temperature or days of rain. Thus the southern boundary of the distribution of the lizard *Amphibolurus ornatus* seems to coincide with 120 wet days per year which is interpreted as the maximum number of days which still allow for the accumulation of enough heat for growth. Those neontologists concerned with numbers of animals in populations frequently demonstrate that unusual weather or climate is associated with changes in population size or structure. On the whole the neontologist works and discerns changes on a scale considerably finer than the palaeontologist.

The animal has no perception of trends but experiences daily, seasonal or other changes in the immediate environment, and demonstrates that the changes have been perceived and are significant by either changing behaviour or responding physiologically or both.

The important difference between what the animal, as contrasted to the palaeontologist, perceives is that without a cultural heritage the animal has no knowledge of past history or trends which might be extrapolated to the future. It is in fact responding to changes and stresses of varying amplitude and frequency of repetition throughout its life. Those animals which survive to reproduce send a new generation into an environment in which life is predictable only if fluctuations remain within the scope of the behavioural and physiological repertoire inherited from the parental generation.

It is only when sufficient animals survive to reproduce that a population can persist. Failure to persist is extinction and the inference when interpreting the palaeontological record is that this is comparable to passing across the geographic boundary of the range of an extant species. In both cases it is assumed that the environmental stresses have passed beyond what can be tolerated. Such an event can be induced by a single environmental oscillation of either great amplitude or duration, so that either organisms are killed or the duration is such that successful reproduction and recruitment of the next generation is not possible before death of the parental generation from physiological old age. Such events would only have to be one or two score years in duration, and if there is no opportunity for re-invasion there is range contraction or extinction.

THE SIGNIFICANCE OF COMPETITION

But the foregoing is too simplistic. On some boundaries of its range a species recognises the presence of other species, and should this recognition be translated into competitive interactions for one of the environmental resources, there will be boundary change depending on the adaptive superiority of one or the other of the competitors. Such competitive changes cannot be interpreted from the fossil record. But the subtlety of the species responses can perhaps be gauged from the case of the western and eastern grey kangaroos.

The distribution of the two species of grey kangaroo poses the sorts of problems discussed by Hutchinson (1958, 1965) where he introduced the concept of the niche as an abstraction of n-dimensional hyperspace. Hutchinson views the wide-ranging, seemingly highly tolerant species - the so-called adaptable species - as those having a large fundamental niche. On the other hand he considered adaptation as possessing inherent competitive powers. Distribution determined by the relative competitive abilities of contiguous species, Hutchinson calls the realized niche.

Since the discoveries of Kirsch and Poole (1972) it has been recognised that there are two species of grey kangaroo. Even as now understood, the eastern grey kangaroo (*M. giganteus*) is exceptionally widespread, occupying habitats over a wide range of latitudes from tropical to temperate. It is thus to be considered an adaptable species in the sense of Hutchinson (1965) as it has a large fundamental niche which might in a way be considered bounded by climate. Nevertheless the western boundary is apparently determined by the occurrence of the western grey kangaroo (*M. fuliginosus*). This boundary is presumably determined by competitive relationships resulting from the adaptations of the two species. Thus each species of grey kangaroo exhibits a realized niche and neither its fundamental niche. Any change of climate across the inter-species boundary which changed the competitive status of either species would change the boundary and hence the distribution, but may not be of a kind that could be detected in any fossil record.

Given that competition is an ecologically significant factor but recognising that climatic changes are much more likely to be interpretable from the fossil record, it is appropriate to list the possible individual and population responses to the various environmental changes that do occur.

From table 9.1 it is possible to reason that climatic events leading to local or regional extinction can be of the following sorts:

(a) seasonal short-term extremes of environment which exceed the animal's capacity to tolerate or behaviourally to avoid (physiological response);

(b) unfavourable environmental fluctuations of amplitude approaching (a) above in which recruitment is low and in which the intervening favourable periods are of such short duration that the population cannot be restored by recruitment from any persistent local stock in refugia;

(c) unfavourable environmental fluctuations which lead to local extinction but followed by favourable environmental conditions so distributed spatially that emigration can restore the former distribution, provided that competitive relationships, where they occur, have not changed;

(d) long-term environmental trends of climatic change, in which the rate or magnitude of change ultimately exceeds the population's capacity to respond by selecting favoured genotypes (evolutionary response).

Table 9.1. Environmental events perceived by animals and their responses to them.

	Daily	Short-term	Annual Weather	Climatic Fluctuations
Characteristic environment	Night - cold Day - warm	As pressure cells pass - rain, wind, sunshine, hot, cold Cold weather on leading edge and hot weather on trailing edge of high pressure cells Duration of temperature regime dependent on rate of passage of cells	Seasonal cycle - summer, autumn, winter, spring Drought possible, i.e. failure of rainfall in any season Temperature associated with pattern of passage of cells	Extended good seasons or Prolonged run of poor seasons, with drought
Animal response	Daily metabolic rhythm and activity	Maintenance of body function modified by behaviour, e.g. in W.A. seek shelter and tend not to emerge on cold days, windy nights	Reproduces recruits to population. Depending on seasons responds physiologically to maintain vital functions, e.g. water and energy balance, ruminant-like forms recycle urea, etc Population movement - nomadic or migratory	Live normal life, breed, recruit; population grows, produces migrants and colonises suitable empty habitats or a) fail to breed b) fail to recruit c) die of drought stress - dehydration, starvation. d) die of physiological old age before environmental amelioration (a) to (d) lead to local extinction

Only (d) of the above are of a sort which can be discerned in our present palaeontological record. Some are extant species known as fossils which have not changed throughout the geological record known to us, or extant species and believed to be descended from distinct but undoubted ancestral forms. Such species are listed in table 9.2. Marshall (1973) and Bartholomai (1975) have both given figures for the reduction in size of these species.

Table 9.2. Species which have experience Late-Quaternary diminution in size.

Fossil Species	Extant Form	Reduction - Author
Macropus titan	*Macropus giganteus*	30% - Marshall (1973)
Osphranter cooperi	*Osphranter robustus*	25% - Marshall (1973)
Macropus agilis siva	*Macropus agilis agilis*	No size difference - Bartholomai (1975)

Others, particularly the so-called mega-fauna are fossil animals, which became extinct. However it must be emphasised that while these large animals became extinct over a very short period of time, the record itself does not distinguish between three possible causes of extinction, namely:

(a) fluctuations of great amplitude which exceed tolerance,

(b) environmental trends whose rate of change or magnitude exceeds the species' capacity to respond - that is, the species loses its adaptation and either becomes extinct or is replaced by a competitively superior (better-adapted) species, and

(c) fluctuations of moderate amplitude but with favourable conditions of too short a duration for adequate population recruitment.

Nevertheless, Bowler et al. (1976) indicate that about the time of the recorded extinction there were great climatic fluctuations.

ECOPHYSIOLOGY

Earlier I distinguished between environmental and ecophysiological studies. In the sense used, ecophysiological studies are broadly of two sorts:

(a) those recording the field conditions experienced by the organism, and

(b) those recording the responses of the organism to laboratory conditions which are believed to resemble in some degree the extreme field conditions.

Laboratory studies tend to relate to levels of achievement, since the organism is given no opportunity to avoid behaviourally and must respond physiologically. Moreover, many laboratory trials involve single factors only (e.g. heat or cold stress, water deprivation) and do not take into account any synergistic effects resulting from interaction between stresses. Thus laboratory trials reflect the organism's capacity to respond to or tolerate specific conditions for specific times, for example, heat stress on the hare wallaby *Lagorchestes conspicillatus* (Dawson and Bennett 1971), or salt water loading on the tammar *Macropus eugenii* (Kinnear, Purohit and

Main 1968; Purohit 1971). Such trials clearly measure survival responses but we do not know how frequently such extreme conditions are experienced in the life of an animal. Radio-telemetry suggests that given adequate cover the hare wallaby need never experience high day-time temperatures (Burbidge and Main 1971); and on neither of two of its island homes have we any evidence of consistent sea-water drinking by the tammar even in severe drought. In general it seems that evaporative cooling, salivation or salt water drinking are responses of the last resort, made when it is almost certain that without them the animal will die from hyperthermia or dehydration; but there is a chance that conditions will ameliorate if death can be so delayed for even a short while.

On natural history grounds it would seem that animals are much more likely to suffer exposure to the more insidious effects of chronic water shortage or food of inadequate quality. These conditions may well be completely within the tolerance of the adult animal. For instance, capacity of the water-conserving mechanism of the kidney allows the production of a concentrated urine and this, along with suitable behaviour, permits the animal to balance its water needs. Alternatively, the animal can recycle urea and thus balance its nitrogen needs or very much reduce the nitrogen deficiency between diet and daily needs, a very considerable achievement which will delay death from starvation. However, these abilities are strictly related to survival of adults, and such conditions as the adults can readily handle may be completely inimical to recruitment because the smaller juvenile animals either lack the full development of physiological abilities or are disadvantaged by their small body size and its metabolic consequences. Without recruitment the population cannot persist but the ecophysiological aspects of environmental stress on the juvenile segment of the population have not been studied, though the work on red kangaroos reported by Newsome (1977) documents the juvenile mortality under severe environmental stresses. Newsome (1977) also points out how unlikely recruitment is in the life-time of a female kangaroo.

The foregoing suggests that in any situation where recurrent drought is interspersed with more favourable periods, organisms will adapt to the intensity of the drought but the population will not persist unless it responds adaptively so that recruitment and population growth can occur in even relatively short periods of favourable environment. One adaptive response to shortened periodicity of favourable/unfavourable environmental oscillations is to increase the number of reproductive events during favourable periods or shorten the period to sexual maturity. The latter is most readily done by maturing at a smaller size. A selection of macropod species is given in table 9.3, and the marked dimorphism shown by some species is consistent with the remarkable similarity in age to first maturity of females of all species.

Macropods only produce a single young, and so their adaptive response is entirely towards smaller size of female at maturity, as suggested by the reduction in female size towards the desert and a breeding biology related to the earliest possible response to favourable conditions when they occur, e.g. continuous breeding of reds, euros, hare wallabies and quokkas (mainland) but not of tammars. Commonly this is related to the so-called unpredictability of the desert. But interpretation revolves around a definition of what is meant by an unpredictable environment. A biologically useful definition must take account of the cues perceived. It is necessary to distinguish between environments that are regular, reliable, predictable, or their opposites.

An event can be regular or irregular in occurrence, for example the seasonal changes in day length or the sequence of seasons. Such events are also reliable since they have an astronomical causation. Depending on the geographic location of any site such events may or may not be good predictors of coming events. In the vicinity of Perth shorter, cooler days in April and May are good predictors of the coming winter rains and are in fact the cues for the breeding of frogs of the genus *Heleioporus* before water is present. In this sense the cues of shorter, cooler days are good cues for predicting

Table 9.3 Sexual dimorphism among macropod marsupials from an array of habitats (after Main and Bakker 1978)

Species	Weight (kg)		Female weight as % of male weight	Earliest recorded age to maturity (in years-months)		Habitat
	Male	Female		Male	Female	
Macropus giganteus, Shaw	*	22.46 ± 0.89	*	4-0	1-6	Woodland and forest.
Macropus fuliginosus (Desmarest)	54.30 ± 2.14	25.5 ± 0.64	46.96	2-7	1-2	Woodland and forest.
Macropus rufus (Desmarest)	42.0 ± 4.81	21.69 ± 0.60	51.64	2-0	1-3 1-6	Mulga plains.
Macropus robustus, Gould	31.0 ± 1.27	14.68 ± 0.27	47.35	1-10	1-6	Rock piles in tussock grassland.
Wallabia bicolor (Desmarest)	16.0	10.1	63.13	-	-	Forest.
Macropus agilis (Gould)	8.29 ± 0.62	7.84 ± 0.24	94.75	1-2	1-0	Riverine forest.
Macropus irma (Jourdan)	7.96 ± 0.10	7.72 ± 0.20	96.98	-	-	Sclerophyll forest.

Species	Weight (kg): Male	Weight (kg): Female	Female weight as % of male weight	Earliest recorded age to maturity (in years-months): Male	Earliest recorded age to maturity (in years-months): Female	Habitat
Macropus eugenii (Desmarest) K.I.	6.3	4.45	70.63	2-0	1-0	Open woodland.
G.I.	5.85 ± 0.65	4.3 ± 0.6	73.5			
Macropus parma, Waterhouse	4.78	3.45	72.18	2-0	1-3	Forest.
Setonix brachyurus (Quoy and Gaimard)	2.91 ± 0.08	2.34 ± 0.12	80.41	1-1	0-8.5	Riparian and swamps.
Lagorchestes conspicillatus, Gould	2.72 ± 0.08	2.37 ± 0.09	87.13	-	-	Tussock grassland.
Lagostrophus fasciatus (Peron)	1.55 ± 0.06	1.66 ± 0.05	107.10	-	-	Scrubland.
Bettongia lesueur (Quoy and Gaimard)	1.145	1.145	100.00	-	-	Scrub and grassland.
Potorus tridactylus (Kerr)	1.35 ± 0.024	1.13 ± 0.026	83.43	0-9.5	0-9.5	Forest.

* Data not available. K.I. = Kangaroo Island. G.I. = Garden Island.

coming events, namely winter rains producing ponds suitable for tadpoles. On the other hand, frogs of the genera *Neobatrachus*, *Notaden* and *Cyclorana* live in arid situations where day length and temperature are good predictors of season but not as cues for successful breeding. Species of these genera are burrowers, and heavy rain which floods them is invariably sufficient to form the ponds in which the larval stages can be completed. In this sense, while rainfall of the places in which the frogs live is both irregular and unreliable, once it does occur it is a good predictor of breeding success insofar as water will be present until the larvae metamorphose, after which the young frogs can burrow and so avoid any subsequent drought.

Red kangaroos and euros live in the same regions as the above frogs but they are large animals which take much longer to achieve independence and reproductive maturity. Moreover, they have only one young at a time, and cues which are useful to burrowing frogs in initiating successful reproduction are of no use to them. In other words, while the environment is regular, it is unreliable and unpredictable insofar as kangaroo breeding is concerned. Clearly for the kangaroo populations to persist some solution independent of the inadequate cues provided by weather is needed, and it is in this light that the breeding biology must be interpreted.

Red kangaroos, euros and hare wallabies breed continuously. Red kangaroos and euros lose many joeys because drought conditions are frequently so bad that reproduction does not lead to recruitment to the population. We can look upon these losses as the costs the population must bear in the absence of a less expensive and more reliable way of anticipating a run of seasonal conditions which will not only permit reproduction but also lead to recruitment of a new generation.

When an event leading to population decline has a probability of occurring at least once in a life-time of residents in a locality, e.g. population decline through successional stage as in the quokka (Christensen and Kimber 1975) or through drought or fire, selection will favour populations producing recruits in excess of the replacement needs of the local population. These excess recruits become available as colonisers of other localities formerly unsuited or from which populations may have been removed because of fire, plant succession or drought. Thus continuous breeding is related on the one hand to the unreliable environment of arid Australia and on the other to the need to maintain a population of such size and age structure as will produce ample numbers of recruits to re-establish populations in other places. The species becomes fugitive in the sense of Hutchinson (1965).

Bartholomai (1975) suggests possible phylogenetic relationships within the euros and wallaroos back to the late Pliocene. This is consistent with their desert occupancy of the rocky broken country where tussock grasslands dominated by *Triodia* reach their fullest development. Under an arid environment the forage varies in water content and nutritional value along with seasonal conditions. The euro has a well-developed set of physiological traits which can be interpreted as being adaptive for such environmental conditions.

In laboratory feeding trials it has been shown that urea, a nitrogen source of no biological value, can replace casein as a source of protein nitrogen (Brown 1968, 1969; Brown and Main 1967). This can be done because when fermentable energy is available the bacteria of the enlarged fore-stomach can incorporate the urea nitrogen into bacterial protein which can then be digested by the host.

In addition to this ability to use urea nitrogen, Brown (1969) has shown that urea fed to euros is not excreted in the urine but is removed from the blood-stream and is presumed to have been utilized by the bacteria of the fore-stomach. Brown and Main (1967) have also shown that urea is actively secreted by the salivary glands. Main (1971) showed that there was a great deal of variability in the urea in the urine of field euros in the Pilbara and that low values were particularly noticeable at the season when euros were eating seed heads of *Triodia*, a good source of fermentable energy. These findings were interpreted as indicating that urea recycling was occurring.

Kinnear and Main (1975) showed that utilization of urea injected into field-caught tammar wallabies was seasonally limited by available energy. Utilization of endogenous urea will be similarly limited and it is likely that all macropods which have the capacity to supplement dietary nitrogen by recycling endogenous urea will have their capacity fully to use this nitrogen source limited by fermentable dietary energy.

Undoubtedly animals living in arid habitats are likely to experience nitrogen deficiency to a more marked degree than animals from mesic habitats, but it is not necessary that all arid land macropods be dependent on urea recycling in order to obtain nitrogen balance. Numerous feeding trials have confirmed that the red kangaroo lacks the euro's capacity to handle a diet high in fibre (Foot and Romberg 1965) and field observations suggest that it is a more selective grazer, eating better quality diet and moving nomadically from poor to better quality pastures (Newsome 1965; Storr 1968).

Feeding trials show that the euro can maintain positive nitrogen balance on a diet of 1% nitrogen (Brown and Main 1967) or 1.0 gm digestible nitrogen per day. These are values which are rarely exceeded in *Triodia* which is a significant dietary component of the forage eaten by the euro (Storr 1968). Storr has also shown that in field-caught animals the daily nitrogen intake agrees closely with Brown's (1964) laboratory determination.

With increasing size dietary needs rise as a function of body weight$^{0.75}$. A consequence is that as animals become larger they can satisfy their needs and remain in nitrogen or energy balance on a poorer quality diet. From this, and knowing that many very large herbivores evolved during the Pleistocene, it is possible to infer that this evolution was the consequence of selection for large size which was advantageous because of a general and long-term decline in the quality of available forage.

The literature reports on the field diet of the euro show that it is of exceptionally poor quality and that seasonally the nitrogen content may fall below the level needed for maintenance of balance in the euro. Consequently, in addition to the euro's well-developed capacity to recycle and utilize urea, we might intuitively think that larger size and consequent reduced needs on a per unit body weight basis would be a reasonable evolutionary response to the environment in which they live.

The euro-wallaroo group is different. It has persisted from the Pliocene to the present but it shows precisely the opposite trend to what we might expect arguing from the metabolic advantages of large body size, namely a diminution rather than increase in body size (table 9.2). A size reduction of 25% (Marshall 1973) might suggest that the needs on a unit body weight basis would be considerably higher than for a larger animal, but the fact that euros are still abundant suggests that reasoning from dietary quality and the metabolic consequences of body size alone fails to accord with the palaeontological record because it has not taken into account some other important biological factor.

The size differences quoted in table 9.2 relate to linear dimensions. What these linear dimensions mean in terms of body size is not clear, although field measurements made on pes length and weight of euros and red kangaroos in the Pilbara (table 9.4) suggest that changes of 25 to 30 per cent in length are equivalent to doubling of body weight. Further understanding of the relationships between linear measurements and bulk and body proportions could perhaps be profitably undertaken using the relationships established by McMahon (1973). Such a study would at least give insight into the body proportions and bulk of the large extinct marsupials.

Calculated water intake of kangaroos from the Pilbara, based on per cent water in dietary plants and data on per cent epidermis of plant species in the faeces, shows seasonal intakes per unit body weight to power 0.82 as in table 9.5 (Main and Bakker 1978). The data of this table indicate that, with respect to water derived from the diet, larger males on a per unit body weight basis are at a disadvantage compared with females, that is they lose the metabolic advantage that large body size confers with respect to energy and nitrogen. Newsome (1977) reports the deficiency of larger and older

Table 9.4. Relationships between pes length and weight in male and female kangaroos. Sample size in parentheses.

Pes length (cm)	Animal weight (kg)			
	Red kangaroo		Euro	
	Female	Male	Female	Male
23			(6) 12.6	
25	(6) 12.6		(40) 14.8	(3) 14.5
26		(3) 11.0		
30	(5) 21.8			(9) 31.7
35		(7) 45.4		

Table 9.5. Changes in dietary water intake of kangaroos at Mundabullanganah during 1961.

Data calculated from Storr's (1968) data on daily dry matter intake (DMI) and the weighted average percent of water in the food eaten as calculated from the percent epidermis of each plant species in the faeces of shot animals (see Storr 1968 for method) and the percent water of the same plant species collected at the same time.

Species, month of collection and sex		Sample size mean weight (kg)	Calculated DMI (gm/day)	Mean percent water in diet	Calculated total daily water intake	Water intake, $ml.kg^{0.82}$/day
Red kangaroo						
Feb.	M.	43.3 (3)	446	75	1338	59.5
	F.	20.4 (14)	250	75	752	63.5
May	M.	34.1 (4)	393	47	349	19.3
	F.	22.7 (15)	277	47	246	19.0
Aug.	M.	28.0 (4)	323	49	311	20.2
	F.	20.6 (11)	258	49	248	20.8
Nov.	M.	23.0 (5)	280	27	104	7.9
	F.	20.2 (14)	255	27	94	8.0
Euro						
Feb.	M.	31.3 (6)	420	59	604	35.9
	F.	14.5 (14)	240	59	345	38.5
May	M.	32.1 (8)	428	49	411	24.0
	F.	15.0 (18)	245	49	236	25.6
Aug.	M.	25.8 (10)	364	44	286	20.0
	F.	15.2 (14)	248	44	195	21.1
Nov.	M.	24.2 (11)	348	27	129	9.5
	F.	15.6 (11)	253	27	93	9.8

males in his population samples of red kangaroos, a circumstance consistant with the above if it is assumed that the water deficiency associated with large body size ultimately reduces the chance of large males surviving drought.

Very large animals will of course have much greater difficulty finding cool refuges, particularly in an environment in which trees and large shade plants are becoming sparser. Should these environmental circumstances force large animals to use water for evaporative cooling, this could easily tip the water balance unfavourably, unless free water was available for drinking; and it is characteristic of much of Australia that rivers and water holes are not common.

Field-caught animals from the arid part of the euro's range show a remarkable size dimorphism between the sexes, with females averaging less than 50% of the weight of males. This figure suggests a very large disadvantage unless females select different dietary items of higher quality than do males, a supposition which we cannot confirm because our data on field diets are not sufficiently extensive.

Size differences between the sexes of macropods from more mesic habitats are not as pronounced as in the euro. Moreover, it is not marked in smaller wallabies whether they come from mesic or arid situations. Small body size suggests an earlier maturity than with large body size, a supposition which tends to be confirmed from the data of table 9.3.

The size reductions shown in the biological record and the reduction of females relative to males in large but not small arid land macropods can be interpreted if it is assumed that selection for early maturity, even if it entails smaller body size and attendant metabolic disadvantages, will allow for population recovery and growth even in the relatively short periods of environmental amelioration between successive droughts.

The foregoing reasoning suggests that very large animals or those unable to undergo a selective diminution in size, especially by precocious maturity of females, would be the ones to become extinct. It is likely that those species which have adapted to modern conditions already exhibited considerable variation in age to sexual maturity and consequent adult size before selection was intensified. It was on the already existing variability that selection operated, and we might ask whether the fossil record confirms this expectation and to what extent the females of extant forms of fossil species retain a greater variability in both age to maturity and size.

DISCUSSION

In terms of the Pleistocene only those forms which are extant were successful. We have no knowledge of what the extinct species could tolerate physiologically, but we do have some data on:

(a) size of fossil species which have persisted (reduced size),

(b) size of extant species without fossil record, especially male/female dimorphism, and

(c) seasonality of breeding of extant species in relation to size and habitat.

Hitherto workers studying ecology and physiology of organisms have not addressed themselves to the problem of why certain forms have persisted since the Pleistocene while others have not, and so it is not surprising that so little evidence can be produced which explains either present-day habitat, occurrence or distribution, or interprets past environments.

I have considered ecophysiology to be the responses to what the organism perceives. It then puts in train certain appropriate physiological mechanisms and these, at different costs, tend to maintain the internal condition of the animal and so ensure its survival. One would not expect the organism to be able to respond effectively to all cues simultaneously, and that this is so is shown by the significance given by the organism to water conservation up to the point when death from hyperthermia is likely, at which stage water conservation is abandoned for temperature regulation by evaporative cooling. Another example is the continuous breeding in females of desert kangaroos, provided it does not jeopardise the survival of the already established female.

I have followed Bartholomew (1958, 1964) in believing that ecophysiological responses alone do not account for present distributions. In particular, assumptions relating to the importance of ecophysiological responses ignore the significance of the outcome of competitive interactions in determining boundaries of distribution of species and their persistence.

The main thrust of the present thesis is that it is unlikely that the same or single factors limit a species on all the boundaries of its range. Likewise it is unlikely that simple trends in either precipitation (towards dryness) or increased temperature have led to extinctions. It is more likely that many factors interacted, and it is suggested that one of the more important is likely to be the amplitude and period of climatic oscillation on which were imposed the long-term trends of aridity. Two lines of evidence favour this:

(a) the size reduction of extant representatives of fossil forms, and

(b) the sexual differences in size of the larger forms from arid situations.

Both these observations are consistent with trends towards earlier sexual maturity of smaller forms, a function which is advantageous when the period favourable for possible population growth is short and also gives competitive advantage. In ruminant-like macropods this reduction in size to obtain advantage of greater reproductive responsiveness has to be compromised with the need to achieve selective advantages relative to metabolic body size when the quality of the diet is low and free water in short supply.

The foregoing has proposed a fluctuating climate as the principal component of extinction in the late Pleistocene. It has made no attempt to interpret the climatic regime under which the large fauna evolved. It is implicit in the reasoning developed earlier in this paper that the metabolic advantages of large body size when dietary nitrogen and energy are low are likely to be the prime basis of selection. Forage and diet would be expected to be poor in an environment that was at least seasonally arid. From Table 9.5 it can be seen that when animals become very large, e.g. male kangaroos, they are at disadvantage with respect to dietary water relative to females of about their weight. From this observation it seems reasonable to suggest that the arid times were cool or cold because in hot dry climates water stresses would have been so marked as to have selected for reduction of the attendant disadvantage of large body size with respect to water metabolism. The reasoning developed earlier then suggests that once the environment changed from a steady or even declining state to an oscillatory one with respect to drought and temperature, biological factors other than large body size began to operate. The result was that extinctions became commonplace for all species which were large and lacked the variability in size or age to maturity to exploit the new competitive ecology that emerged.

ACKNOWLEDGEMENTS

Research funds for a continuing programme on arid land adaptations of the Australian fauna have come from the Australian Research Grants Committee, C.S.I.R.O. and the University of Western Australia Research Grants Committee, and this support is gratefully acknowledges. Drs H.R. Bakker and J.E. Kinnear read and commented on the manuscript during preparation.

REFERENCES

Bartholomai, A. 1975 The genus *Macropus* Shaw (Marsupialia: Macropodidae) in the Upper Cainozoic deposits of Queensland. Mem. Qd. Mus., 17, 195-235, pl. 7-26.

Bartholomew, G.A. 1958 The role of physiology in the distribution of terrestrial vertebrates. pp 81-95 in Zoogeography (ed. C.L. Hubbs), Publ. 51 A.A.A.S., Washington.

Bartholomew, G.A. 1964 The role of physiology and behaviour in maintenance of homeostasis in the desert environment. Symp. Soc. Exp. Biol., 18, 7-29.

Bowler, J.M., Hope, G.S., Jennings, J.N., Singh G. and Walker D. 1976 Late Quaternary climate of Australia and New Guinea. Quaternary Research, 6, 359-394.

Brown, G.D. 1964 The nitrogen requirements of macropod marsupials. Ph.D. Thesis, University of Western Australia.

Brown, G.D. 1968 The nitrogen and energy requirements of the euro (*Macropus robustus*) and other species of macropod marsupials. Proc. Ecol. Soc. Aust., 3, 106-112.

Brown, G.D. 1969 Studies on marsupial nutrition. VI. The utilization of dietary urea by the euro or hill kangaroo, *Macropus robustus* Gould. Aust. J. Zool., 17, 187-194.

Brown, G.D. and Main, A.R. 1967 Studies on marsupial nutrition. V. The nitrogen requirements of the euro *Macropus robustus*. Aust. J. Zool., 15, 7-27.

Burbidge, A.A. and Main, A.R. 1971 Report on a visit of inspection to Barrow Island, November 1969. Report No. 8, Department of Fisheries and Fauna, W.A.

Christensen, P.E. and Kimber, P.G. 1975 Effects of prescribed burning on the flora and fauna of south-west Australian forests. pp 85-106 in Managing Terrestrial Ecosystems (eds J. Kikkawa and H.A. Nix), Proc. Ecol. Soc. Aust., 9.

Dawson, W. and Bennett, A. 1971 Thermoregulation in the marsupial *Lagorchestes conspicillatus*. Journal de Physiologie, 63, 239-241.

Foot, J.Z. and Romberg, B. 1965 The utilization of roughage by sheep and the red kangaroo *Macropus rufus* (Desmarest). Aust. J. Agric. Res., 16, 429-435.

Hutchinson, G.E. 1958 Concluding remarks. Cold Spring Harbour Symposia on Quantitative Biology, 22, 415-427.

Hutchinson, G.E. 1965 The ecological theatre and the evolutionary play. Yale Univ. Press, New Haven.

Kinnear, J.E. and Main, A.R. 1975 The recycling of urea nitrogen by the wild tammar wallaby (*Macropus eugenii*) - a ruminant-like marsupial. Comp. Biochem. Physiol., 51A, 793-810.

Kinnear, J.E., Purohit, K.G. and Main, A.R. 1968 The ability of the tammar wallaby (*Macropus eugenii*, Marsupialia) to drink sea water. Comp. Biochem. Physiol., 25, 761-782.

Kirsch, J.A.W. and Poole, W.E. 1972 Taxonomy and distribution of the grey kangaroo *Macropus giganteus* Shaw and *Macropus fuliginosus* (Desmares and their subspecies (Marsupialia: Macropodidae). Aust. J. Zool., 20, 315-3

McMahon, T. 1973 Size and shape in biology. Science, 179, 1201-1204.

Main, A.R. 1971 Measures of wellbeing in populations of herbivorous macropod marsupials. pp 159-173 in Proceedings of the Advanced Study Institute on Dynamics of Numbers in Populations (Oosterbeek 1970) (eds P.J. de Boer and G.R. Gradwell), Centre for Agricultural Publishing and Documentation, Wageningen, Netherlands.

Main, A.R. and Bakker, H.R. 1978 Adaptations of macropods to aridity. Biogeography and Ecology in Australia (ed A. Keast), Junk, The Hague (in press).

Marshall, L.B. 1973 Fossil vertebrate faunas from the Lake Victoria region, S.W. New South Wales, Australia. Mem. Nat. Mus. Vic., 34, 151-171, pl. 14-16.

Newsome, A. 1965 The distribution of the red kangaroo *Megaleia rufa* (Desmarest) about sources of persistent food and water in Central Australia. Aust. J. Zool., 13, 268-299.

Newsome, A. 1977 Imbalance in the sex ratio and age structure of the red kangaroo *Macropus rufus* in Central Australia. pp 221-233 in The Biology of Marsupials (eds B. Stonehouse and D. Gilmore), McMillan, London.

Purohit, K.G. 1971 Absolute duration of survival of common wallaby (*Macropus eugenii*, Marsupialia) on sea water and dry food. Comp. Biochem. Physiol., 39A, 473-481.

Storr, G.M. 1968 Diet of kangaroos (*Megaleia rufa* and *Macropus robustus*) and Merino sheep near Port Hedland, Western Australia. J. Roy. Soc. W.A., 51, 25-32.

10. DISPERSAL AND ESTABLISHMENT OF ANIMALS

by J.H. Calaby

SUMMARY

All animals have a dispersal phase in their life cycle, e.g. the desiccated resting egg in some aquatic invertebrates, small larvae in many sedentary marine invertebrates and independent juveniles in warm-blooded vertebrates.

The various means by which animals disperse include passive dispersal, migration and nomadism. Some small light insects and the resting eggs of some aquatic Crustacea are examples of animals adapted for passive dispersal by air currents and winds. Other animals dispersed passively include rodents and reptiles accidentally carried across marine barriers on rafts of vegetation. Migration is defined as the regular annual movements carried out by a large number of species between areas in which they breed and other places, which may be many thousands of kilometres away, where they spend the non-breeding season. The phenomenon is very common in birds but is found also in other groups including insects and mammals. Nomadism is common in animals and is usually correlated with availability of food. It is particularly associated with species inhabiting zones of high climatic variability. The independent juveniles of birds and mammals, the dispersal phase in these classes, may wander very widely searching for places in which to settle.

The geographical ranges of animals have unstable boundaries because of variability in climatic factors and short and long term changes in climate. Spread of animals into newly available suitable habitat is extremely rapid.

It does not seem likely that a knowledge of animal dispersal and establishment will lead to much improvement in the interpretation of Quaternary environments.

INTRODUCTION

It is perhaps rather trite to begin by saying that animals differ from plants in that they are mobile and can walk, fly, swim or otherwise get from one place to another. Most plants disperse in the form of seeds but to establish an animal species in a new area a group of adults or independent juveniles that will reach breeding age is necessary in the great majority of cases.

In this review the various kinds of dispersal in animals, and some of the mechanisms and ecological background involved, are discussed. It should be remembered, however, that animals display an almost infinite variety of environmental adaptations that are relevant to dispersal and establishment. The review is restricted to land (including fresh water) animals although it should be noted that remains of marine animals such as mollusc shells and vertebrate bones can yield important information on the environment during past times.

ECOLOGY OF ANIMAL DISTRIBUTION

The distribution of animals is dependent on many factors of which climate, especially temperature and rainfall, and correlated vegetation are of primary importance. In the tropics and at least warm temperate areas, animal distributions can usually be defined quite well in terms of rainfall isohyets; for example the red kangaroo *(Megaleia rufa)* is more or less confined within the area enclosed by the 500 mm isohyet. Within this broad general picture a variety of other factors may be involved. The nature of the substrate may be crucial. Soil type usually has an important bearing on the nature of the vegetation cover, and its physical properties may influence the distribution of small burrowing mammals, other vertebrates and numerous invertebrates. A great variety of animal species is found only in rocky places. Many animals are restricted in their food habits to only one or a few related species of plants. Such monophagy or oligophagy is common in insects but is found even in higher vertebrates, for example the koala *(Phascolarctos cinereus)* and greater glider *(Schoinobates volans)* live largely on species of *Eucalyptus*. The variety of mammalian anteaters, monotreme, marsupial and eutherian, have anatomical specializations for dealing with their chief food source, the ants and/or termites, social insects that live in large colonies and are a rich food source.

The distribution of aquatic invertebrates may depend on the amount and chemical composition of dissolved salts, pH, turbidity, temperature, depth of water, vegetation and other factors.

Other ecological factors involved in animal distribution include competition for necessary requisites with other species.

In summary, although animal distribution can be broadly defined in ecological terms, complete details are not known for any species as all have unique adaptations to their environments. The ideal of a complete ecology of any species is approached only in the case of a few insects (Clark et al. 1967).

DISPERSAL

Dispersal includes a number of kinds of movements both within and outside the normal geographical range. All animals, even the most sedentary, have a dispersal phase in their life history. The various kinds of dispersal in animals are described below.

Passive dispersal.

This is a common method of dispersal of seeds of plants; it is very common in marine animals but rare in land animals. The majority of bottom-living marine invertebrates produce free-swimming planktonic larvae that may be carried by currents for long distances before settling. By this means littoral species have spread and colonized remote oceanic islands.

Characteristics of land and aquatic animals that are best adapted for passive dispersal have been discussed by Simpson (1952). These include small size, low specific gravity, resistant exterior and a dormant phase. Among the best known examples are rotifers, some groups of Crustacea and tardigrades. It is not surprising that these groups include a number of species that are cosmopolitan in distribution. Rotifers are minute aquatic animals whose resting eggs are thick-shelled, resistant to temperature extremes and hatch only after a dormant period. The eggs of cladoceran crustaceans have generally similar characteristics. The eggs of notostraca Crustacea are very resistant to long periods of dry conditions and are blown about by wind (Williams 1968). They are very common in ephemeral pools in the inland deserts. The tardigrades are minute animals that live in aquatic or damp terrestrial situations. They are resistant to unfavourable environmental conditions and can withstand extremes of desiccation and temperature.

The insects provide some well-known examples of passive dispersal. Some species of aphids disperse in large numbers, sometimes for long distances, during periods of favourable weather conditions. The bushfly *(Musca vetustissima)* has a poor cold tolerance and cannot persist over winter in southern Australia. It breeds all year round in northern Australia and, as spring advances, bushflies are blown southwards in large numbers by favourable winds (Hughes et al. 1972). Other insects that are widely dispersed as adults by winds include species with aquatic larval stages such as midges (Chironomidae), mosquitoes (Culicidae) and biting-midges (Ceratopogonidae). Murray (1970) studied a major outbreak of the cattle disease ephemeral fever which began in northern Australia in September 1967 and within a few months was co-extensive with cattle herds throughout northern and eastern Australia. He was able to show a very close correlation of the spread of the disease with surface wind patterns. The disease is known to be carried by small biting insects and there can be no doubt that the epidemic was caused by passively transported insect vectors. Some strong-flying insects such as locusts are sometimes caught up and carried along passively by winds.

The juveniles of many species of spiders are adapted for dispersal often for great distances by winds and air currents. The spiderlings produce gossamer threads which are caught and drawn out by air currents, and carry off the young spiders. The phenomenon is known to arachnologists as ballooning (Main 1976).

Small molluscs, including juveniles of large species such as the freshwater mussels, are thought to be carried from place to place on the feet of water-birds and in wind-storms. Many species of molluscs in inland water are resistant to desiccation and survive for considerable periods in dried-out waterholes.

Passive over-sea dispersal on rafts of vegetation has been the means by which land vertebrates such as some rodents and reptiles and other animals including land-snails have colonized new islands and isolated continents. The ancestors of the Australian rodents and several reptile groups must have arrived in this fashion. Catastrophic events such as violent storms and hurricanes would facilitate the carrying-off of such castaways. Even in birds, passive dispersal was probably involved in the colonization of remote oceanic islands.

Expansion and contraction of range.

The boundaries of species distributions are constantly changing because of variability in temperature and rainfall together with short and long term changes in climate. Usually such changes in range are slow and more or less imperceptible. Expansion of range of this nature may conveniently be called spread.

The reproductive capacity of animals is such that many more individuals are produced than the habitat can carry and the newly independent juveniles are footloose and disperse widely, searching for places in which to live. Hence if suitable new areas of habitat become available they are quickly colonized. Areas of forest cleared or partially-cleared for farming are soon colonized by birds typical of woodland or open country, such as the Australian magpie *(Gymnorhina tibicen)*, kestrel *(Falco cenchroides)*, stubble quail *(Coturnix novaezealandiae)*, etc. Unsuitable habitat is as much a barrier to spread as a stretch of ocean. The rapidity of colonization of newly available areas of suitable habitat is shown in spectacular fashion by the invasion of Australia by introduced species such as the rabbit *(Oryctolagus cuniculus)*, house mouse *(Mus musculus)*, feral domestic cat *(Felis catus)* and fox *(Vulpes vulpes)*.

Nowak (1975) has provided a detailed review of the spread of native and introduced animals in Europe in historical times. The ranges of only a small percentage of native species are expanding or have expanded appreciably in recent times. He could offer no explanation for the cause of spread of several species and in other cases the explanation involved the activities of man. However, higher average temperature in northern Europe during the past 100-150 years was apparently the chief reason for the northward spread of some species such as the polecat *(Mustela putorius)*. Kalela (1942, 1944) has given details of many species of mammals, birds and other animals that have expanded their breeding ranges northward in northern Europe, and of northern species whose ranges have contracted from the south, in the same period.

Some cases of explosive spread have been documented, in which species whose ranges have been stable for centuries have suddenly and rapidly colonized huge areas thought to be unsuitable for them. Two recent examples studied in some detail are the serin finch *(Serinus serinus)* and collared turtle dove *(Streptopelia decaocto)* in Europe (Mayr 1963). In about 1800 the serin finch was found in Spain, Italy, part of Switzerland and the Balkans. After 1800 it spread fairly rapidly throughout Europe and at the present day it is found as a breeding species as far north as southern Sweden, Finland and England. The dove is found as a breeding species in virtually the whole of Europe which it has colonized from the Balkans in about 40 years (Nowak 1975). There is no ready explanation for this phenomenon. Mayr (1963) suspects that some genotypic change occurred in a peripheral population that resulted in a change in climatic or habitat tolerance permitting a subtropical species to expand into the temperate zone.

Studies of cave and human occupational deposits in Europe, America and Australia have shown contractions in ranges of mammals in late Pleistocene and Recent times. Whilst most of these are doubtless due to climatic changes, disappearance from islands that were connected to continents in the late Pleistocene should not necessarily be interpreted in this way. Extinctions on such islands may also be due to competitive exclusion or loss of genetic variability in small populations. The island may be too small to support a viable breeding population of some species (Main and Yadav 1971).

<u>Migration</u>.

Dispersal movements of animals of any sort, especially large-scale ones, are commonly referred to as migration. Here the term is strictly defined to mean the regular annual movements of species from areas in which they breed to other areas where they spend the non-breeding season, and back again. The best known examples are found in the birds.

Summer in the higher latitudes of the northern continents is mild and daylength is long, and birds are able to breed there in abundance. However winters are long and severe, and the migrant species leave the breeding areas and spend the winter in places better suited to support them. Such movements may be of great length, for example from northern Europe to southern Africa, or eastern Siberia to Tasmania. The most spectacular movement is that of the Arctic tern *(Sterna paradisaea)* which breeds in high latitudes as far north as 82° and spends the off-season among the Antarctic pack-ice as far south as 74°, a latitudinal range of roughly 18,000 kilometres. Migration may also be partial. Some populations of a species may be sedentary, others may be nomadic in the non-breeding season, while others again may be true migrants. A good discussion of bird migration is given by Dorst (1974).

Because of Australia's mild climate, migration in birds is not very noticeable here. Some species leave the cooler parts of the southern States after breeding and spend the winter further north. Few species leave Australia for islands to the north.

Some mammals are considered to be migratory although their movements are not so regular as those of birds and might better be described as nomadism. The caribou *(Rangifer tarandus)* is a gregarious species almost continually on the move, and must cover enormous distances in the course of the year. For the tundra subspecies the summer tundra range and winter forested range may be 1300 kilometres apart (Banfield 1974). The American bison *(Bison bison)*, another gregarious ungulate, moves around a great deal in the course of its grazing and other activities. It is believed that all the herds make annual southward movements for 300 kilometres or so to more favourable winter pasture (Banfield 1974).

Migratory behaviour is also known in insects. The best known examples are found in the Lepidoptera. The monarch butterfly *(Danaus plexippus)* breeds in temperate North America but no stage of its life cycle can withstand the freezing winter temperatures. In autumn the adult insects congregate and migrate to overwintering sites in Florida, California and especially some areas in the Mexican mountains. At these sites they cluster together in enormous numbers (Brower 1977). The species is self-introduced in Australia and was first seen in numbers about 1870. It is found wherever its food plants, the introduced milkweeds *Asclepias* and *Calotropis*, occur. The extensive migration and overwintering behaviour is not a feature of its biology here but winter aggregations do occur in South Australia and New South Wales (Common 1970). Another well-known migrant is the Bogong moth *(Agrotis infusa)*. The larval feeding areas are on the lowland plains west of the Dividing Range in New South Wales and southern Queensland. Conditions there are too extreme for larval survival in summer. In early summer the adult moths leave the breeding areas and migrate to mountain peaks above 1500 metres in the Australian Alps. There they aestivate clustered together in rock crevices. During autumn the adult moths disperse to the breeding grounds to mate and lay eggs in the pastures where dicotyledonous annuals are available as larval food in place of the unsuitable grasses that dominated the pastures in summer (Common 1954).

Nomadism.

Nomadic movements are common in mammals and birds. Many species of birds hold territories only during the breeding season and for the rest of the year wander about within their habitats, usually in groups. This nomadism is presumably related to local variations in food supply. Young birds are usually nomadic and may travel long distances until they either become established or die. Some nectarivorous and frugivorous species follow the blossoming and fruiting of their food trees. Among such nomads in Australia are the flying foxes *(Pteropus* spp.*)* (Nelson 1965).

A high degree of nomadism is characteristic of some species inhabiting low-rainfall country. The best-known species in Australia is the budgerigar *(Melopsittacus undulatus)*, a gregarious bird sometimes found in the inland in enormous flocks. The nomadism is doubtless an adaptation to the vagaries of the inland weather and the birds' needs for seeds and water. A study of marked emus *(Dromaius novaehollandiae)* that breed in inland Western Australia has shown that they leave the inland in cold dry winters and move south into the winter rainfall zone for distances up to 500 kilometres. They move north-east again when cyclonic rains fall in the mid-central inland. These extensive movements are largely correlated with availability of food. The movements are regular only because the rainfall pattern is regular (Davies 1976).

Some inland water-birds have evolved a high degree of nomadism. The spectacular Australia-wide movements of banded grey teal *(Anas gibberifrons)* during a drought in 1957 following a population increase during floods of the previous year has been documented by Frith (1977). There was a "massive influx" into New Zealand at that time.

The red kangaroo is the most mobile of the Australian mammals (excluding bats) but its movements are not at all remarkable when compared with those of birds or gregarious ungulates of other continents. Frith (1964) studied it during a non-drought period in north-western New South Wales, and in four years following the marking of some individuals the greatest distance from the marking place an animal was recovered was 32 kilometres. In the same general region Bailey (1971) recorded movements of up to 220 kilometres in later years that included a drought period.

The wanderings of most nomadic species largely takes place within the borders of their established ranges. Juvenile or vagrant individuals or small groups of very mobile species are fairly commonly reported outside their normal ranges. The literature of ornithology is replete with such occurrences.

There are a small number of species that suddenly appear at intervals in large numbers well outside their normal ranges. Such movements are known as irruptions. The well-known periodic irruptions of lemmings *(Lemmus lemmus)* in Scandinavia are described in Elton's (1942) classic book. The lemming normally lives in the subalpine zone of the mountains. Its populations increase until the numbers are so high that the habitat can no longer carry them. They then leave their normal habitat in large numbers and invade the lowlands, crossing rivers and lakes and passing through towns in their path. The survivors that reach the coast swim out to sea. Irruptions of lemmings are fairly regular and occur in cycles of about four years.

The best Australian example is the long-haired or plague rat *(Rattus villosissimus)*; it is normally a relatively rare species found on the Barkly Tableland, in western Queensland and in north-eastern South Australia. Heavy or prolonged rains induce lush persistent growth of ground vegetation and this in turn leads to very large population increases, followed by dispersal in large numbers away from the breeding areas. During the irruption of 1967-70, the most widespread ever recorded, rats were found up to 800 kilometres from the breeding areas (Newsome and Corbett 1975).

Drought in areas of reliable rainfall may lead to irruptions. Little rain fell in northern Australia in the summer of 1951-52. In the autumn and winter of 1952 southern Australia was invaded by large numbers of water-birds and a couple of species of hawks that are normally restricted to northern Australia (Serventy and Whittell 1976).

An important point about irruptions is that the emigrants are surplus to what their native habitat can carry and, in the case of mammals at least, probably none return. A few may breed for a time in the invaded areas but all die fairly soon and no extensions of range result.

RATES OF SPREAD OF VERTEBRATES

No specialized adaptations are involved in spread. The animals merely discover and occupy all suitable habitat by moving around in search of food and other requirements. Juveniles are much more prone to wandering than are established adults and it is no doubt chiefly these that colonize newly available areas of suitable habitat. In nomadic species there is considerable individual variation in degree of nomadism as studies of marked animals have shown. In the populations containing marked individuals, of red kangaroos studied by Frith (1964) and Bailey (1971) and of grey teal observed by Frith (1977), some individuals moved long distances while others never went far from the marking place. It would be expected that the chief colonizers of new areas would be the more adventurous ones. Selection for nomadic characteristics would confer an advantage on land and aquatic species adapted to low-rainfall habitats with high climatic variability.

Except for the giant toad *(Bufo marinus)*, introduced into Australia, I know of no figures on rates of spread of poikilothermic vertebrates; and few are available for mammals and birds. Species boundaries normally change slowly and there is usually no continuity of observation over a sufficiently long period.

Serventy and Whittell (1976) record that a few species of birds such as the galah *(Cacatua roseicapilla)* and crested pigeon *(Ocyphaps lophotes)* characteristic of the drier inland of Western Australia have penetrated into the southwest as breeding species during the present century. The major factor involved was usually thought to be the provision of new suitable habitats resulting from European settlement. However, Serventy and Whittell believe that the increases in range are probably a consequence of decreases in rainfall over the period. Some locality records and dates are given and it is clear that some of the range extensions are of the order of several hundred kilometres.

Nowak (1975) has analysed available records of range extensions of native and introduced animals in Europe. Some of his figures for representative native species are given in Table 10.1.

Table 10.1 Rates of spread of mammals and birds in Europe in recent times.

Species	Period	Newly settled area sq km x 1000	Newly settled area sq km/year
Polecat	1909 - 1919	19.5	1940
	1919 - 1929	39	3870
Elk *(Alces alces)*	1850 - 1928	820	10,500
	1928 - 1958	1840	61,300
Syrian woodpecker	1901 - 1930	352	12,000
(Dendrocopos syriacus)	1931 - 1960	same	same
Yellow-brested bunting	1825 - 1850	511	20,400
(Emberiza aureola)	1900 - 1925	864	34,600
Serin finch	1800 - 1875	520	6900
	1925 - 1968	1200	27,900
Collared turtle dove	1929 - 1931	6	2000
	1958 - 1961	ca.500	167,000

As mentioned previously, the polecat has increased its range northward as a result of increases in temperature in recent times. The elk is returning to the range it occupied several centuries ago and from which it was eliminated by hunting. Protection and rational use have ensured its increase in numbers. Most of the increase in range has been due to natural spread but it has been helped to some extent by planting of shelter belts and elimination of wolves. The serin finch and the dove are the examples of explosive spread given previously. The woodpecker has slowly increased its range into south-eastern Europe from Asia Minor. The range of the bunting expanded from Asia across European USSR to Finland throughout the 19th and part of the 20th centuries; it is no longer spreading.

The northern elephant seal *(Mirounga angustirostris)* that bred on islands and some mainland sites on the Californian and Lower Californian coasts was hunted almost to extinction for its oil during the last century. Between 1865 and 1880 only a few stragglers were seen. None was recorded from 1884 to 1892. In that year a collecting expedition from the Smithsonian Institution saw a group of eight on Isla de Guadalupe and promptly killed seven of them for the national collection! It is believed that less than 100 were in existence at that time and the present population is derived from that remnant. In 1922, 264 were counted on Isla de Guadelupe. The Mexican Government immediately granted the species complete protection, as did the United States Government when individuals began to appear off the Californian coast. In 1957 the total population was estimated at 13,000 and is probably now about 50,000. Breeding colonies are established on several islands spread along about 1400 kilometres of the Californian and Lower Californian coasts (Le Boeuf 1977).

Information on spread of animals in suitable habitat is also given by records of introduced species. After establishment in 1859 the rabbit spread over southern and eastern Australia at the rate of about 110 kilometres per year (Ratcliffe and Calaby 1958). Within half a century it had virtually reached its present distribution limits where it occupies almost four million square kilometres (Myers 1971). By contrast the rabbit's conquest of Europe has been much slower. Most of Europe was less suitable as rabbit habitat than Australia because of low winter temperatures and the density of the forest cover. Initially the rabbit was limited to the Iberian Peninsula. Spreading presumably began about the sixth century A.D. and in 14 centuries it has colonized about 2.6 million square kilometres. Its spread has been helped by introductions and man-made changes in the forest cover (Nowak 1975).

A small group of Bali cattle *(Bos javanicus)* were introduced to the Cobourg Peninsula, Northern Territory, presumably in 1849 (Calaby 1975). They were present on the Peninsula in numbers in the 1870's. They are common over the 1920 square kilometres of the Peninsula but are confined there because of unsuitable habitat on the 5 kilometre wide neck of the Peninsula. The water buffalo *(Bubalus bubalis)* was introduced to the same area at various times between 1827 and 1849 and by the 1880's had occupied all suitable wetland habitat in the north-eastern part of the Northern Territory. Stragglers from this area, usually mature bulls, are found hundreds of kilometres from the main breeding concentrations, and have been recorded in Queensland over 1000 kilometres away.

Elton (1958) has discussed in detail the spread of many introduced species of animals, plants and pathogens and illustrated his book with maps showing the spread of a variety of organisms.

It is clear from the data available that the spread of animals into newly available suitable habitat is extremely rapid, even in large mammals that produce at most one young per mature female per year. In the case of a species whose habitat is a complex climax community, spread could be a relatively slow process. As an example, the greater glider is an inhabitant of tall open eucalypt forest, and also needs over-mature tall trees with hollow limbs or trunks for shelter. It would take a considerable time, probably hundreds of years, for such a habitat to develop following an appropriate increase in rainfall and/or decrease in temperature.

ANIMAL DISPERSAL AND QUATERNARY STUDIES: SOME DIFFICULTIES

It is difficult to see that a knowledge of animal dispersal and establishment can be of much help in Quaternary studies. Caution would need to be observed when basing interpretations on species known to be very nomadic or subject to irruption, or those that disperse passively. Without other evidence it might be difficult to sort out whether remains of species subject to irruption were those of refugees or animals in their breeding range when the climate was different from that of the present day. It could not be deduced from fossils that a species experienced an explosive spread and the change in environmental tolerances which that implies. Such events are so rare, however, that they could safely be ignored. Fossil remains in island sites might have resulted from a waif colonization that persisted for a number of generations and then died out for genetic reasons unrelated to environmental conditions.

One of the more difficult problems in studies of mammal-bearing deposits is that involving layers of sediments containing two or more species, apparently in close juxtaposition, that are not found in the same habitats nor even in close geographic proximity in modern times. Wakefield (1967) reported one such deposit in McEachern's Cave, south-western Victoria, where he found mammals that at the present day are characteristic of forest habitat mixed with species of drier woodlands and grasslands. It is possible that there were minor climatic fluctuations that allowed different species to occupy the area for varying periods, but these were relatively short and do not show in the stratigraphy. Animals, even large mammals, colonize suitable habitat so rapidly that only very slow changes over long periods or rapid changes at long intervals could be deduced from the fossil record. It is also possible and perhaps even likely, as Dr. J.H. Hope suggested in discussion, that during the Quaternary there existed communities of plants and animals that have no counterparts in the modern biota.

REFERENCES

Bailey, P.T. 1971 The red kangaroo, *Megaleia rufa* (Desmarest), in north-western New South Wales. I. Movements. CSIRO Wildl. Res., 16, 11-28.

Banfield, A.W.F. 1974 The mammals of Canada. University of Toronto Press, Toronto.

Brower, L.P. 1977 Monarch migration. Nat. Hist., 86, 40-53.

Calaby, J.H. 1975 Introduction of Bali cattle to northern Australia. Aust. Vet. J., 51, 108.

Clark, L.R., Geier, P.W., Hughes, R.D. and Morris, R.F. 1967 The ecology of insect populations in theory and practice. Methuen, London.

Common, I.F.B. 1954 A study of the ecology of the adult bogong moth, *Agrotis infusa* (Boisd.) (Lepidoptera: Noctuidae), with special reference to its behaviour during migration and aestivation. Aust. J. Zool., 6, 69-88.

Common, I.F.B. 1970 Lepidoptera (moths and butterflies). pp 765-866 in The insects of Australia. Melbourne University Press, Melbourne.

Davies, S.J.J.F. 1976 The natural history of the emu in comparison with that of other ratites. pp 109-20 in Proceedings of the 16th International Ornithological Congress (eds H.J. Frith and J.H. Calaby), Australian Academy of Science, Canberra.

Dorst, J. 1974 The life of birds. Weidenfeld and Nicolson, London.

Elton, C. 1942 Voles, mice and lemmings: problems in population dynamics. Clarendon Press Oxford.

Elton, C.S. 1958 The ecology of invasions by animals and plants. Methuen, London.

Frith, H.J. 1964 Mobility of the red kangaroo, *Megaleia rufa*. CSIRO Wildl. Res., 9, 1-19.

Frith, H.J. 1977 Waterfowl in Australia. A.H. and A.W. Reed, Sydney.

Hughes, R.D., Greenham, P.M., Tyndale-Biscoe, M. and Walker, J.M. 1972 A synopsis of observations on the biology of the Australian bushfly (*Musca vetustissima* Walker). J. Aust. ent. Soc., 11, 311-31.

Kalela, O. 1942 Die Ausbreitung der kulturbedingten Vogelfauna als Glied der spätquartären Faunengeschichte Europas. Ornis Fenn., 19, 1-23.

Kalela, O. 1944 Zur Frage der Ausbreitungstendenz der Tiere. Annales Zoologici Societatis Zoologicae Botanicae Fennicae 'Vanamo', 10, 1-23.

Le Boeuf, B.J. 1977 Back from extation? Pacific Discovery, 30, 1-9.

Main, A.R. and Yadav, M. 1971 Conservation of macropods in reserves in Western Australia. Biol. Conserv., 3, 123-133.

Main, B.Y. 1976 Spiders. Collins, Sydney.

Mayr, E. 1963 Animal species and evolution. Harvard University Press, Cambridge, Mass.

Murray, M.D. 1970 The spread of ephemeral fever of cattle during the 1967-68 epizootic in Australia. Aust. Vet. J., 46, 77-82.

Myers, K. 1971 The rabbit in Australia. pp 478-506 in Dynamics of Populations. Proc. Adv. Study Inst. Dynamics Numbers. Popul. (Oosterbeck, 1970) (eds P.J. den Boer and G.R. Gradwell), Centre Agr. Publ. Document, Wageningen.

Nelson, J.E. 1965 Movements of Australian flying foxes (Pteropodidae: Megachiroptera). Aust. J. Zool., 13, 53-73.

Newsome, A.E. and Corbett, L.K. 1975. Outbreaks of rodents in semi-arid and arid Australia: causes, preventions, and evolutionary considerations. pp 117-53 in Rodents in desert environments (eds I. Prakash and P.K. Ghosh), W. Junk, The Hague.

Nowak, E. 1975 The range expansion of animals and its causes (as demonstrated by 28 presently spreading species from Europe). Nat. Tech. Inf. Serv., U.S. Dept. Commerce, Springfield, Va.

Ratcliffe, F.N. and Calaby, J.H. 1958 Rabbit. Aust. Encyclopaedia, 7, 340-47. Angus and Robertson, Sydney.

Serventy, D.L. and Whittell, H.M. 1976 Birds of Western Australia. University of W. Aust. Press, Perth.

Simpson, G.G. 1952 Probabilities of dispersal in geologic time. Bull. Amer. Nat. Hist., 99, 163-76.

Wakefield, N.A. 1967 Preliminary report on McEachern's Cave S.W. Victoria. Vict. Nat., 84, 363-83.

Williams, W.D. 1968 Australian freshwater life. The invertebrates of Australian inland waters. Sun Books, Melbourne.

11. DETERMINANTS OF ENVIRONMENTAL TOLERANCE LIMITS IN PLANTS

by H.A. Nix

SUMMARY

The environmental factors which can influence a plant's physiological response are enumerated and the notion of a minimum set of estimates sufficient to predict such response discussed. The difficulties of assessing the determinants of tolerance limits through a range of areal scales are considered.

INTRODUCTION

'Every plant is a measure of the conditions under which it grows' (F.E. Clements (1920))

That plants provide a measure of environmental conditions in general and of climatic conditions in particular is not disputed. The use of plant species or vegetation types as a proxy for environmental conditions is as well sanctioned in the applied sciences (agriculture, forestry, rangeland and catchment management) as it is in the more esoteric pursuits of palaeobotany and palaeoclimatology. But the basic question confronting us here is the kind of measure that plants and plant fossils can provide. Which environmental variables have been measured and with what degree of precision and level of accuracy? In what form is this information encoded and how may it be retrieved and deciphered?

The specific technical problems of sampling, preparation and analysis of pollen and plant macrofossil materials are reviewed by Davis and Watts respectively in this volume. It is clear from these two reviews and others (e.g. Walker 1971) that the basis for palaeoenvironmental reconstruction from plant fossil data rests upon two primary assumptions:

(a) that modern taxa and their fossil antecedents are equivalent in terms of their physiological and ecological responses, and

(b) that environmental domains can be defined using knowledge of present-day plant/environment and vegetation/environment interactions.

If we accept the uniformitarian stance adopted, then it is clear that progress hinges upon further development of our understanding of ecophysiological and genecological responses of plants to their total environment. Fortunately, such understanding is also a primary objective of the applied sciences in agriculture and forestry. In this case however, the objective is to 'fit' or 'tailor' a specific plant to a particular environment, or, to use environmental data to specify which economic plants can be grown where and how well. Formerly, developments in the more fundamental and the applied branches of plant science proceeded more or less independently if in parallel, but a scan of recent literature indicates that a convergence in basic philosophy and methodology is taking place.

Before proceeding to consider some of the basic concepts and methods in plant/environment analysis and synthesis, the problem of reconstructing environments from limited plant sample data is restated in a different context. Consider an experienced plant ecologist who is presented with the task of reconstructing a present-day environment (or environments) given a sample data set with the following properties.

(a) A list of taxa present, identified in some cases to species level, but in others only to generic or even higher taxonomic rank; ecotypic variation is unknown.

(b) Absence of a taxon may be real but can be an artefact of sampling, since the sample is known to be biased through the operation of a selective filtering process which is only partially understood.

(c) A measure of abundance of individual taxa. However the measure is imprecise as an estimator of population size, particularly with respect to comparisons between taxa.

(d) The centroid of the sampled area is precisely located, but the outer perimeter remains undefined. Indeed, the sample area is known to vary with individual taxa. The taxa listed are not necessarily associated in space in the community sense and more than one vegetation type or plant association may be represented.

(e) The time frame of the sample lies between 10 and 100 years, but could be more. Thus the taxa listed are not necessarily contemporaries.

Accepting all the limitations of the data, I think it likely that the experienced ecologist could make an intuitive but reasonable assessment of the environmental domains represented. Initially perhaps he would attempt to match the assemblage of taxa with one, or more, known plant communities and on the basis of their known distribution, infer some of the properties of the macroclimate and the physical substratum. Inferences of a more specific nature would be derived from autecological knowledge of individual taxa. Although species provide the most explicit information with respect to environmental gradients, higher taxonomic ranks can still provide useful information (Dale and Clifford 1976; Del Moral and Denton 1977; Walker and Guppy 1977).

Although it is climate that receives the major emphasis in the reconstruction of environments, the physical substratum, possible or probable vegetation associations and even the putative presence of associated biota, also receive attention. In each case the assemblage of taxa represented does provide a context, but it is easy to agree with both Davis and Watts (this volume) that autecological studies of component taxa will yield the most definitive information. Given the time and effort required for a thorough autecological analysis of a single species it is obvious that selection of species for study warrants critical attention.

SOME BASIC CONCEPTS OF PLANT/ENVIRONMENT RELATIONSHIPS

In attempting to define the environmental domain of any given species, the concept of limiting factors is of fundamental importance. Its development can be traced back to Liebig's 'Law of the Minimum' which, as originally formulated in 1840, related only to mineral elements then known to be essential for plant growth. Plant response was held to be dependent on that essential element in minimum supply. Blackman (1905) extended the concept to other environmental factors and clarified the form of plant response along an environmental gradient by demonstrating a maximum as well as a minimum level of tolerance. Thus, an organism can be shown to have a characteristic physiological response curve to an environmental gradient with minimum and maximum limits to tolerance and an optimum range between these values.

It is necessary to distinguish between the physiological response to an environmental variable in the absence of competition and the ecological response in the presence of competition for the same environmental resources. Studies under controlled environment conditions show that these two responses are rarely if ever identical. The basic features of these two responses and their significance for environmental interpretation are presented graphically and economically by Davis (figure 2.1, this volume).

If we consider all the environmental factors that can influence physiological response, then the concept of niche as developed by

Hutchinson (1958, 1965) provides a theoretical basis for further development and quantification of environmental domains. The niche is "the sum of all the environmental factors acting on an organism; the niche thus defined is a region of n-dimensional hyperspace". It is an abstract environmental volume within which a species' tolerance limits lie. This concept however merely defines the outer boundaries of a 'potential' niche. Wuenscher (1969, 1974) has suggested modifications of the basic Hutchinsonian concept which will permit evaluation of 'optimum' niche as defined by differential response to each environmental gradient and 'realised' niche as defined by the hypervolume actually occupied by the species when it is limited by competition.

A complete specification of a species' niche (i.e. every response to every environmental variable) may be a desirable objective in theory, but is rarely necessary in practice, since a limited number of variables exert a major influence on plant response. More precise specification of the environmental gradients can reduce very markedly the number required to define a niche. Thus, a measure of the soil water-balance is more directly related to plant response and, moreover, is less location-specific in action than, say, separate axes for precipitation, potential evaporation and soil depth. A more difficult conceptual problem relates to the time integration of plant response to environmental factors. One solution is to use dynamic modelling and simulation techniques and these will be considered in a later section.

The total response of a plant to its environment is the result of its genetically programmed morphology and physiology interacting with available environmental resources, but modified by competition and predation. The genetic program is the product of evolutionary history and changes through time at varying rates (Franklin, this volume). Physiological ecology (or ecophysiology) and genecology are the primary disciplines involved in the search for the physiological and genetic bases of plant response to environment.

ENVIRONMENTAL VARIABLES

A comprehensive list of environmental variables that have been considered in vegetation studies and their principal effects on plants has been developed by Billings (1952, 1974) and modified by Scott (1974) (Table 11.1). Many of the environmental variables do not act directly, but through other variables that are more closely linked to plant response. Also, many of them are unlikely to be limiting except in very specific localized situations. Analysis of causal pathways and linkages suggests that a functional grouping of environmental variables in terms of the major regimes (radiation, thermal, water, mineral element and biotic) that modulate plant response is most useful.

The key role of climate and weather in conditioning biological and physical processes and in determining basic strategies of plant growth, development, reproduction and survival needs no emphasis. While existing climatic station networks may provide an adequate coverage for analysis of plant distributions and identification of major limiting factors at global and continental scale, they often prove completely inadequate at regional and more local scales. There is a pressing need for development of methods for mesoscale extrapolation and interpolation and the generation of synthetic climatic data. Useful progress has been made in some areas, such as estimation of solar radiation in complex terrain where slope angle and aspect significantly modify radiation receipts (e.g. Holland and Steyn 1975).

As an example of the inadequacy of existing climatic station networks, I recently examined the possibility of a bioclimatic analysis of the distributions of *Araucaria* and *Nothofagus* species in Australia, New Guinea, New Caledonia and South America. Despite an apparently reasonable geographic network of climate stations, closer inspection reveals that

TABLE 11.1

Aspects of environmental variables that have been considered in vegetation studies and their principal effects on plants (modified from Billings, 1952).

A. Radiation regime; B. Temperature or energy exchange regime; C. Nutrient regime; D. Water regime; E. Living regime in time or space or factors controlling plant responses.

(After Scott 1974)

Variables	Aspects	Plant regime
CLIMATIC		
Radiation:		
solar	Spectral composition, intensity, direction, duration, periodicity	A,B,D,E
terrestrial	Intensity, integration	B
Temperature:		
air	Degree, periodicity, integration, lateral & vertical variation	B,D,E
soil	Degree, periodicity, lateral & vertical variation, freeze - thaw phenomena	B,C,D,E
Water:		
vapour	Amount or densities, vapour pressures & VP deficits, evaporation, transpiration	D,B
condensed water	Cloud, fog, dew, salt spray	D,C
precipitation	Types, amounts, frequencies, droughts, snow cover, snow abrasion	D,E,C,B
soil water	Content, tensions, supply rates, soil aeration	D,B,C,E
Atmospheric gases:		
composition	Oxygen & carbon dioxide contents and partial pressures, ozone, pollutant gases	C,E
pressure	Weather phenomena, gas partial pressures	A,B,D,C
wind	Frequency, force, direction, evaporation, abrasive agents	E,D

Table 11.1 (cont'd)

Variables	Aspects	Plant regime
EDAPHIC		
Parent material:	Minerals present, structure, weathering susceptibility	C,D
Soil:		
physical properties	Profile, structure, texture, mechanical analysis, soil moisture, stability	C,D,B
chemical properties	Clay mineralogy, organic compounds & breakdown, base exchange properties, pH macro- & micro-nutrients, toxicity elements	C
biotic properties	Soil flora & fauna, litter & humus, breakdown rates, antibiotic effects	C,E
GEOGRAPHIC		
Position:		
latitude, longitude, etc.	Work through other factors, e.g. climate, erosion	A,B,C,D
Topography:		
slope, aspect, altitude	Work through other factors, e.g. climate, erosion	A,B,C,D
Vulcanism:		
thermal effects		B
mechanical	Ash cover, lava flows	E
PYRIC		
Climatic:	Temperature (air & soil), intensity & duration, post-burning micro-climatic effects	B
Edaphic:	Organic matter destruction, mineral release, erosion	C,D,E
Biotic:	Community composition	E
BIOTIC		
Other plants:		
competition	For light, water & nutrients, anti-biotic & autotoxic effects	A,B,C,D,E
dependence	Litter & humus, physical & chemical effects, cover	A,B,C,D,E
Animals:		
destructive	Grazing, feeding, etc., effect on soil	E
beneficial	Nutrient cycling, fruit & seed dispersal	C,E
Man:	Can change almost any factor, at least locally	A,B,C,D,E

remarkably few of them represent the environments in which the target species grow. Commonly the climate station is located in a broad open valley and the target species occur on adjacent mountain slopes and ridges which are known to have significantly different radiation, thermal and water regimes.

Recent developments in the U.S.A. (J.T. Ritchie, pers. comm.) indicate that reliable estimates of solar radiation can be generated on a 0.8 kilometre grid using hourly cloud data from geostationary satellites calibrated against ground-based radiation measurements. Since solar radiation data provide necessary input to the more general and reliable predictive equations for potential evaporation (e.g. Penman 1948), this new technique offers prospects of much finer spatial resolution of key components in the energy and water balances.

Although the choice of environmental variables to be used in plant-environment analysis must appear very wide, it is possible to nominate some key environmental variables which are necessary for any adequate explanation of plant responses and tolerance limits. The concept of a minimum data set necessary for explanation of the phenomena studied is a useful one. The minimum data set for analysis of plant response would include measures or estimates of global solar radiation, maximum and minimum temperature, precipitation, potential evaporation and soil extractable water content (a resultant of soil physical properties and plant root distribution). From these data it is possible to calculate plant response to the primary light, thermal and water environments. This is not to deny the significance of mineral element status (whether in deficiency or excess), but at present the analytical methods available do not have sufficient generality for our purpose. In the absence of direct experimentation, the best that they can do is indicate the possibility of deficiency or excess. In broad geographic studies the use of qualitative rating scales for inferred soil nutrient status can prove useful (e.g. Specht and Brouwer 1975).

Identification of the more important environmental factors is one thing, but what of their spatial and temporal variation? The degree of resolution sought and the amount of data reduction considered necessary will depend on the scale and purpose of the study. Data reduction (and expansion) can be a particular problem with climatic and weather data. The time scale of analysis will dictate whether hourly, daily, weekly, monthly or even annual integrals are most appropriate. Some of the problems of reduction of climatic data in relation to ecological analysis have been considered by White and Lindley (1976). An example of data expansion is provided by Fitzpatrick and Nix (1970) who used harmonic analysis techniques to derive estimates of weekly values from published monthly climatic normals.

METHOD IN PLANT/ENVIRONMENT ANALYSIS AND SYNTHESIS

Although complete explanation of plant/environment interactions and functional relationships must be an ultimate goal, for the present we must be satisfied with the lesser goal of prediction. Acceptable levels of prediction do not depend on a complete explanation of relationships between variables. Our present concern is to make predictions about the nature of the biophysical environment from knowledge of the physiological and ecological responses of plant species. The converse of this is of particular importance in applied ecology (e.g. agriculture and forestry) where we wish to make predictions about the potential distribution and performance of an economic plant given a set of biophysical environment data. In each case, our ability to make useful predictions hinges upon adequate description and quantification of the relationships that couple the plant to its environment.

According to Scott (1974) the presence of a plant demonstrates, firstly, that none of the environmental variables acting alone or in combination have exceeded its tolerance range at any stage during its life and, secondly, that the plant has made net growth. On this basis he defines two kinds of analytical problems:

(a) defining tolerance limits, and

(b) defining performance under fluctuating environmental conditions.

Although both tolerance limits and relative performance are included in the concept of the physiological response curve along an environment gradient, the proposed subdivision provides a useful basis for discussion of methods.

Defining tolerance limits.

Much of the early evidence for establishment of tolerance limits in various species and vegetation types came from studies of plant distribution and this tradition continues. Particular emphasis is given to environmental conditions prevailing at the margins of distributions and to study of distributions along steep environmental gradients. Where one or other environmental variable completely dominates plant response, reasonable inferences may be drawn. Thus Dahl (1951) demonstrated a strong coincidence between the ranges of alpine plants and various isotherms for more than one hundred species in Scandinavia. However, such simple correlations are the exception rather than the rule because of the various interactions and substitutions between environmental variables over the whole range of the plant species.

The experimental basis for determination of tolerance limits to particular environmental variables really goes back to Liebig's studies in the mid-nineteenth century, but major impetus has come from development of controlled environment facilities during the last two decades (Went 1957). The role of the 'phytotron' in research and some of the problems of extrapolating to the field has been reviewed by Evans (1963, 1976).

Although examples of the use of controlled environment facilities abound in the literature, most are concerned with general plant growth and development responses (particularly temperature and photoperiod) rather than with specific definition of tolerance limits. However, the screening of plant species in the search for ecotypes with wider tolerances to specific limiting environmental variables has been a consistent theme in the applied spheres of agriculture and forestry. Thus, for example, Hacker et al (1974) use controlled environmental facilities to simulate frosting a number of tropical pasture species and cultivars.

Obviously, definition of tolerance limits, taking into account all the major environmental variables and interactions between them and for different stages of the plant growth cycle, is a major undertaking for even one plant species. In fact, the nearest to satisfying such a requirement is the work of Forman (1964) who grew the moss *Tetraphis pellucida* under various combinations of temperature, relative humidity, pH and substrate conditions. Tolerance limits were determined for several stages of the life history of this cryptogam. The results were used, in combination with environmental data, to explain the distribution of the moss at scales ranging from global down to the level of a population on a stump.

Forman's work also provides a classic illustration of the fact that different environmental variables and combinations of variables can

(a) assume importance at different scales, and

(b) become limiting in different parts of the total distribution of a species.

Thus, with *Tetraphis pellucida* at global scale, the climatic variables of temperature and relative humidity were dominant, while at the level of local stream-bank populations light intensity and pH were most significant. In North America, the total distribution was limited by low temperature in the north, low relative humidity in the southwest and high temperature in the southeast.

Finally, we must not ignore the reality of ecotypic variation, since tolerance limits and also the optimum range may vary in space and time within a species. The classical example is provided by crop species where such variations are the deliberate goal of the plant breeder. In plant species with extensive distributions such variation has been demonstrated many times. A classical example is the work of Mooney and Billings (1961) in their study of arctic and alpine populations of *Oxyria digyna*. An Australian example is provided in a study of the tree species *Acacia harpophylla* by Coaldrake et al (1971).

Defining performance under fluctuating environmental conditions.

While tolerance limits may determine presence/absence of a species in terms of physiological response, it is the relative performance (i.e. productivity, dry matter production) in the response range between tolerance limits that may determine survival under competition. Growth and development responses to the major light, thermal water, nutrient and associated biotic regimes have been a major preoccupation of plant physiologists, ecophysiologists and agricultural scientists for most of this century.

Growth response, dry matter production and standing biomass provide a measure of the relative importance of a species at a given location. Both standing biomass and the structural form or physiognomy of vegetation are broadly correlated with major environmental gradients of temperature and water regimes. Recognition of such broad categories led to the first attempts at climatic classification and the general climatic classifications of Köppen (1913) and Thornthwaite (1931, 1948) are still used today. These more rational and systematic classifications play an important role in codifying climatic information at global and perhaps continental scales, but are of limited utility for study of specific genotype/environment interactions.

The Köppen and Thornthwaite classifications of climate do relate to some major zonal changes in vegetation, but at a scale and level of precision which is not very useful. A much more explicit classification for vegetation/climate at global scale has been erected by Holdridge (1947, 1966). A graphical representation of this system provides a fascinating exercise in trapezoidal symmetry with major axes representing effective mean air temperature, annual precipitation and a potential evaporation surrogate. The vegetation types are structural/physiognomic. If these types can be recognised, if their floristics can be inferred and if the defined relationships between climate and vegetation class are shown to be stable and consistent, then the Holdridge system may offer some possibilities in palaeoclimatic reconstruction.

Comprehensive classifications which directly link climate and vegetation types can have value, since in many cases it is sufficient to show that major shifts in zonal vegetation or plant formations have taken place. Thus Maley (1977), in reconstructing palaeoclimates of the central Sahara during the early Holocene, interpreted his pollen core data in terms of four major vegetation zones of the Lake Chad basin. Pollen grains were classified according to present geographic distribution of the taxa to which they belong and the most typical were placed in four phytogeographical elements.

At the other extreme from the widely known general purpose, multi-attribute, classifications of climate are those based on genotype/environment interactions (i.e. actual plant response to environment). Logically, such an approach requires a special classification for each genotype (or group of genotypes with similar ecophysiological responses) and, in consequence, tends to be rejected on grounds of impracticality. But advances in ecophysiological knowledge on one hand and in computer based storage on the other, make such an approach entirely feasible.

The beginnings of such a physiological response based approach are evident in the work of Livingston (1916a,b) who developed temperature indices and a combined temperature/moisture index for analysis of plant growth in relation to climatic conditions. These were later used in a classical study of the distribution of vegetation in the United States (Livingston and Shreve 1921). The climatic classification of Papadakis (1961, 1970) is based on broad tolerance limits to temperature and water regimes of ecologically similar crop species.

Direct coupling of the forcing functions of climate and weather with known physiological responses (modified by the effects of physiographic and edaphic conditions if available data permit these to be specified) offer prospects of much more explicit analysis of plant/environment interactions and resultant plant distributions. Fitzpatrick and Nix (1970) derived separate light, thermal and moisture indices for each of three major pasture groups with distinct physiological responses. The indices transform the non-linear responses of plants to major light, thermal and water regimes on to a linear scale with zero indicating completely limiting conditions and unity indicating optimum conditions. These indices, together with a multiplicative growth index, have been used in a bioclimatic analysis of the grassland ecology of Australia.

Of more direct interest in this volume is the use of these indices in an analysis of limiting values for the major structural and floristic zones in northern Australia and New Guinea (Nix and Kalma 1972) and the results used in simulation of vegetation changes in the region during the past 30,000 years. More precise specification of the physiological responses of selected taxa, offers prospects of more detailed analysis in autecological studies. One example of such an approach is provided by the analysis of distribution and environment of mulga *(Acacia aneura)* in Australia at two different scales, one continental and one regional (Nix and Austin 1973).

The need for a holistic approach to problems of plant/environment interactions has long been recognised (e.g. Billings 1952) but until recently the conceptual and computational tools were simply not available. Systems analysis and simulation techniques offer prospects of much better understanding of the dynamics of growth, development, reproduction and survival under varying environmental conditions. Basic concepts and techniques have been discussed by Watt (1966, 1968) and Dale (1970).

At the single species level most of the dynamic growth models so far developed apply to economic plants (crop, forage, forest) and progress is presented in a comprehensive review edited by Evans (1975). As yet, very few really comprehensive models have been developed and many show gross imbalance in that certain processes are emphasized and others, equally important, are neglected. Thus, process models for the basic photosynthesis process dominate while equally important processes such as respiration, translocation and morphogenesis languish for want of the same attention. Despite these criticisms, it seems clear that systems analysis and simulation techniques do offer the best prospects for a more complete understanding of complex plant/environment interactions.

CONCLUSIONS

That the coupling of plants and environment involves exceedingly complex interactions should need no emphasis. Complete explanation of these interactions must remain a goal for the distant future. However, basic understanding of the more important ecophysiological processes does exist and is developing rapidly. We will have to wait for really comprehensive dynamic models of plant growth and development, but meanwhile the simpler, semi-empirical, models should prove useful in the exploration of plant species distribution in space and time.

The capacity to reconstruct palaeoenvironments from interpretation of plant fossil data will depend on increasing our knowledge of the physiological and genetic characteristics of the species concerned. Thorough and comprehensive autecological studies, involving comparative experiments in the field and under controlled environment conditions, will be essential. The effort required demands that very careful attention be given to the selection of species (and ecotypes), species groups and perhaps higher taxonomic groupings.

REFERENCES

Billings, W.D. 1944 Environment Concept and Reality. pp 9-35 in Handbook of Vegetation Science, Part VI, Vegetation and Environment (eds B.R. Strain and W.D. Billings), Junk, The Hague.

Billings, W.D. 1952 The environmental complex in relation to plant growth and distribution. Quarterly Rev. Biol., 27, 251-265.

Blackman, F.F. 1905 Optima and limiting factors. Ann. Bot., 19, 281-95.

Clements, F.E. 1920 Plant Indicators. Carnegie Inst., Washington.

Coaldrake, J.E. 1971 Variation in some floral, seed and growth characteristics of *Acacia harpophylla* (brigalow) Aust. J. Bot., 19, 335-352.

Dahl, E. 1951 On the relation between summer temperature and the distribution of alpine vascular plants in the lowlands of Fennoscandia. Oikos, 3, 22-52.

Dale, M.B. 1970 Systems analysis and ecology. Ecology, 51, 7-16.

Dale, M.B. and Clifford, H.T. 1976 On the effectiveness of higher taxonomic ranks for vegetation analysis. Aust. J. Ecol. 1, 37-62.

Del Moral, R. and Denton, M.F. 1977 Analysis and classification of vegetation based on family composition. Vegetatio, 34, 155-165.

Evans, L.T. 1963 Extrapolation from controlled environments to the field. pp 431-437 in Environmental Control of Plant Growth (ed L.T. Evans), Academic Press, New York.

Evans, L.T. (ed) 1975 Crop physiology - some case histories. Cambridge University Press, Cambridge.

Evans, L.T. 1976 The role of phytotrons in agricultural research. pp 11-27 in Proc. of Symposium on Climate and Rice. IRRI, Los Banos, Philippines.

Fitzpatrick, E.A. and Nix, H.A. 1970 The climatic factor in Australian grassland ecology. pp 3-26 in Australian Grasslands (ed R.M. Moore), ANU Press, Canberra.

Forman, Richard T.T. 1964 Growth under controlled conditions to explain the hierarchical distributions of a moss *Tetraphis pellucida*. Ecol. Monogr., 34, 1-25.

Hacker, J.B., Forde, B.J. and Gow, M. 1974 Simulated frosting of tropical grasses. Aust. J. Agric. Res., 25, 45-47.

Holdridge, L.R. 1947 Determination of world plant formations from simple climatic data. Science, 105(2727), 367-368.

Holdridge, L.R. 1966 The life zone system. Adansonia, 6, 199-203.

Holland, P.G. and Steyn, D.G. 1975 Vegetational responses to latitudinal variations in slope angle and aspect. J. Biogeog., 2, 179-183.

Hutchinson, G.E. 1958 Concluding remarks. Cold Spring Harbor. Symp. Quat. Biol. 22, 415-427.

Hutchinson, G.E. 1965 The ecological theatre and the evolutionary play. Yale Univ. Press, New Haven.

Köppen, W. 1931 Grundriss der Klimakunde. de Gruyter, Berlin.

Livingston, B.E. 1916a Physiological temperature indices for the study of plant growth in relation to climatic conditions. Phys. Res., 1, 399-420.

Livingston, B.E. 1916b A single index to represent both moisture and temperature conditions as related to plants. Phys. Res., 1, 421-440.

Livingston, B.E. and Shreve, F. 1921 The distribution of vegetation in the United States, as related to climatic conditions. Carnegie Institution Publ. 284, Washington.

Maley, J. 1977 Palaeoclimates of Central Sahara during the early Holocene. Nature, 269, 573-577.

Mooney, H.A. and Billings, W.D. 1961 Comparative physiological ecology of arctic and alpine populations of *Oxyria digyna*. Ecol. Monogr., 31, 285-308.

Nix, H.A. and Austin, M.P. 1973 Mulga : A bioclimatic analysis. Trop. Grasslands, 7, 9-21.

Nix, H.A. and Kalma, J.D. 1972 Climate as a dominant control in the biogeography of northern Australia and New Guinea. pp 61-91 in Bridge and Barrier : The natural and cultural history of Torres Strait (ed D. Walker), Department of Biogeography and Geomorphology, Publication BG/3. Australian National University, Canberra.

Papadakis, J. 1961 Climatic tables for the world. Published by the author, Buenos Aires.

Papadakis, J. 1970 Climates of the world : their classification, similitudes, differences and geographic distribution. Published by the author, Beunos Aires.

Penman, H.L. 1948 Natural evaporation from open water, bare soil and grass. Proc. Roy. Soc. A., 193, 120-45.

Scott, D. 1974 Description of relationships between plants and environment. pp 49-69 in Handbook of Vegetation Sciences, Part VI, Vegetation and Environment (eds B.R. Strain and W.D. Billings), Junk, The Hague.

Specht, R.L. and Brouwer, Y.M. 1975 Seasonal shoot growth of Eucalyptus spp. in the Brisbane area of Queensland (with notes on shoot growth and litter fall in other areas of Australia). Aust. J. Bot., 23, 459-474.

Thornthwaite, C.W. 1931 The climates of North America according to a new classification. Geogrl. Rev., 21, 633-655.

Thornthwaite, C.W. 1948 An approach toward a rational classification of climate. Geogrl. Rev., 38, 55-94.

Walker, D. 1971 Quantification in historical plant ecology. pp 91-104 in Quantifying Ecology (ed H.A. Nix), Proc. Ecol. Soc. Aust., 6, Canberra.

Walker, D. and Guppy, J.C. 1976 Generic plant assemblages in the highland forests of Papua New Guinea. Aust. Journ. Ecol., 1, 203-212.

Watt, K.E.F. (ed) 1966 Systems analysis in ecology. Academic Press, New York.

Watt, K.E.F. 1968 Ecology and resource management. McGraw-Hill, New York.

Went, F.W. 1957 The experimental control of plant growth Chronica Botanica, Waltham.

White, E.J. and Lindley, D.K. 1976 The reduction of climatological data for ecological purposes : a preliminary analysis. Journal of Environmental Management, 4, 161-182.

Wuenscher, J.E. 1969 Niche specification and competition modelling. J. Theoretical Biol., 25, 436-443.

Wuenscher, J.E. 1974 The ecological niche and vegetation dynamics. pp 39-45 in Handbook of Vegetation Science, Vegetation and Environment, Part VI (eds B.R. Stain and W.D. Billings), Junk, The Hague.

12. DISPERSAL AND ESTABLISHMENT OF PLANTS

by J.M.B. Smith

SUMMARY

Dispersal mechanisms and range, and some factors affecting establishment, are briefly reviewed. No adaptations are specifically for dispersal over very long distances, this being rare though important and probably achieved by birds. Dispersal and establishment over more than short distances are most likely for species of pioneer strategy, and least likely for "climax" species. Rapid migrations are therefore restricted to pioneers, which are the first species to benefit from climatic amelioration. Especially in temperate zones, the vegetation of Quaternary interglacial periods is successional, "climax" species migrating too slowly to occupy their full potential ranges and many probably having become extinct. Detailed autecological research is needed to further understand the plant migration process.

INTRODUCTION

The topic of dispersal and establishment of plants has a large and varied literature which, however, permits few useful generalizations. This is largely because each species is unique in its dispersal and establishment characteristics, and will have different potential for spread and migration in different habitats or parts of its actual or potential range. It is also due to the operation of chance factors, especially where dispersal and establishment over relatively long distances are considered; such factors may have low probabilities but operate potentially over long periods of time during which environments and their inhabitants are constantly changing.

In this paper I consider not only the range of characteristics which may be of significance to the spread of the plants possessing them, but also environmental and evolutionary factors which influence plant migration. I will attempt to simplify the discussion slightly by considering only phanerogams (seed plants). Cryptogams, to judge by their generally wider global distributions, may have a greater capacity for dispersal and establishment. This may be partly due to dispersal of their spores by wind over long distances and to their greater ecological tolerance (lack of specialisation) especially in the cases of many non-vascular thallophytes (algae, fungi, lichens and bryophytes). I will also largely ignore the important role of man, considered by Wace (this volume).

ADAPTATIONS FOR SEED DISPERSAL

Many plants produce seed or fruit with mechanisms which can readily be interpreted as favouring dispersal away from the parent plant. A similar number has no such discernible mechanisms, but in many cases these plants appear to be dispersed just as efficiently. Dispersal mechanisms have been extensively reviewed elsewhere (Ridley 1930; van der Pijl 1969) and will only be treated briefly here.

The principal modes of dispersal for which structural mechanisms can be recognised are listed below.

Wind Dispersal.

Hairs or plumes of various forms and morphological origins form part of the seeds or dry fruits of many plants at the time of their liberation from the parent plant. In many cases these are efficient aids to dispersal. Van Steenis (1967) reports a probable single dispersal flight by *Crassocephalum crepidioides* of 100-300 kilometres during the spread of this alien species in Java. During colonisation of drained polders in Holland, Bakker (1960) reports flights of up to 4 kilometres by *Tussilago farfara* fruits.

Some trees have winged fruits or seeds, the wing(s) slowing their fall from the parent allowing some lateral movement by wind. Where the seed is small and the wing papery this may be effective in dispersal over moderate distances, e.g. up to 5 km for *Pinus radiata* and 15 km for *P. contorta* in New Zealand (Bannister 1965). Where the seed is larger and the wing more robust, dispersal is rarely for more than a few hundred metres; for example Dipterocarpaceae in Borneo are unable to cross "all but the narrowest dispersal barriers" (Ashton 1969).

Very small seeds may be readily carried by wind. This is no doubt of importance for example to epiphytic orchids' seed dispersal to branches other than those of the tree on which the parent is growing.

Internal dispersal by animals.

Fleshy fruits are eaten by birds, bats or other animals, which may move some distance before regurgitating or defaecating the indigestible seeds. Waterbirds may ingest seeds, sometimes as parts of whole shoots which are eaten, and act as dispersal agents for aquatic or marsh plants. To what extent seeds may be carried internally over long distances by birds is unclear (Cruden 1966; DeVlaming and Proctor 1968) but numerous observations confirm the efficiency of this mechanism over at least short distances.

In cases where the seed is the animal attractant, rather than the fruit, it is usually killed during the digestive process. Beans *(Vicia faba)* have been found germinating in the decomposing crops of dead pigeons, however, and this may have realistic dispersal significance when it is considered that falcons and other predators may kill, carry and eat large numbers of birds whose crop contents are liable to be scattered at the site where the prey is dismembered. Of probably greater significance, at least for some species, is the hoarding by certain vertebrates of seeds not all of which are later found and eaten. Holtmeier (1972), for example, records the dispersal of seeds of *Pinus cembra* in the European Alps by nutcrackers, usually to unforested areas at higher altitudes than the parent tree. Jays have been seen to carry acorns for over 4 km in Germany (Jones 1959). Ants may carry large numbers of seeds, and bury them, but only over short distances.

External dispersal by animals.

A diversity of hooks, barbs and similar structures, and glutinous substances, may cause seeds or fruits to attach to the fur, plumage, etc. of animals coming into contact with them. Few studies appear to have been undertaken with wild animals. In one, Agnew and Flux (1970) found in Kenya that 160 of 369 hares examined carried 810 disseminules of 17 species, demonstrating the importance of this means of dispersal, at least in some environments for some species and over short or moderate distances.

Water.

Seeds of many wetland and seashore plants are undamaged by water and may float for lengthy periods. 50% of those of the strand plant *Crambe maritima*, for instance, remain viable and afloat for 14 days (Scott & Randall 1976) permitting dispersal to distant shoreline sites. Seeds of the small tree *Sophora microphylla* gathered from beaches on the Kermadec Islands and found to be viable must have floated about 800 kilometres from New Zealand (Sykes and Godley 1968). Observations by Praeger (1913) and others on seeds of many species suggest that those of about 10-18% can float for at least a week, some for many months. Local down-river dispersal of seeds probably occurs in many species not necessarily specifically adapte to it.

Self-dispersal and rain splash dispersal.

Some plants exhibit devices whereby seeds are thrown from the plant; such dispersal, of itself, is effective over only a few metres. The same is true of seed scattering by the impact of raindrops, for which the hygroscopic "splash-cups" of a few species appear adapted (Smith 1977).

THE GEOGRAPHICAL RANGE OF DISPERSAL

The great majority of seeds of almost all species are dispersed only a few metres from the parent plant, and indeed most seeds may fall actually beneath the parent. This may ensure that these seeds arrive in a favourable habitat for that species. On the other hand the developing seedlings will suffer from intense intraspecific competition with each other, and (if it continues to live) their parent, although in some species seedlings are apparently favoured in their early growth by close proximity of siblings (Linhart 1976).

However, seeds suffer heavy predation, especially where they are shed in large numbers, so that commonly very few survive to germinate. Seed predators are most abundant near to seed sources, and may consume all seed falling near the parent plant. Janzen (1972) has demonstrated this for *Sterculia apetala* in Costa Rica, where predation by bugs is total near the seeding tree and tapers off to zero at 30-60 metres distance, so that only those few seeds dispersed by monkeys and squirrels beyond this radius are able to germinate. Where trees grow in nearly pure stands, predators may destroy nearly all their seeds (e.g. in *Fraxinus excelsior* woodlands in England; Gardner 1977) unless mast fruiting occurs in which massive seedfall only every few years satiates seed predator populations, allowing the surplus seeds to germinate. Especially in very mixed stands, such as lowland tropical forests, plants often produce seeds containing toxic materials, which are eaten largely by specialised animal species which take seed of only one or a few plant species. In such cases dispersal away from the parent tree is unlikely to be to the vicinity of a reproductive individual of the same species, and will therefore probably result in escape from the rather specific seed predators. This has a strong bearing upon the species richness of tropical forests (Janzen 1970) and perhaps upon the ease with which invading species may occupy such a habitat. Seed predation has been reviewed by Janzen (1971). Predation of vegetative parts is also likely to be heaviest where individuals of the same plant species are near each other.

In many cases, then, we can conclude that very local dispersal, though most common, is often ineffective in reproduction. It can also be only slowly effective in extending the range of a species, and not at all if such range extension involves the crossing of even small barriers. We are therefore led to consider the important minority of seeds which are dispersed over longer distances. Unfortunately such dispersal is not easy to observe directly; especially for those rare but important dispersal events over distances of more than several kilometres, our data are largely circumstantial and anecdotal. Most such data relate to the occurrence of seeds or plants in environments distant from possible seed-sources, such as on islands; the actual dispersal event is not observed, but inferred from its result. Plant migration can also be deduced from paleoecological data though in these cases the assumptions implicit in conclusions relating to dispersal are even greater.

The floras of islands which have never been connected to mainland areas, or whose vegetation was destroyed after such connection by volcanic or other catastrophe, provide lists of taxa which have succeeded in dispersing across long distances. Such lists of course provide minimum numbers of species; many will have established themselves for a period to become extinct later. Nor must evolution, which may be rapid in small island populations, be ignored. Several species may have evolved from a single colonising ancestor, and evolution may result in different characteristics from those of the initial colonist. Carlquist (1966) has postulated an evolutionary loss of dispersal mechanisms in some Pacific Island Compositae, and similar observations have been made by Guppy (1906) and Ridley (1930) for other island plants.

The floras of remote islands have long been regarded as disharmonic, being small and atypical by comparison with continental floras. While some disharmony can be explained by low species equilibrium levels (MacArthur and Wilson 1967; Simberloff 1974), much is also due to the varying dispersal "abilities" (or chance events affecting dispersal) of plants in the pool of potential colonists. It is probably no accident that island equilibrium theory derives from studies of birds and insects, whose over-water dispersal abilities appear to be generally greater than those of seed-plants.

The most revealing island situations are those which have floras of recent immigration. From these cases, it appears that dispersal by wind over distances exceeding some tens of kilometres is rare, and that by flotation in sea water is effective for usually only a small number of specialised littoral plants. Dispersal by birds (and bats) may account for the bulk of immigrant plants. For example,Jarak Island (Straits of Malacca, over 500 kilometres from the nearest land) has 93 seed plants of which 6 or 7 are sea-dispersed, 3 are wind-dispersed, 2 introduced by man and 81 probably carried by birds or bats (Wyatt-Smith 1953). The flora of the island appears to be youthful, and probably entirely post-dates a heavy tephra fall which is likely to be only a few millennia old.

Further examples are provided by subantarctic islands, at least some of which were probably buried by ice during the last glacial period and whose floras have therefore immigrated in postglacial time and over long distances. On Macquarie Island, Taylor (1954) found an unidentified but not native seed in a lake inhabited by petrels, and seeds adhering to the feet of albatrosses by a hard layer of dried regurgitated food material. Much of the island's flora of mostly widespread species clearly owes its presence to bird dispersal.

Dispersal by wind, though not actually random, may be effectively so for seeds, for which the difference between two sites only metres apart may be the difference between a small chance and no chance for establishment. On the other hand dispersal by animals is to a degree guided dispersal; animals choose where they go, or at least settle, and select sites suitable to them, likely also to be suitable for the plants whose seeds they carry. A duck bearing seeds from one swamp or lake will fly, not to a forest or ridge, but to another wet environment.

However, there remain in most insular floras a residue, sometimes large, of species quite lacking apparent dispersal mechanisms, but which have nevertheless succeeded in crossing wide dispersal barriers. An example here is the flora of the high mountains of New Guinea, the bulk of which probably immigrated from temperate zone regions (in some cases no doubt via "stepping-stone" peaks) in Quaternary times. Of 100 native species in the tropicalpine flora of Mt Wilhelm which were examined, 44 had no discernible dispersal mechanism. Type or absence of dispersal mechanism showed no correlation with time or direction of migration deduced from other data. 14 of the 27 of these species which were conspecific with populations outside New Guinea, and which therefore may be expected not to have evolved substantially since migration, lacked dispersal mechanisms (Smith 1977). Dahl (1963) concluded from an analysis of dispersal mechanisms in terrestrial floras in the north Atlantic region "either that the adaptations favouring long-distance dispersal are of no biological significance, or that the problem of Amphi-Atlantic plant distribution is not a matter of long-distance dispersal".

When dispersal over very long distances is considered it therefore appears that most seed and fruit mechanisms are rarely effective. Carlquist (1965) questions whether any species are specifically adapted to it. Those plants which succeed in crossing large barriers do so in many and probably most cases by means not facilitated by special adaptations. A category of "occasional" dispersal agents can be imagined to account for long distance dispersal: cyclones, jet streams, natural rafts and prehistoric human transport have all been suggested, and may all have acted occasionally. Bird dispersal over long distances has often been discounted on the grounds that, prior to migration, birds empty their guts and carefully preen their plumage. However, Taylor's (1954) albatrosses would no doubt find it impossible to clean the hard, seed-containing material from their feet, and the coincident distributions of birds and plants in New Zealand quoted by Falla (1960) also suggest that accidental dispersal by migrating birds may occur. Furthermore birds, even weakly flying non-migratory forms like the Rallidae, not infrequently are blown very long distances by strong winds, and presumably such involuntary migration is not preceded by internal and external cleansing. One of the most wide-ranging groups, Charadriiformes, consists of birds usually regarded as carnivorous; but the finding of DeVlaming and Proctor (1968) that the Killdeer eats a variety of seeds (many not normally considered as showing adaptations to internal bird dispersal), and may retain these in viable condition for up to 120 hours, strengthens the suggestion that most long distance dispersal is by birds.

The great problem, of course, is that it is almost impossible to observe such a cryptic event as seed dispersal if it occurs with very low frequency. Fosberg (1963) has estimated that the Hawaiian islands were colonized by only 272 ancestral seed-plant species, these by evolution leading to the entire native flora of 1729 species and varieties. If we assume a conservative age for the archipelago of 10 million years, this means that only one successful dispersal and establishment each 37,000 years need be postulated to account for the flora. No doubt several extinctions have taken place, and many successful dispersals did not result in establishment at all. Nevertheless, it is clear that dispersal over very long distances need only be a very infrequent event to account for the cases in which it has evidently occurred.

FACTORS AFFECTING ESTABLISHMENT

In most species under natural conditions each successive generation consists of approximately the same number of individuals. It follows that the vast majority of seeds will fail to develop into mature plants, in other words to effect establishment. The role of seed predators has already been mentioned. Even of those seeds which by good dispersal and good fortune evade predation, only a few will develop into mature plants. Harper (1965) considers that for invading species it is in the phases of germination and seedling establishment that their success or failure is most critically determined. Factors affecting the establishment process are grouped below under the three headings: seed, site and plant characteristics. Only a few of the most important factors are considered even briefly; a complete account of establishment, from seed to seeding plant, would be virtually a full account of the autecology of that species.

Seed Characteristics.

Aspects of seed biology have been reviewed extensively elsewhere (e.g. Heydecker 1973; Kozlowski 1972; Stebbins 1971; Wareing 1966), and the subject will be touched upon cursorily here.

Size. Seed size determines the amount of food reserves available to the developing seedling, and hence its potential for rapid early growth in the period before the plant need depend solely upon its own photosynthate for growth. Large seeds produce large seedlings which may initially compete more effectively. Large seeds are produced particularly by species of forest environments where the seedling needs to become established in a poorly lit environment (Ashton 1969) and often from within a layer of deep or bulky litter and in dry environments where the rapid development of a deep root system may save the seedling from an early death by drying, (Baker 1965).

Seed size and number are to a degree inversely related, since only a finite proportion of available energy can be invested by a plant in seeds. This proportion varies, tending to be lower in plants of more mature ecosystems, which are also those plants tending to have larger seeds. Large seeds, with generally poor dispersal ability, are therefore those produced in smallest numbers relative to the total biomass of the parent plant (Harper, Lovell and Moore 1970).

Shape. The shapes of seeds are often interpreted as having dispersal significance, but may also be of great though less obvious importance in positioning the seed in micro-sites favourable to germination (Harper, Lovell and Moore 1970; Sheldon 1974). Some idea of the range of microsites on a bare soil surface, and the different responses to them by germinating seeds of related species, is provided by the work of Harper, Williams and Sagar (1965). Some species exhibit seed polymorphism, producing seeds of two or more shapes and sizes (Harper, Lovell and Moore 1970).

Dormancy. Different species produce seeds with different degrees of dormancy which, when it occurs, may be broken by various external or internal stimuli. In many species, of which many weeds are examples, a single plant produces seeds with differing dormancy characteristics. While dormant, seeds may be resistant to a variety of more or less extreme external conditions, such as high or low temperatures, dryness, contact with sea water, etc. Dormant seeds are therefore better able to survive the possible vicissitudes of dispersal, and may delay their germination after dispersal until this is triggered by the onset of favourable conditions. Plants with seeds lacking dormancy may in general be less capable of dispersal, and subsequent establishment, over more than short distances.

Site characteristics.

Physiochemical. Climatic and edaphic factors are of importance to any plant. They are especially critical to the seedling, which is less robust and tolerant, and more easily killed, than a larger plant. If the attempted establishment is in a site distant from that in which the parent grew, it is likely that some of these factors will prove to some degree unfavourable.

Competition. The effects of competition by plants already present at a site may nearly or entirely prevent establishment of seedlings. Schuster (1976) believes long-distance dispersal to be much commoner than often assumed, but to lead only rarely to establishment in relatively closed communities. Germination even a few days earlier than later emerging competitors confers a lasting advantage upon seedlings (Ross and Harper 1972). This is further illustrated by several experimental attempts to establish or augment plant populations in apparently suitable vegetation types (e.g. Holt 1972; Miles 1972; van Steenis, Hamzah and Toha 1972). Such attempts generally fail unless accompanied by physical or chemical disturbance of the soil, reducing competition for nutrients and perhaps water and light, or unless mature individuals are transplanted to the site so bypassing the critical early stages of seedling establishment.

This phenomenon is part of what has been described as vegetation inertia by Pearsall (1959). The crucial role of competition in producing such inertia is underlined by Miles' (1975) demonstration that alternative perennial vegetation, including both woody and herbaceous species, can be established (at least for several years) after, but not without, removal of the previously present vegetation.

Predation and parasitism. In its native environment a species is likely to have collected during its long local history an associated suite of more or less specialised predators (of both vegetative and reproductive organs) and parasites. Although the odds are heavily stacked in other ways against the chances of successful establishment following dispersal over a long distance, with regard to predators and parasites the seed at great distance from its parent is often more favourably placed than one falling nearer to hand. The case with seed predators has already been mentioned. The unaccustomed vigour of many species after being given a helping human hand across barriers shows how important is the stress imposed by predators and parasites of the vegetative plant in the same species' native location. Such vigour in numerous examples has been reduced when some of these consuming organisms, or alternative ones, have been deliberately or unwittingly introduced to the immigrant host population.

Plant characteristics.

A seed may be successfully dispersed, germinate and grow into an immature plant, but establishment implies also the successful and continued reproduction of this plant and its descendant population. The whole gamut of factors impinging upon the relationship between the species and its complex environment(s) comes into play, and will of course be different in every instance. Considering for the present only the plant derived from a distantly dispersed seed, two reproductive factors threatening its future establishment of a population are self-fertility and reproductive capacity.

Following long distance dispersal the immigrant is unlikely to find itself within pollination range of another of the same species. Should the plant not be self-fertile, therefore, it can only at best establish a genetically uniform and slowly spreading population by vegetative means. Discussing this, Baker (1955) mentions the case of the usually dioecious (and hence self-sterile) *Coprosma*, of which an immigrant population on Macquarie Island is atypically monoecious.

Even if the individual in question is able to scatter seed in its new location, too poor a seed output and too slow a population growth may sound its death-knell. Any small population is liable to extinction through local chance events. The importance of reproductive capacity (rate of seed production, Salisbury 1942) is emphasised by Harper's (1966) study of five closely related species of *Papaver*, whose abundance was closely related to reproductive capacity.

DISPERSAL AND ESTABLISHMENT AS PART OF OVERALL ECOLOGICAL STRATEGY

Dispersal and establishment strategies and the range of external factors affecting them are almost infinite in their variety. However, some generalisations are possible if the problem is considered in the context of overall ecological strategy.

Plant (as well as animal) species can be crudely aligned along a gradient from those inhabiting and adapted to very transient environments to those of very stable environments. The former class may variously be described as fugitives, pioneers or adventives, or in the terminology of MacArthur and Wilson (1967), r-species, while the latter are the species of "climax" (very mature) ecosystems or K-species. Pianka (1970) has tabulated the climate, population, competition, evolution and phenology characteristics of species of each type, and the environments they inhabit. Briefly, r-species inhabit variable and unpredictable environments; they have variable populations subject to catastrophic, density-independent mortality; they experience low competition pressure; they exhibit rapid development with an early and single reproductive phase; and they are small and short-lived. K-species show opposite characteristics.

As Harper (1961) has pointed out, this gradient is maintained by opposing selection pressures upon a plant in a successional situation: the plant may either evolve characteristics enabling it to efficiently escape from the maturing ecosystem and colonize other more or less distant open sites; or it may evolve towards survival in the closed, more stable vegetation of the more mature ecosystem.

The fugitive species, fleeing the closing vegetation, are also the pioneering colonists, establishing new populations in other open areas. This strategy involves efficient dispersal of seeds but also great wastage of seeds, many of which fail to arrive in the small (before man) open sites created by geomorphic processes. For both these reasons seeds are small. It also involves a syndrome of characters better fitting the establishing plant to its transient but less competitive niche: rapid growth rate with prodigal use of nutrients, early reproduction and rapid life-cycle, and perhaps above all a phenotypic and physiological plasticity enabling survival and reproduction in a range of more or less favourable sites. Weeds of tillage, though many show particular adaptations in addition, are paramount examples of good pioneer species (Baker 1972, 1974; Ehrendorfer 1965, Hill 1977). While they may have evolved further in response to man's peculiar disturbances of the environment, many were preadapted to the weed niche by evolution under conditions of natural disturbance, as clearly indicated by the presence of many British "weeds" in early postglacial deposits there (Godwin 1975).

The other end of the spectrum is occupied by "climax" species, whose environment, though stable so that dispersal over long distances is an unnecessary luxury, involves intense competition. Here the stress is not upon abundant and quick reproduction but upon the relentless battle for resources (including light) over lengthy life-cycles. Seeds tend to be large and few, vegetative reproduction is often important, and ecological specialisation is precise. While competitive ability is great under a limited set of environmental conditions, "climax" species are unable to cope with the wider range of conditions tolerated by pioneers.

I choose to loosely define for present discussion purposes a third, intermediate, strategic category. This I call nomad, borrowing a term from van Steenis (1958) who recognised such plants as forming the "scab" over gaps in the forests of southeast Asia. While not excluding herbs and even animals, most of his examples are woody plants, and I restrict the term to these in the present context. Nomads succeed the herbaceous pioneers in previously disturbed sites, and precede the "climax" species of the later, longer-lived and more mature ecosystems. In stature and life-form, they compare more closely with "climax" species, but as van Steenis stresses, there is a sharp distinction in ecological strategy, which in nomads is more of the pioneer type. They produce an abundance of well dispersed seed, are ecologically rather tolerant, quick-growing and of short life-cycle by comparison with other trees, and are unable to establish themselves in closed woody vegetation.

Van Steenis cites *Pinus merkusii* in Sumatra as one of his examples, and it appears that most species of *Pinus* adopt this strategy. Other clear temperate zone examples include *Betula* in North America and Eurasia and some species of *Eucalyptus* in the wetter parts of eastern Australia. Dobzhansky (1959) suggested that species selection is for efficiency in the relatively stable tropics, where ecological controls are biotic, while in the more variable temperate zones species selection is for fecundity. The relative instability of temperate environments can perhaps be seen on a geological as well as a seasonal timescale. It may be that the bulk of temperate zone trees may best be regarded as nomads rather than as "climax" species.

Generalising yet further, pioneers, nomads and "climax" species differ both in their dispersal abilities and in their chances of establishment following dispersal over more than short distances.

All strategies will involve very common local dispersal, say to distances of less than a kilometre; relatively few seeds of any type are dispersed beyond this range. Similarly very long distance dispersal (say beyond 100 km) will be very rare in almost all cases. The differences between strategies will mainly appear at intermediate dispersal ranges; pioneers may commonly be dispersed to distances exceeding 10 km but such dispersal of nomads may be rare, and of climax species virtually non-existent.

If we consider how well these categories of plants may become established after dispersal, pioneers and to a lesser degree nomads are most favourably placed. They are preadapted by broad ecological tolerance and phenotypic and physiological plasticity to conditions differing somewhat from those at the parental location, and different conditions are increasingly likely to be encountered at increasing dispersal distances.

So it should not be surprising that the species which best succeed in colonising distant insular sites (e.g. MacArthur and Wilson 1967, Smith 1977) and which have been the first to take advantage of newly favourable territory following climatic amelioration (e.g. Godwin 1975, Iversen 1954) are pioneers (including aquatics). In some cases they have subsequently evolved towards a "climax" strategy, part of the taxon cycle as identified in Polynesian ants by (Wilson 1965; Wilson and Taylor 1967). But even after evolutionary change, the very widespread affinities of such immigrant floras (e.g. Smith 1977, Taylor 1954), including a large proportion of genera which are nearly cosmopolitan and which commonly include weed species, attest to their pioneer origins.

At the other end of the strategy gradient, "climax" species pursue their migrations not (or very rarely) by leaps across barriers, but by progressive spread through continuous vegetation. *Nothofagus* provides a good example of this strategy, an ancient genus whose distribution is so constrained by barriers that it provides a partial map of Gondwanaland fragments which were still in contact at the time of its initial Cretaceous evolution (Smith 1974). Ecological data from New Zealand confirm that *Nothofagus* species have poor dispersal ability (Preest 1963) and that they (and some other species) have not yet had sufficient time since the end of the last glacial period to spread to the limits of their ecologically potential ranges (Holloway 1954, Wardle 1963, 1964). Paleoecological data suggest other examples (e.g. *Castanea* in North America; Davis 1976) but we must beware of circular argument in deducing strategy to explain migration from such data.

CHANGES IN GEOGRAPHICAL RANGES OF PLANTS

Range expansion by pioneers.

Mayr (1965), in summarising the proceedings of a conference on colonising species, suggested that there are three kinds of successful colonisers: pioneers invading newly created or disturbed environments, species crossing barriers, and species invading new ecological niches after evolution of appropriate characters. The nature of pioneers, as discussed above, clearly makes them the best fitted species for succeeding in the first two categories. Their relatively rapid evolution compared with that of climax species (Allard 1965, Margalef 1968) also makes them the best candidates, at least over relatively short time-spans, for the third category (exemplified by the delayed but sudden spread of *Senecio squalidus* in England; Turril 1948). As discussed by Watts (1973) there are also other types of range expansion, following natural environment change (climatic change, soil development), and although these are not restricted to pioneers, plants of this strategy are likely to be the first to benefit from some of these changes.

Many cases of rather rapid range expansion by pioneers can be quoted from modern studies and inferred from paleoecological data. The spread of exotic weeds, though their dispersal is no doubt in part due to man, shows their great powers of establishment in a range of disturbed environments. The migration of *Crassocephalum crepidioides* all over southeast Asia to as far as Queensland within 30 years of its introduction in India or Sri Lanka is one example (van Steenis 1967).

Following climatic amelioration, formerly glaciated and periglacial areas in north temperate regions were quickly colonised by herbaceous plants of pioneer strategy, many of them now considered weeds (Godwin 1960, 1975). Amongst this immigrant flora are aquatics which show by their sexual reproduction that warm conditions prevailed. This suggests (Iversen 1954, West 1964) that the pioneer flora was present due to its superior migrational abilities, the less able woody nomads and climax species which could also have occupied the region migrating more slowly.

Range expansion by nomads and "climax" species.

During a local succession, pioneers precede nomads, which are finally followed by "climax" species, although in marginal habitats distortion imposed by extreme stress may make the distinction between plants of these differing strategies unclear. Traditionally the succession has been seen as being controlled by the difficulty experienced by the later species in establishing in open conditions, they requiring a vegetation cover below which their seedlings may become established. On a larger scale dispersal difficulties are also involved. For example, lowland rainforest in Malaysia can only regenerate (in the short term) beyond a nomad stage if there is a nearby seed source for "climax" species.

Davis (1976) has emphasised that in north temperate areas, the limited duration of interglacial periods is inadequate for the migration of many species here called nomad and "climax" to occupy their full ecologically potential ranges. In different interglacials, species have arrived in different orders, due perhaps to differing chance long distance dispersal events, so that in any given region the vegetation has varied between interglacials in a partly random way. Succession on this large scale, then, beginning with a quick immigration of pioneers, does not proceed to its end-point before the onset of the ensuing glacial period. This had been recognised earlier in the case of the present interglacial (postglacial) in New Zealand from ecological and biogeographical data (Holloway 1954, Wardle 1963, 1964). Davis suggests that most species of north temperate deciduous forest whose ecology permits only slow migration through rather mature vegetation ("climax" species) may have become extinct.

Lowland tropical environments may have experienced as great a temperature change between glacial and interglacial periods as that demonstrated for temperate and tropical highland environments. However this would not have involved deforestation, and for the many plants whose altitudinal range exceeds several hundred metres may not even have necessitated migration of a part of their populations. Dense forests also control to a degree their own subcanopy environment, and the distance of any enforced migration would in any case have probably been less than that imposed upon north temperate plants. To this is probably related the larger proportion of "climax" species to those of other strategies in the lowland tropics. Indeed it may in part explain the richness of tropical vegetation generally, as biotic diversity appears to be closely related to environmental stability (Colinvaux 1973, Slobodkin and Sanders 1969).

Contraction of geographical range.

Plant species may lose as well as gain ground. In so doing they may retain a continuous but smaller range, or survive as a series of relict populations not in genetic contact with each other. Especially in the case of species of "climax" strategy, these relict stands may survive for long periods under sub-optimal conditions and possibly as sexually sterile small populations, as may be the case with tundra outliers of *Picea glauca* and *P. mariana* in Canada (Nichols 1976) and very small stands of *Nothofagus moorei* in Queensland. Such cases may be possible both through the long life of individuals, and their ability to reproduce vegetatively (by layering or coppicing, for example) in situations where seed production is impossible.

When conditions change again, relict populations may be allowed to expand and coalesce, restoring or even enlarging the plant's former range. A case in point is that of many pioneer species which were widespread and abundant in late glacial times in Britain, survived subsequent forest periods tenuously on dunes, cliffs, etc., and with human clearance of forests became widespread and common once more (Godwin 1975). Other interesting relicts survive (or did so until recent human influence) in isolated species-poor environments, where presumably they can exist under conditions of lower competition than in other places. Arborescent

Compositae are (or were) conspicuous on some tropical mountain-tops and oceanic islands; they are primitive and appear to have retreated to these peculiar sites from formerly larger ranges (Mabberley 1974). These "decadent" plants, of "climax" strategy, cannot be expected ever to expand their ranges, unlike adaptable pioneers which are quick to spread where disturbance (natural or human) permits it.

A plant's range may contract through one or several of a variety of causes. Any assemblage of organisms is in a continuing dynamic internal struggle, every species being directly or indirectly influenced by all others. A plant's evolution in this struggle will improve its position vis-a-vis its competitors, predators and parasites, but its failure to evolve as fast or as effectively as its congeners may result in its decline and eventual extinction in that ecosystem, and hence in loss of part or all of its range. This has been highlighted in the case of seed predators by Janzen (1970, 1971) as previously discussed, and there is no reason to suppose that a plant species' relations with other parts of its biotic environment are any less dynamic or critical.

More abrupt than evolutionary changes may be the immigration of new species to the community, whose impact may be too rapid to allow an effective evolutionary response by previously present species. The devastation and probable extinction of *Castanea* in North America by an immigrant fungal parasite is an extreme example.

Physical environmental changes may also lead to range contraction, just as they may do to range expansion. Such changes may be climatic, edaphic, or catastrophic such as the result of volcanic activity. It should be remembered, however, that most physical parameters controlling geographical ranges usually only act directly at the harsher edge of a species' range, if at all. The commoner situation is for biotic controls to become more or less acute along a physical gradient, so that the proximal cause of a limit of distribution is biotic, and only the distal cause physical. Since plants inhabit different communities in different parts of their ranges; this implies that the coincidence of limits with critical physical conditions may not be constant in different regions, even if the species itself does not display geographical variation. Davis (this volume) points out that the ecological reponses of species may differ geographically with their varying suites of competitors.

Range contraction is the other side of the coin from range expansion which results from dispersal and establishment. Its consideration is valuable, if only to further bring to attention some of the many factors apart from initial dispersal which may prevent establishment of a species beyond its previous range.

CONCLUSIONS

The Quaternary has been a period of unusual environmental instability, and hence of unusually high levels of plant migration, evolution and extinction. A variety of studies have resulted in an emphasis upon the variable geographical characteristics of Quaternary vegetation types, even during the past few millennia in which climates have not changed greatly (e.g. Davis 1976; Holloway 1954; Leak and Graber 1974; Tallantire 1977; Wardle 1963; West 1964).

The different extents to which environments support communities of "climax" strategy plants are related to their age and stability, not only in the short term but also over millennia, and perhaps even longer periods as implied by Dansereau (1963) and Smith (1977). Less mature ecosystems of more unstable or youthful environments, viewed on any of these scales, include species of more pioneering strategy, and are "unsaturated" and invasible. Even rather maturer ecosystems, at least at temperate latitudes, may offer little initial resistance to invasion, the invader's population apparently often being controlled by intra- rather than inter-specific competition (Watts 1973). The species of these ecosystems, albeit of varying strategy, are all presumably adapted to penetration of closed communities, in order to have survived the forced marches of the Quaternary (Davis 1976).

From the preceding discourse, five broad conclusions of relevance to paleoecologists can perhaps be drawn.

(1) Mechanisms for seed dispersal are not effective over very long distances; such dispersal is rare and by occasional "chance" processes often involving birds.

(2) Dispersal over more than short distances, and establishment after such dispersal, is most likely for seeds of pioneer species and least likely for those of "climax" species.

(3) The species most likely to establish themselves in locations distant from the parent plants, and so to be the first to avail themselves of new geographical opportunities such as provided by climatic amelioration, are therefore those of pioneer strategy.

(4) Much Quaternary vegetation change is best explained by its being large scale and never completed succession, whose details are determined in large part by dispersal and establishment abilities and chances.

(5) Because of the migration difficulties encountered by "climax" species, most of these have become extinct in those parts of the world where Quaternary environmental changes have had the widest geographical impact. The most mature ecosystems in such regions (e.g. cool temperate North America and Europe) are dominated mainly by nomads.

Palaeoecologists may then ask what is known of the processes of dispersal and establishment of plants today, for their own data demand explanation and analysis based on more than circular argument.

Webb (1966) concluded that we know little of these processes. Actually we know a great deal, but understanding of the totality of plant migration is elusive in its complexity. Though a wealth of facts has been accumulated, and we have learnt to look at plants in the light of their overall strategies, we have not advanced far beyond the bleak situation outlined by Webb thirteen years ago. Biogeographical and palaeoecological data tell us that migration has occurred, the result of dispersal and establishment. However, actual observation of the important but rare dispersal events over longer distances is nearly impossible, and their inference from the discovery of distant individuals must usually be clouded by doubts over human seed transport. Study of a species' establishment is nearly synonymous with that species' entire autecology. And, anyway migration today need not be under conditions comparable with those in the past, even under "natural" circumstances which are themselves becoming rarer.

To proceed beyond the stage of generalisation exemplified by the present paper, further detailed autecological studies are needed (West 1964) which pay proper attention to factors influencing dispersal and establishment chances. Since the number of plants of paleoecological interest is much larger than the number of ecologists prepared to study them, this is a tall order. Perhaps if autecological studies can at least elucidate the range of combinations of conditions under which particular key species cannot (instead of usually) establish themselves, a realm of impossibility can be defined in a past situation. It is within the remaining area of improbability that rare but important events may and frequently did happen to advance migration. After all, it was the one acorn in tens of thousands carried 25 km, there to develop against the odds to maturity, and not the multitude dropped near the parent tree, that enabled the oak to spread across postglacial Europe.

A difficulty here lies in distinguishing the impossible from the highly improbable, especially in the largely unknown conditions of the past.

Another possible approach is to assess for individual species a range of characters possibly facilitating dispersal and establishment, such as self-fertility, reproductive capacity, dispersal mechanisms, seed size, seed dormancy, phenotypic plasticity and ability to tolerate a wide range of conditions. An aggregate score for such characters could be used to assign to species positions along a gradient of migration ability. In this way an objective if imprecise measure of the degree to which a species is likely to indulge in rapid (or slow) migration can be arrived at. But whether this potential is realised will depend also upon other factors, intrinsic, environmental and chance.

Palaeoecological data reveal an imperfect record of the changing plant distributions of the past. These patterns result from a combination of biotic, physical and historical constraints, including failure to establish in distant but otherwise suitable sites. Different combinations of constraints will operate at different margins of species' ranges. There is no short cut to understanding plant migration processes and past distributions. Perhaps the best that can reasonably be aimed at is increasingly informed intuitive judgements based on greater understanding of autecology through detailed and extensive field observations.

REFERENCES

Agnew, A.D.Q. and Flux, J.E.C. 1970 Plant dispersal by hares *(Lepus capensis* L.*)* in Kenya. Ecology, 51, 735-7.

Allard, R.W. 1965 Genetic systems associated with colonizing ability in predominantly self-pollinated species. pp 49-75 in The Genetics of Colonizing Species (eds H.G. Baker and G.L. Stebbins), Academic Press, New York.

Ashton, P.S. 1969 Speciation among tropical forest trees: some deductions in the light of recent evidence. Biol. J. Linn. Soc., 1, 155-96.

Baker, H.G. 1955 Self-compatibility and establishment after "long-distance" dispersal. Evolution, 9, 347-9.

Baker, H.G. 1965 Comment in discussion. pp 399-400 in The Genetics of Colonizing Species (eds H.G. Baker and G.L. Stebbins), Academic Press, New York.

Baker, H.G. 1972 Migrations of weeds. pp 327-347 in Taxonomy, Phytogeography and Evolution (ed D.H. Valentine), Academic Press, London.

Baker, H.G. 1974 The evolution of weeds. Ann. Rev. Ecol. & Syst., 5, 1-24.

Bakker, D. 1960 A comparative life-history study of *Cirsium arvense* (L.) Scop. and *Tussilago farfara* L., the most troublesome weeds in the newly reclaimed polders of the former Zuiderzee. pp 205-222 in The Biology of Weeds (ed J.L. Harper), Blackwell Sci. Publs., Oxford.

Bannister, M.H. 1965 Variation in the breeding system of *Pinus radiata.* pp 353-374 in The Genetics of Colonizing Species (eds H.G. Baker and G.L. Stebbins), Academic Press, New York.

Carlquist, S. 1965 Island Life. Natural History Press, New York.

Carlquist, S. 1966 The biota of long-distance dispersal. II. Loss of dispersibility in Pacific Compositae. Evolution, 20, 30-48.

Colinvaux, P.A. 1973 Introduction to Ecology. Wiley, New York.

Cruden, R.W. 1966 Birds as agents of long-distance dispersal for disjunct plant groups of the temperate western hemisphere. Evolution, 20, 517-32.

Dahl, E. 1963 Plant migrations across the North Atlantic ocean and their importance for the paleogeography of the region. pp 173-178 in North Atlantic Biota and their History (eds A. Löve and D. Löve), Pergamon, Oxford.

Dansereau, P. 1963 Naturalization hazards in the Pacific. pp 293-297 in Pacific Basin Biogeography (ed J.L. Gressitt), Bishop Mus. Press, Honolulu.

Davis, M.B. 1976 Pleistocene biogeography of temperate deciduous forests. Geoscience and Man, 13, 13-26.

DeVlaming, V. and Proctor, V.W. 1968 Dispersal of aquatic organisms: viability of seeds recovered from the droppings of captive killdeer and mallard ducks. Amer. J. Bot., 55, 20-26.

Dobzhansky, T. 1950 Evolution in the tropics. Amer. Sci., 38, 209-21.

Ehrendorfer, F. 1965 Dispersal mechanisms, genetic systems, and colonizing abilities in some flowering plant families. pp 331-351 in The Genetics of Colonizing Species (eds H.G. Baker and G.L. Stebbins), Academic Press, New York.

Falla, R.A. 1960 Oceanic birds as dispersal agents. Proc. Roy. Soc., B 152, 655-9.

Fosberg, F.R. 1963 Plant dispersal in the Pacific. pp 273-281 in Pacific Basin Biogeography (ed J.L. Gressitt), Bishop Mus. Press, Honolulu.

Gardner, G. 1977 The reproductive capacity of *Fraxinus excelsior* on the Derbyshire Limestone. J. Ecol., 65, 107-18.

Godwin, H. 1960 The history of weeds in Britain. pp 1-10 in The Biology of Weeds (ed J.L. Harper), Blackwell, Oxford.

Godwin, H. 1975 The history of the British Flora. A Factual Basis for Phytogeography (2nd Ed.). University Press, Cambridge.

Guppy, H.B. (1906) Observations of a Naturalist in the Pacific. 2. Plant Dispersal. Macmillan, London.

Harper, J.L. 1961 Approaches to the study of competition. Symp. Soc. Exp. Biol., 15, 1-39.

Harper, J.L. 1965 Establishment, aggression and cohabitation in weedy species. pp 243-265 in The Genetics of Colonizing Species (eds H.G. Baker and G.L. Stebbins), Academic Press, New York.

Harper, J.L. 1966 The reproductive biology of British poppies. pp 26-39 in Reproductive Biology and Taxonomy of Vascular Plants (ed J.G. Hawkes), Pergamon, Oxford.

Harper, J.L., Lovell, P.H. and Moore, K.G. 1970 The shapes and sizes of seeds. Ann. Rev. Ecol. & Syst., 1, 327-356.

Harper, J.L., Williams, J.T. and Sagar, G.R. 1965 The behaviour of seeds in soil. 1. The heterogeneity of soil surfaces and its role in determining the establishment of plants from seed. J. Ecol., 53, 273-86.

Heydecker, W. 1973 Seed Ecology. Pennsylvania State Press.

Hill, T.A. 1977 The Biology of Weeds. Arnold, London.

Holloway, J.T. 1954 Forests and climates in the South Island of New Zealand. Trans. Roy. Soc. N.Z., 82, 329-410.

Holt, B.R. 1972 Effect of arrival time on recruitment, mortality and reproduction in successional plant populations. Ecology, 53, 668-73.

Holtmeier, F.K. 1972 The influence of animal and man on the alpine timber-line. pp 93-96 in Geoecology of the High-Mountain Regions of Eurasia (ed C. Troll), Steiner, Wiesbaden.

Iversen, J. 1954 The late-glacial flora of Denmark and its relation to climate and soil. Danmarks Geologiske Undersögelse, 2, 80, 87-119.

Janzen, D.H. 1970 Herbivores and the number of tree species in tropical forests. Amer. Nat., 104, 501-28.

Janzen, D.H. 1971 Seed predation by animals. Ann. Rev. Ecol. & Syst., 2, 465-92.

Janzen, D.H. 1972 Escape in space by *Sterculia apetala* seeds from the bug *Dysdercus fasciatus* in a Costa Rican deciduous forest. Ecology, 53, 350-61.

Jones, E.W. 1959 Biological flora of the British Isles. *Quercus* L. J. Ecol., 47, 169-222.

Kozlowski, T.T. 1972 Seed Biology. Academic Press, New York.

Leak, W.B. and Graber, R.E. 1974 A method for detecting migration of forest vegetation. Ecology, 55, 1425-7.

Linhart, Y.B. 1976 Density-dependent seed germination strategies in colonizing versus non-colonizing plant species. J. Ecol., 64, 375-80.

Mabberley, D.J. 1974 Pachycauly, vessel-elements, islands and the evolution of arborescence in "herbaceous" families. New Phytol. 73, 977-84.

MacArthur, R.H. and Wilson, E.O. 1967 The Theory of Island Biogeography. Princeton Univ. Press.

Margalef, R. 1968 Perspectives in Ecological Theory. Univ. of Chicago Press.

Mayr, E. 1965 Summary. pp 553-562 in The Genetics of Colonizing Species (H.G. Baker and G.L. Stebbins), Academic Press, New York.

Miles, J. 1972 Experimental establishment of seedlings on a southern English heath. J. Ecol., 60, 225-34.

Miles, J. 1975 Performance after six growing seasons of new species established from seed in Callunetum in northeast Scotland. J. Ecol., 63, 891-901.

Nichols, H. 1976 Historical aspects of the northern Canadian treeline. Arctic, 29, 38-47.

Pearsall, W.H. 1959 The ecology of invasion: ecological stability and instability. New Biology, 29, 95-101.

Pianka, E.R. 1970 On r- and K- selection. Amer. Nat., 104, 592-7.

Praeger, R.L. 1913 On the bouyancy of the seeds of some Britannic plants. Scientific Proc. Roy. Dublin Soc., 14, 13-62.

Preest, D.S. 1963 A note on the dispersal characteristics of the seed of the New Zealand podocarps and beeches and their biogeographical significance. pp 415-424 in Pacific Basin Biogeography (ed J.L. Gressitt), Bishop Mus. Press, Honolulu.

Ridley, H.N. 1930 The Dispersal of Plants throughout the World. Reeve, London.

Ross, M.A. and Harper, J.L. 1972 Occupation of biological space during seedling establishment. J. Ecol. 60, 77-88.

Salisbury, E.J. 1942 The Reproductive Capacity of Plants. Bell, London.

Schuster, R.M. 1976 Plate tectonics and its bearing on the geographical origin and dispersal of Angiosperms. pp. 48-138 in Origin and Early Evolution of Angiosperms (ed C.B. Beck), Columbia Univ. Press.

Scott, G.A.M. and Randall, R.E. 1976. Biological flora of the British Isles. *Crambe maritima* L. J. Ecol., 64, 1077-91.

Sheldon, J.C. 1974 The behaviour of seeds in soil. 2. The influence of seed morphology and the behaviour of seedlings on the establishment of plants from surface-lying seeds. J. Ecol., 62, 47-66.

Simberloff, D.S. 1974 Equilibrium theory of island biogeography and ecology. Ann. Rev. Ecol. & Syst., 5, 161-82.

Slobodkin, L.B. and Sanders, H.L. 1969 On the contribution of environmental predictability to species diversity. pp 83-93 in Diversity and Stability in Ecological Systems, Brookhaven Symposia in Biology, 22.

Smith, J.M.B. 1974 Southern biogeography on the basis of continental drift. A review. Austr. Mammalogy, 1, 213-230.

Smith, J.M.B. 1977 Origins and ecology of the tropicalpine flora of Mt Wilhelm, New Guinea. Biol. J. Linn. Soc., 9, 87-131.

Stebbins, G.L. 1971 Adaptive radiation of reproductive characteristics in angiosperms, 2: seeds and seedlings. Ann. Rev. Ecol. and Syst. 2, 237-60.

Sykes, W.R. & Godley, E.J. 1968 Transoceanic dispersal in *Sophora* and other genera. Nature, Lond., 218, 495-6.

Tallantire, P.A. 1977 A further contribution to the problem of the spread of spruce (*Picea abies* (L.) Karst.) in Fennoscandia. J. Biogeogr., 4, 219-227.

Taylor, B.W. 1954 An example of long distance dispersal. Ecology, 35, 369-72.

Turrill, W.B. 1948 British Plant Life. Collins, London.

van der Pijl, L. 1969 Principles of Dispersal in Higher Plants. Springer, Berlin.

van Steenis, C.G.G.J. 1958 Rejuvenation as a factor for judging the status of vegetation types: the biological nomad theory. pp 212-218 in Study of Tropical Vegetation, Proc. Kandy Symp. 1956, UNESCO, Paris.

van Steenis, C.G.G.J. 1967 Notes on the introduction of *Crassocephalum crepidioides* (Bth.) S. Moore in Indo-Australia. J. Ind. Bot. Soc. 46, 463-9.

van Steenis, C.G.G.J., Hamzah, A. and Toha, M. 1972 The Mountain Flora of Java. Brill, Leiden.

Wardle, P. 1963 Evolution and distribution of the New Zealand flora, as affected by Quaternary climates. N.Z. J. Bot., 1, 3-17.

Wardle, P. 1964 Facets of the distribution of forest vegetation in New Zealand. N.Z. J. Bot., 2, 352-66.

Wareing, P.F. 1966 Ecological aspects of seed dormancy and germination. pp 103-120 in Reproductive Biology and Taxonomy of Vascular Plants (ed J.G. Hawkes), Pergamon, Oxford.

Watts, W.A. 1973 Rates of change and stability in vegetation in the perspective of long periods of time. pp. 195-206 in Quaternary Plant Ecology (eds H.J.B. Birks and R.G. West), Blackwell Sci. Publs., Oxford.

Webb, D.A. 1966 Dispersal and establishment: what do we really know? pp 93-100 in Reproductive Biology and Taxonomy of Vascular Plants (ed J.G. Hawkes), Pergamon, Oxford.

West, R.G. 1964 Inter-relations of ecology and Quaternary palaeobotany. J. Ecol. 22, (suppl.), 47-57.

Wilson, E.O. 1965 The challenge from related species. pp. 7-24 in The Genetics of Colonizing Species (eds H.G. Baker and G.L. Stebbins), Academic Press, New York.

Wilson, E.O. and Taylor, R.W. 1967 An estimate of the potential evolutionary increase in species diversity in the Polynesian ant fauna. Evolution, 21, 1-10.

Wyatt-Smith, J. 1953 The vegetation of Jarak Island, Straits of Malacca. J. Ecol., 41, 207-25.

13. HUMAN MODIFICATION OF THE NATURAL RANGES OF PLANTS AND ANIMALS

by N.M. Wace

SUMMARY

Naturalised species may tell a lot about human cultures but it is doubtful whether the past and present ranges of such species have much to say about former climate. Except in the grossest climatic terms, volunteer species extending their ranges as a result of human activities have rather little value as indicators of past climatic conditions. The reasons for this may be summarized as follows.

(a) The success and abundance of naturalized species depends in large part upon the competitive context into which they are introduced. This context will vary greatly from one part of the world to another, and in ways that are difficult to define, or to quantify, because of widely differing native biotas.

(b) The conditions under which the remains of exotic organisms are fossilized will be determined largely by contemporary human activities. The species represented will reflect firstly the mobility of man and his tools and domesticated and feral animals, and only secondarily the nature of the overall physical environment. The dispersal and deposition of propagules need not imply former establishment, or even the climatic suitability of the region in which they are found.

(c) The uniformitarian assumptions which must underlie any environmental interpretation of fossil data, are harder to sustain for most volunteer man-adapted species than for those from stable climax communities. Most naturalized species probably have shorter generation times, show greater morphological plasticity and have geographical ranges which fluctuate more than native species from more stable habitats. They are therefore less likely to be good indicators of past environmental conditions than native species.

INTRODUCTION

A few millennia hence, paleontologists will perhaps view the last thousand (and certainly the last few hundred) years as the great age of stratigraphic confusion; as a time during which much of the earth's terrestrial flora and fauna began to undergo massive rearrangements which could not be directly related to, nor explained by, secular changes in the physical environment. They will note the sudden appearance of rabbit bones and pine pollen in Australian and New Zealand sediments, and of gum-nuts and possum bones in New Zealand deposits. They will note quite sudden declines of the pollen of trees such as some of the chestnuts in North America or elms in Europe in sediments of similar age, or the slower, but accelerating, decline in the remains of all large carnivores, and all but a very few species of large ugulates over most of the earth's grasslands.

As they look at the situation more closely, and try to account for these abrupt changes in the distribution of the inhabitants of the earth's land surfaces, our future paleontologists might notice several striking features that seemed to be associated with them, and their differences from previous changes in the distribution of terrestrial organisms as revealed in the older fossil record. They would first of all be struck by the suddenness of many of the changes, but more particularly by the fact that the most rapid changes of all seemed to have taken place in those lands most isolated by sea barriers from the larger continents in the northern hemisphere. They would note that the species which began to extend their ranges seemed mostly to have spread from these large northern lands to the more isolated southern ones, rather than vice-versa. Above all they would note with perplexity that although many of the newly-dispersed species which had so suddenly increased their ranges were plants and animals of unstable or transient habitats, it was difficult to relate the phenomenon as a whole to any conceivable secular changes in the physical environment, such as climatic changes, shifts in sea levels, or other major events in the history of the earth's crust.

These rearrangements of the world's terrestrial biota by humans and their activities have been in progress as long as man himself has existed as a toolmaking animal. His culture has from the start been based on the manipulation of plants and of other animals for his own use, and on the movement with him of the species that he domesticated. With him on his travels have gone the weeds and pests of his crops and other domesticates, together with a less specialized retinue of camp-following plants and animals. Biogeography, or the study of the distribution of plants and animals over the earth, is traditionally interpreted and understood in terms both of the impact of present environments, and of the history of both organisms and environments stretching back into the distant past. The appearance of man as the ecologically dominant species over much of the earth has accentuated the importance of the last two million years, as compared to those of earlier times, in interpreting the distribution of an increasing number of species. Alterations in the natural ranges of plants and animals by man have introduced extra complications into the paleoecologists' task of reconstructing the terrestrial environments of the Quaternary.

NATIVE AND INTRODUCED SPECIES AND THE NATURAL RANGES OF PLANTS AND ANIMALS

Unless we know what the natural ranges of plants or animals were before the onset of human activities, we are not going to be able to address the question of the human modification of their natural ranges at all. In a careful analysis of the status of plants inhabiting Clare Island, off the west coast of Ireland, Praeger (1911) stressed that "in attempting to express with any accuracy the present standing of species as ... native or introduced, we are really trying to express their past history". He held that *"a species is native in a natural locality to which it has spread by a natural means from a natural source"*. Leaving aside the implication that there is something inherently "unnatural" about man himself and his activities, three criteria are thus to be applied to determine the status of any species: its habitat, its means of dispersal, and its immediate source. Under such a rigid definition, few species in settled countries would qualify as fully native everywhere within their range, because there are now few habitats or sources of specie[s] unaffected by past or present human activities. Praeger pointed out that definitions of "native" as applied to species (and therefore to their ranges) are based almost entirely on negative criteria, in that there is no evidence in the case of a fully native species that man has had anything to do with its source, dispersal or present habitat: *"a native plant means to us a plant the record of whose introduction is lost"*.

In this paper, a less rigid view is taken of what constitutes a native species. A species is considered to be introduced to an area (and its range therefore extended) if its dispersal has been intentionally or unintentionally aided by man, or by the movement of his tools or his domesticated animals. Introduced species may or may not be naturalized; a naturalized species is one that grows and reproduces for several generations at least, without the direct aid of man, outside its native range.

In those parts of the earth's surface longest inhabited by man, especially in the tropics of Asia and Africa, he has been present for so long that it may be difficult to determine whether a particular species is truly native in any given region in the absence of fairly complete fossil and archaeological records. In the intensively-studied regions into which Paleolithic and Neolithic peoples were moving following the retreat of the Pleistocene ice, we may have a more certain base for deciding whether a species is truly native, or whether it owes its presence to human activities. The confident assertion "native" included against the names of 1702 species (or 72%) of the 2375 species of vascular plants thought to be native or naturalized in the British Isles (Clapham, Tutin and Warburg 1962) is based very largely on subfossil evidence. In so far as it depends on that evidence alone, it suffers from the uncertainties of interpretation described by Davis and Watts (this volume), including the possibilities of dispersal of fossils beyond the range of the species concerned and differential preservation, and also ignorance about the degree to which human activities created suitable habitats outside the range within which they might be thought of as truly "native". It has been suggested (Salisbury 1932, Godwin 1960, 1975) that many of the weeds of early Neolithic cultivation in northwest Europe may have moved into man-made situations from the open habitats of glacial moraines and outwash deposits created during late-glacial times. It may thus be difficult to distinguish the impact of early human societies in creating new habitats for colonizing species, from the purely natural creation of new habitats, whether or not these result from climatic changes. If such movements have taken place, the species concerned would only doubtfully be "native" in Preager's (1911) sense.

Local extensions of range due to human activities are well documented, and probably very frequent. Again, subfossil evidence is crucial in establishing such local movements. It has long been accepted from subfossil evidence that the European beech *(Fagus sylvatica)* although native in southern England and parts of Wales, is not so in northern England, Scotland or in Ireland. It has been planted in these latter places. Even in such an intensively botanized area as the British Isles, it is now impossible to be sure of the precise range of many species which are accepted on good fossil evidence as having dispersed there independently since the last retreat of the ice, because of the long history of management of the landscape and the likelihood of human dispersals, intentional or accidental over this time. Similar uncertainty may prevail with many vertebrate animals: if we did not know of the hunting activities of the Normans and of their introduction of the rabbit to Britain (Thompson and Worden 1956) it would surely be accepted as native there, although not perhaps in Ireland (see also Calaby, this volume).

In lands which are traditionally thought to have been more isolated from the main centres of human dispersals and cultural evolution in Pleistocene times, such as Australia, we may perhaps ascribe the status "Native" or "Introduced" with more confidence. But even here, we can scarcely do more than guess at the number of plant species which have been brought in by Aboriginals and other pre-European visitors to Australia. Golson (1971) showed that a number of genera of plants used for food were shared between Arnhem Land and Cape York on the one hand, and Malaysia on the other. He argued that the exploitation of plants by Australia's earliest settlers may have been patterned by traditions established in Malaysia, although he did not suggest that the species involved had necessarily been introduced from Malaysia to Australia (or vice-versa). In the absence of analyses of more than a few subfossil deposits in the Australian tropics, and of scant knowledge of the ecological behaviour of the plants which are likely to be involved in the traditional human economies of the peoples who first reached Australia, it remains impossible to do more than hint at species which may be part of an "Aboriginal Element" in the exotic flora of Australia - plant equivalents of the dingo. Despite their lack of a system of settled agriculture, some plants must have been brought to Australia by pre-European peoples during the 40,000 years or more that they have been present in the continent. Especially in Australia, archaeological work is of the greatest value in establishing such an "Aboriginal Element", the existence of which would be extremely interesting because of its bearing on the earliest stages of plant domestication, which in Australia never developed as far as settled agriculture (but see Jones 1975).

Because of the distinctiveness of much of the Australian flora from that in other parts of the world, and the more distant geographical origins of the first agricultural settlers in Australia, (who introduced species often quite unlike those already here) it has been thought possible to make some estimates of the numbers of species naturalized in Australia during the last two centuries. But such estimates are little more than guesses at present, due to lack of published taxonomic knowledge of both the native and naturalized floras of much of the continent, especially those parts likely to contain species introduced from the north (Burbidge 1963, p. xvi). Working from State floras, published lists such as Anderson (1939), and the unpublished lists of W. Straatmans (1962) and G. Scurfield (1958), I estimated in 1972 that there were about 1500 species of flowering plants naturalized in Australia (Wace 1972, 1973). Following later work in the Melbourne Herbarium, I considered that this figure was probably an underestimate, but that extensive field and herbarium work, as well as taxonomic studies were necessary before any more accurate figure could be derived. Moore & Perry (1973, p.73) considered that there were about 1300 introduced plants (? vascular plants) naturalized in Australia, but did not define what they meant by "naturalized".

In New Zealand, with its smaller and more intensively studied flora, Godley (1975, quoting A.J. Healey) estimated that 560 species of vascular plants had become naturalized since European contact and settlement, compared to a total native vascular flora of 1996 species. Some 22% of the vascular flora of New Zealand is thus thought to consist of naturalized species, compared to about 28% for the British Isles. But even in New Zealand, the status of some weedy species such as *Cotula australia* and *Sonchus asper* remains doubtful, and they may owe their presence there to Maori introduction (Allan 1961). In other cases, such as *Oxalis corniculata* and *Calystegia sepium* and its allies (Ogden, in press), native and alien taxa may have hybridized. Amongst the vertebrate land animals, as in Australia, the distinction between native and exotic species is clearer than amongst the plants. Only one well-attested contribution to the now-naturalized fauna was made by pre-European inhabitants, the Polynesian rat (Gibb and Flux 1973). But the depredations of the Maoris themselves, as well as their rat and their dog, reduced to the point of extinction numbers of native birds, including many moas (Williams 1973).

Floristic and faunistic data of these kinds, although of some interest to those concerned with the degree of human disturbance of different parts of the earth's surface, tell little of the processes by which introduced species extend their ranges.

In Australia, with its spectacular and well-remembered examples of rabbits, cane toads and prickly pears from other continents, we often forget how many of our native species have extended their range within the continent as a direct result of human activities. One of the commonest wattles around Canberra, *Acacia baileyana* has a restricted natural range near Cootamundra, yet it seeds freely and is commonly regarded as "native" here, (and is sold as such to gardeners). It also hybridises with the local native wattles (Burbidge and Gray 1970). The crested wattle *(Albizzia lophantha)* on the New South Wales coast is no more native there than the sycamore *(Acer pseudoplatanus)* in England, or the Monterey cypress *(Cupressus macrocarpa)* in Florida.

Long distance dispersal of both native and introduced species independently of man continues into the present, although the likelihood of successful establishment may be markedly influenced by human alteration of environments into which the naturally dispersing species are spreading. The unaided arrival and establishment in New Zealand during the last 150 years of some ten species of land birds native to Australia and the spontaneous spread of eleven species of European passerines naturalized in New Zealand to its outlying islands (Williams 1973) are good examples of these continuing processes which may or may not be linked to past or present human activities. Because of the ubiquity of these activities, any new arrival of organisms which are more or less passively dispersed tends now to be attributed directly or indirectly to man.

The foregoing discussion is intended to show that defining the status of species as "native" or "introduced" is by no means a simple matter, even with ample subfossil historical and ecological information. A species is here considered to be introduced if there is fairly clear evidence (either direct or circumstantial) that it has been transported by human activities outside the region where it appears to have evolved and become distinct. The kind of habitat in which it grows as an introduced species is a part of that circumstantial evidence, but a native species may grow in a man-made environment without being considered as introduced (cf. Praeger 1911). Domesticated species are those in which man has consciously manipulated their genetic nature (usually by selective breeding) to such an extent that they now differ from their wild relatives. Such domesticated species can scarcely be said to have a native range in the same sense as non-domesticated species because they owe their origin as species to man himself. They are not considered here. Naturalized species are those which have been introduced, and which maintain themselves indefinitely (or at least through several generations) in the wild or in man-made habitats without feeding or cultivation.

HUMAN CULTURES AND THE MODIFICATION OF THE NATURAL RANGES OF PLANTS AND ANIMALS

Ecosystems are altered by human activities both by the direct impact of such physical processes as burning and vegetation clearance, as well as by the introduction of new species from elsewhere. Alteration and introduction in these senses are sometimes regarded as more or less distinct processes, but they are so closely connected that they can scarcely be considered in isolation from one another. Competition ensures this. Any plant or animal is a part of the environment of its neighbour, so that the arrival of a new species on the scene (or even an alteration in the density of the same species) changes both the environment and the terms of competition which may determine ecological success or failure and hence the range of a species.

These processes, both of physical alteration of the environment and of the introduction of new genetic material, are clearly related to the type of culture of the human community involved. But whatever his culture, man operates fundamentally as a mediator of the terms of competition between species, whether they be native or introduced, domesticated or wild, cultivated or naturalized.

Each human culture has its own characteristic assemblage of plants and animals whose ranges are extended as a result of the type and intensity of human activities characteristic of that culture. At each cultural stage, man also tends selectively to destroy some wild organisms, because they are of use to him, or because they get in the way of his activities, or merely as a side effect of some of those activities. For these he thereby diminishes their natural range.

The species whose ranges are extended or diminished as a result of human activities will vary from place to place, and will be determined obviously by the available biota. It is an important biogeographic feature of human cultural change that in the sequence hunter-gatherer → nomadic pastoralism → settled agriculture → urban-industrialism, in which man has increasing effects on the environment due to the impact of his activities, he also increases his overall mobility. The biota available for distribution is thus greatly enlarged. Rapid intercontinental transport now makes possible the equally rapid transport of a very large number of species from one part of the world to another. Although a sparse population of hunter-gatherers may be locally more mobile than the far more numerous modern city dwellers, the possibilities of rapid long-range movement of species by members of modern urbanised communities are immeasurably greater than they have ever been before.

Hunter - gatherer cultures.

At this stage, man may live as a more-or-less generalized nomadic herbivore on a wide range of plant foods; or as a generalized and nomadic, or specialized, selective and more sedentary carnivore. Or his culture may include a mixture of these characteristics, often arranged on a seasonal basis, according to the availability of food. The range of human activities which have impacts upon the environment and its biota at this cultural stage are very great even within Australia (McCarthy 1957; Lawrence 1971). But whatever the details of the hunter-gatherer economy, it seems often to have reduced to the point of extinction the ranges of many of the larger animal species on which humans depended for food, both in Australia (Jones 1975) and elsewhere (Martin 1967), although the relative importance of fluctuating climates (Main, this volume) and human predation in causing these Pleistocene extinctions is debated.

Hunter-gatherers have generally extended the range of some scavenging animals such as the dog and the pig and the rat, and probably also plants which inhabit his middens and other sites with enriched soils near his campsites. Animals and plants which have thus moved in on him and exploited his scraps and campsite soils may be amongst his earliest domesticates (Zeuner 1963). Plants possibly falling into this category in various hunter-gatherer cultures are:

BRITAIN
(Salisbury 1961)

Aphanes arvensis
Atriplex patula
Chamaenerion angustifolium
Centaurea cyanus
Cerastium holosteoides
Cirsium arvense
Cirsium vulgare
Euphorbia helioscopia
Galium aparine
Linaria vulgaris
Plantago lanceoleta
Plantago major
Polygonum lapathifolium
Polygonum persicaria
Ranunculus repens
Rumex acetosella
Rumex crispus
Rumex obtusifolius
Senecio sylvaticus
Silene vulgaris
Solanum nigrum
Sonchus arvensis
Stellaria media
Taraxacum officinale
Tripleurospermum maritimum
Tussilago farfara
Viola tricolor agg.

NORTHERN AUSTRALIA
(Jones 1975)

Cycas media
Dioscorea transversa
Eugenia suborbicularis
Oryza fatua

CENTRAL AUSTRALIA
(Cleland 1966)

Amaranthus mitchellii
Cyperus victoriensis
Dactyloctenium radulans
Ipomaea calobra
Panicum decompositum
Pimelia microcephala
Polygonum glabrum
Polygonum plebejum
Portulaca oleracea
Rumex crystallinus
Sida corrugata
Solanum ellipticum
Tragus australianus
Tribulus occidentalis

In Northern Australia the seeds from gathered fruits were often discarded intentionally on to midden heaps, an almost 'agricultural' process (Jones 1975).

The possible effects of pre-agricultural man in profoundly affecting the vegetation cover, and therefore the ranges, of many plants and animals, by the use of fire has generally been discounted when compared to man's later use of fire to clear forest and shrublands for agriculture. But since charcoal is frequently associated with the encampments and overall environment of people who did not practice agriculture, such hunter-gatherers must have favoured the spread of plants which could tolerate wildfires. The results of such firing should be particularly obvious in subfossil records of vegetation which is not naturally fire-prone.

Smith (1970) argued that the abundance of hazel *(Corylus avellana)* pollen in the early post-glacial vegetation of north-west Europe may be related to the effects of firing by Mesolithic peoples. In the far more fire-prone climates of much of Australia, any such human effects in increasing the frequencies of fire may have been very important in determining the extent of different types of vegetation and therefore the ranges of many species. The clearest evidence for any such man-induced changes would be found in changes in the position of the boundaries between fire-tolerant and fire-sensitive vegetation, as between eucalypt and rain forests (e.g. Gilbert 1959). But even with such fluctuating boundaries explanations based on climatic, rather than purely man-induced, effects are difficult to separate from one another.

Although it is hard to judge how far the burning of vegetation by pre-agricultural man was important in altering the local ranges of plant species, it is striking that the invasion of recently fired vegetation in Australia by species introduced from outside the continent appears to be far more important in the dry sclerophyll woodlands of the winter rainfall regions than in the heaths, mallees and arid vegetation. But again, it is difficult to compare the degree to which these communities are subjected to different carriage of plant propagules and to direct human effects in altering the environment, apart from increasing the fire-frequency itself (Aschmann 1973).

Nomadic-Pastoralist cultures.

Once a human community concentrates its activities on herding animals, a whole new range of selective processes is brought into play which begins radically to alter the competitive balance between different species in the region grazed by its herds. The concentration of grazing pressure on vegetation, together with localized effects at herding and slaughtering sites and at watering points, accentuates and extends the disturbance-plus-eutrophication effect on soils. "Piospheres" are thus set up (Lange 1969) in which there is rigorous selection for plant species which are unpalatable to stock, and able to tolerate trampling and dunging. Combined with this, the mass carriage of plant propagules by the herded stock, together with the repetitive creation of similar environments by the activities of the stock themselves, ensures the spread of any species adapted to these conditions, often from one piosphere to another. The nomadic herding of stock thus imposes on a small scale, a set of conditions favourable for the dispersal of some plants, and for the parasites or other close biological associates of the stock themselves. This situation is analagous to that later created by man on a vastly bigger scale by the establishment of cities.

The management of grazing animals in different parts of the world must have led to the selection from native floras, and to the spread outside their native ranges, of species adapted to animal dispersal. Burr-dispersed plants are perhaps the most obvious candidates for such early spread. It is perhaps no accident that many of the most troublesome burrs in wool belong to species which are native to Mediterranean lands and South-West Asia where nomadic pastoralism has been long established. Salisbury (1961), using the data of Dony (1953), listed the flowering plants imported with wool (mostly from Australia) found as introduced plants in Bedfordshire, England. The categories I to V are arranged in order of decreasing frequency of the species in England: only those in category I were considered to be "definitely naturalized". Most of the species include the Mediterranean - South-West Asia region in their supposedly native ranges, although the original range of some species is in doubt.

I -	*Acaena anserinifolia*	S.E. Australia, N.Z.
	Melilotus indica	Mediterranean, S.W. Europe
	Sisymbrium orientiale	Mediterranean
	Vulpia bromoides	Mediterranean

II -	*Medicago minima*	Europe
	Medicago arabica	Mediterranean
	Erodium cicutarium	S. Europe
	Erodium moschatum	S. & W. Europe
	Xanthium spinosum	S. America
	Datura stramonium	Eurasia &? America
	Amaranthus chlorostachys	Tropical America
	Medicago laciniata	N. Africa, S.W. Asia
	Erodium botrys	Mediterranean
	Polypogon monspeliensis	Mediterranean
III -	*Erodium cygnorum*	Australia
	Trifolium subterraneum	Mediterranean, W. Europe
	Phalaris minor	Mediterranean, S.W. Asia
	Tagetes minuta	S. America
	Hordeum hystrix	Mediterranean, S.W. Asia
	Hordeum leporinum	Mediterranean, S.W. Asia, India
	Trifolium angustifolium	Mediterranean
	Carduus tenuiflorus	W. Europe & Mediterranean
	Physalis ixacarpa	S. America
	Bromus uniloides	S. America
IV -	*Medicago praecox*	Mediterranean
	Agrostis lacnantha	
	Hordeum glaucum	Mediterranean, S.W. Asia
	Amaranthus thunbergii	
	Chenopodium probstii	Europe
	Echinochloa crus-galli	Pantropical
	Agrostis avenacea	Australia
	Carthamnus lanatus	Mediterranean
	Sisymbrium irio	Mediterranean
	Erigeron bonariensis	S. America
V -	*Setaria verticillata*	Subtropical Asia & Africa
	Calotis cuneifolia	Australia
	Bidens pilosa	S. America
	Bidens bipinnata	S. America
	Lepidium hyssopifolia	Australia
	Senecio inaequidens	S. Africa
	Ammi majus	S. Europe
	Rumex brownii	Australia
	Brassica tournefortii	Mediterranean
	Chenopodium giganteum	N. India

Many burred plants and other pasture weeds which were thus spread to distant parts of the world (where nomadic pastoralism was never established), with the later expansion of commercial agriculture, may therefore have started their itinerant careers before the advent of tillage. Thistles may be another fairly well-defined group of plants whose spread is historically related to the spread of pastoralism.

Sauer (1952) challenges the view that nomadic pastoralism necessarily preceded agriculture, but since the method of dispersal, as well as the range of habitats for exploitation were greatly extended when settled grazing became established, it would be difficult to distinguish as a group those species which were dispersed by nomadic pastoralists from those which originated at the times of settled pastoralism and agriculture.

Agriculture.

The advents of shifting and then of settled agriculture greatly increased the opportunities to expand their ranges for plants and animals that learnt to accommodate to these activities. In addition to the communal middens, the burned areas, and those affected by concentration of livestock already mentioned, which were all maintained or extended by agricultural peoples, three radically new types of habitat were created: tilled land, cities and highways.

Tillage at once presented a new type of environment for volunteer plants and animals to exploit, but one closely similar in some respects to early seral stages in some natural situations, especially where sudden events or where high energy inputs from natural processes are involved (e.g. landslides, exposed coasts). Opportunities for colonizing species and pests of cultivated crops became as diverse as the crops themselves and their methods of cultivation. The movement of tools domesticated animals (used as a source of power or fertilizer) and of men themselves, provided a variety of dispersal methods for many plants and crop pathogens.

The built environment of cities again presents a variety of new habitats for volunteer plants and animals to exploit. Cities usually include middens (rubbish tips) and cultivated areas, along with more or less disciplined habitats such as mown, paved and trampled surfaces, which acquire characteristic assemblages of plants and animals. However, the very high population densities in ancient cities (Mumford 1956) may have left little space on the ground within their walls for the variety of volunteer plants characteristic of modern cities. The process of building itself ensures that much of the land in any city is more or less continuously in process of disturbance, thus presenting permanent opportunities here and there for ruderal plants.

Highways are a third major kind of environment which first became important with the development of the settled life of agriculture and cities. As a strip of land with more or less untended verges bordering on the man-made agricultural and urban environment, the highway must have been from its earliest times an important corridor for the movement of species carried by people and animals, and the temporary sojourn of many species in passage along its length. Highways became of greater importance with the diversification of transport methods (rail, automobile) and with the rise of urban-industrial settlements.

The Neolithic revolution thus set up, with tillage, cities and highways, a system of replicated habitats presenting similar opportunities for adapted plants and animals, together with opportunities for the movement and dispersal of propagules between them.

Mobile, urban-industrial societies.

Trade between Europe and the Orient by way of the Mediterranean, and the far-ranging movements of seagoing peoples such as the Polynesians, Vikings and Phoenecians, must have provided for some movement of plants and animals before the expansion of the maritime Europeans across and out of the confines of the Atlantic beginning some 500 years ago. But this expansion initiated a new phase in the mixing and rearrangement of the world's floras and faunas that is still in progress. The long distance transport of both people and their goods by sea after this time, including the carriage of domesticated plants and animals, for almost the first time provided a means of transport of large numbers of species between climatically similar parts of the world in different continents. The temperate regions of the northern and southern hemispheres, and the tropics of Africa, Asia and the Americas began to exchange both domesticated and volunteer species in large numbers. This process of the breakdown of former biotic isolation continues today, as more and more people and goods are moved about the world more and more rapidly, both in the course of international trade and for pleasure. In the face of mass tourism quarantine controls are probably quite unable to prevent the arrival of the plant and animal propagules, except in the case of a few pests, and pathogens of man's domesticates (e.g. Laird 1961).

The industrial revolution has itself had a profound effect upon the movement of species about the world over the last 250 years, because of the enormous demands it has generated for food and raw materials from parts of the world formerly sparsely inhabited and comparatively lightly used by indigenous peoples with traditional economies. The Europeans established colonies in the temperate latitudes of the Americas, Southern Africa, Australia and New Zealand based on crop plants and livestock most of which were domesticated in the Mediterranean or South-West Asia

(Sauer 1952). Their agricultural and pastoral activities thus reconstructed in different parts of the world habitats for volunteer species (including many which were seen as weeds and pests) similar to those in lands with thousands of years of such land use.

More recently, the sprawl of large conurbations, with their mixture of industrial, residential and commercial activities, with the simultaneous adoption of mechanised high-energy agriculture using similar tools and similar techniques in widely scattered parts of the world, has again replicated environments for volunteer species adapted to these activities. Specifically, the invention of new, biologically active, molecules used as pesticides and weedicides in agricultural and pastoral systems, has led to a closer definition and replication of both agricultural, industrial and urban-residential environments. Species which have become adapted to such environments in one part of the world are readily dispersed by man to similar environments in other continents.

Plant species which are native to certain types of natural habitat, such as fresh water lakes and coastal dunes, have a wider geographical range than those of more stable terrestrial climax communities (Ridley 1930; Good 1953). Such lake and coastal sites share with man-made habitats a scattered or discontinuous distribution linked by ready and reliable means of dispersal between them (birds, sea-flotation). The species which comprise the man-adapted flora of cities, agriculture and highways are achieving similar cosmopolitan distributions.

This historical view of the modification of the ranges of organisms by human activities suggests that his construction of similar habitats in distant parts of the world, together with the provision of the means of dispersal between them, is resulting in the erosion of the biotic individuality of all the inhabited parts of the earth. The timing of the arrival of volunteer organisms, as well as domesticates, may thus be useful to indicate the spread of particulr kinds of economic activity, where no historical records exist. They may also be used as stratigraphic markers, where there are historical records, as with pine pollen in swamps and bogs of southeast Australia (Singh, personal communication).

BIOLOGICAL STRATEGIES IN THE BIOTA SPREAD BY MAN

The range of human activities, and the opportunities thus created for other organisms to exploit, are so various that their adaptive responses can only be typified at the broadest level of generalization. The ecological and reproductive strategies of almost all the plants and animals that exploit both the mobility and the range of habitats created by man, are above all adapted to disturbance and continuing change.

Amongst the species whose ranges are extended by man, the only exceptions to this occupancy of changing environments are those heterotrophic organisms which are biologically adapted to man himself, or to his domesticates, as parasites or pathogens, and those few species living within the almost totally controlled environments of his own habitations. If their remains survive, such close associates of man may provide valuable information about his conditions of life to later investigators. Fossil remains of alligators in the sewers of New York, or tropical rainforest plants in public buildings in London, would tell future archaeologists something of energy use in the built environments of big cities. Similarly, surviving evidence of pathogens of man or his domesticates may be of interest to anthropologists and historians of disease, but can tell us little of overall environments. These species which are thus obligatorily associated with man will not be further considered here.

The great majority of plants and animals whose ranges are extended by man are opportunist species, which have been selected for their ability to disperse readily, to move into unoccupied habitats, and which invest a large proportion of their energy in reproduction. They are r-selected species (MacArthur and Wilson 1967) which are "continually colonizing habitats of a temporary nature ... and they are exposed to selection at all population densities. Their strategy is basically opportunistic, 'boom and bust'. Migration will be a major component of their population process, and may even occur every generation" (Southwood 1976). The principal characteristics of some of this volunteer man-adapted biota (other than the pathogens and parasites of man himself) as they are displayed in the dispersal, ecesis, and reproduction of some of the species involved, are discussed below.

Dispersal.

Ridley (1930) discussed in detail "the methods by which alien plants are commonly dispersed by man". He thought that the most important of such methods were:

(a) as impurities with cereals, vegetables and birdseed

(b) as attachments to fleeces, hides etc of domesticated animals

(c) in animal fodder

(d) in ships' ballast and exported soil

(e) in packing materials

(f) as escapes from cultivation

Of these methods, perhaps only (b) would select for seeds (or other propagules) having macroscopic properties readily identifiable as aids to dispersal.

Propagules dispersed as impurities amongst the seeds of domesticated plants may tend to be similar to those of the crop plants amongst which they occur. This tendency to propagule mimicry is probably widespread amongst crop weeds, and may also include other phases in the life cycle of the weed whose range is extended along with the crop. The domestication of cereal weeds such as oats and rye probably arose through such mimicry of the prime domesticates, wheat and barley, as the crops and their weeds were together moved from warm temperate Mediterranean to the wetter and cooler climates of northwest Europe (Sauer 1952). In this case, there was selection in the weeds mimicking the initial crop species for timing of maturation, grain retention at harvesting, and probably other features; but seed cleaning methods then were primitive or non-existent. Seed cleaning is becoming more sophisticated, relying on specific gravity, magnetic and electrostatic separation as well as on the size, shape and density of seeds usually employed in sieving systems, to separate weed from crop seeds. As such sophisticated methods are more widely employed, along with mechanized sowing and harvesting and the use of new weedicides, selective pressures on weeds to adapt to agricultural tools and technology and therefore to mimic their crops more closely will increase. Changes in seed cleaning methods have greatly influenced the abundance of some crop weeds: the wheatfield weed *Agrostemma githago* has become rare in Britain since the adoption of standard mechanized seed cleaning (Salisbury 1961), but it is said to be naturalized as an occasional weed of wheat in SE Australia (Willis 1972). If seed cleaning technology had developed to its present stage when seed wheat was being brought from Europe to Australia, it is likely that *Agrostemma* would be unknown as a weed here. Technological change as well as human mobility controls the degree to which species are spread about by man. Studies of weed seed size and uniformity might thus be used as a measure of the effort employed in handling and cleaning crop seeds in the past.

Burrs, and other plant propagules which are adapted to external animal adhesion, constitute a geographically widespread group in the exotic element of all terrestrial floras, and have excited interest because of their economic importance as vegetable fault in wool and hides (Snelson 1977). Man himself also transports burrs on his clothing and footwear (van der Pijl 1969). In southern Australia, a wide range of both native and exotic species are commonly found as "sockplants", including some which may have no clear morphological adaptation to external adhesion of their seeds or fruits:

Acaena anserinifolia
Acaena ovina
Carrichtera annua
Cryptostemma calendula
Cynodon australe
Daucus glochidiatus
Medicago polymorpha
Plantago lanceolata
Rumex brownii
Bromus unioloides
Danthonia spp.
Hordeum leporinum
Phalaris aquatica
Vulpia bromoides
Stipa spp.

A smaller, but troublesome, group of plants with spiny fruits which are often loosely referred to as "bindy-eyes" are commonly carried about on human footwear and motor tyres in Australia, and are no doubt dispersed thus. They also include both native and introduced species:

Bassia quinquecuspis
Bassia birchii
Calotis hispidula
Emex australis
Soliva pterosperma
Tribulus terrestris

Plants dispersed by the other four methods listed by Ridley mostly have no obvious morphological adaptations for dispersal of their seeds or fruits. The lack of obvious adaptations for dispersal is typical also of the native floras of remote islands whose native plants must have reached them by long distance dispersal (Wace and Dickson 1965; Carlquist 1965). Success in reaching remote islands by natural means, as well as extension of range by dispersing with man, is a matter of total reproductive and biological strategy. Harper et al. (1970) have stressed that although variations in seed (i.e. propagule) size and shape have been interpreted almost entirely as adaptations for dispersal, many other selective forces such as predation, mode of nutrition, status in seral succession and even shape and density of packing within the ovary, may exert a marked influence on seed morphology.

Since Ridley's review of plant dispersal, human mobility has greatly increased. Other modes of dispersal, in addition to those that he considers, have become important. Mechanized transport by fast land and air vehicles is now widespread. In developed industrial societies the numbers of road vehicles are very large. In the United States there are now half as many vehicles as human beings. Throughout the world, there were in 1975 some 324 million motor vehicles in use, which probably travelled as much as 518.4^{10}km each year (International Road Federation 1976). The potential of this gigantic activity for moving organisms around has never been assessed. Some attempt to measure the potential of motor cars for moving plants around, by sampling the viable botanical content of carwash sludge has recently been carried out in Canberra (Wace 1978, in press). The propagule characteristics of plants thus dispersed by cars, sampled in carwash sludge and germinated within six months are of some interest. Few propagules show any clear morphological adaptation to dispersal by any particular means. Those producing the most numerous seedlings (table 13.1) belong to taxa having rather small seeds or fruits (Ridley 1930). The only clearly identifiable groups in respect of propagule type are the berries (e.g. *Pyracantha*, *Cotoneaster*) which are presumably picked up easily with their squashed fruits on roads, and the grasses with intricate wind-blown inflorescences (*Chloris truncata*, *Eragrostis trachycarpa*, *Panicum capillare*) which collect in car radiator grilles.

The motor car is a new human tool. There is not yet an identifiable car-adapted flora, as there is a flora variously adapted to animal dispersal. The motor car simply picks up and spreads around the propagules of all sorts of plants indiscriminately: trees, shrubs and herbs, annuals and perennials, natives and aliens, domesticated, cultivated and naturalized species. Although it has a clear relationship to the local flora, the assemblage of species germinated from the contrived sedimentary situation which is set up in the carwash sludge tanks seems to be an

Table 13.1 Numbers of seedlings of species germinated from Canberra carwash sludge between June 1973 and August 1975. The small numbers on the left are the number of seedlings in control trays. Species native to Australia are underlined. (From Wace 1978, in press.)

	1922	Unidentified grasses
	1671	*Polygonium aviculare*
	1473	*Betula alba/pendula/verrucosa*
1+	883	*Eleusine tristachya*
	870	*Poa annua*
2+	693	*Phalaris aquatica*
	565	*Trifolium subterraneum*
	544	*Trifolium glomeratum*
1+	512	<u>*Cynodon dactylon*</u>
	485	<u>*Eragrostis trachycarpa*</u>
	477	*Lolium perenne*
	462	*Paspalum dilatatum*
50+	452	*Oxalis corniculata*
3+	409	*Echinochloa crus-galli*
	288	*Digitaria sanguinalis*
	252	*Agrostis tenuis*
1+	244	*Trifolium* spp.
	236	*Digitaria ciliaris/parviflora*
	236	*Holcus lanatus*
	229	Unidentified dicots
	228	<u>*Chloris truncata*</u>
	200	*Poa pratensis*
	197	*Panicum laevifolium*
	195	<u>*Lythrum Hyssopifolium*</u>
	172	*Buddleia davidii*
	171	*Plantago lanceolata*
	169	*Rumex acetosella*
3+	160	<u>*Chenopodium pumilio*</u>
12+	158	*Conyza bonariensis/floribunda*
	157	*Eragrostis cilianensis*
	151	*Trifolium dubium*
	150	*Platanus orientalis*
	130	<u>*Alternanthera denticulata*</u>
	125	*Vulpia myuros*
	120	<u>*Casuarina cunninghamiana*</u>
	115	*Lolium* sp. (annual)
	105	*Eragrostis benthamii*
	103	<u>*Eragrostis parviflora*</u>
	95	*Digitaria ischaemum*
	92	*Eleusine indica*
	90	*Trifolium repens*
	85	*Chenopodium album*
	84	*Eragrostis* sp.
1+	80	Unidentified sedges
	79	*Cyperus eragrostis*
	69	*Trifolium arvense*
	68	*Setaria geniculata*
	56	*Rumex crispus/conglomeratus*
	52	*Amaranthus hybridus/retroflexus*
	50	*Trifolium campestre*
	49	*Petrohagia nanteuilli*
	48	<u>*Solanum nigrum*</u>
	47	<u>*Juncus bufonius*</u>
	46	*Modiola caroliniana*
	42	*Panicum effusum*
	41	*Vulpia bromoides*
	40	<u>*Cyperus sanguinolentus*</u>
	37	*Anagallis arvensis*
	37	*Digitaria ischaemum/sanguinale*
1+	37	*Sonchus oleraceus*
1+	35	*Sonchus asper*
	33	*Sophora japonica*
	33	*Ulmus parviflora*
	31	<u>*Danthonia* spp.</u>
	30	*Pyracantha yunanensis*
	29	*Fraxinus velutina*
	29	*Vulpia megalura*
	26	*Hirshfeldia incana*
	25	*Medicago lupulina*
	24	*Medicago polymorpha*
	24	*Cirsium vulgare*
	23	*Hydrocotyle pilulifera*
	23	*Ligustrum ovalifolium*
	24	*Acer negundo*
	22	*Cotoneaster pannosa/franchettii*
	22	<u>*Cotula australis*</u>
	21	*Bromus mollis/molliformis*
	21	*Taraxacum officinale*
	19	<u>*Agrostis avenacea*</u>
	19	*Hypochoeris radicata*
4+	19	<u>*Scirpus subtilissimus*</u>
	18	*Axonopus affinis*
	18	*Conyza canadensis*
	17	*Ceratochloa uniloides*
	17	*Lolium multiflorum*
	17	*Lycopersicum esculentum*
	16	<u>*Bothriochloa macra*</u>
	16	<u>*Cyperus rotundus*</u>
	16	<u>*Dichondra repens*</u>
	16	*Malva parviflora*
	16	*Vicia disperma*
	16	*Vulpia* or *Festuca* sp.
	15	*Aster subulatus*
	15	*Dactylis glomerata*
	14	*Briza minor*
	14	*Cynosurus echinatus/cristatus*
	14	*Pyracantha angustifolia*
	14	<u>*Stipa* sp.</u>
	13	<u>*Centipeda cunninghamii*</u>
	13	*Cotoneaster* spp. ined
	13	<u>*Dianella* sp.</u>
	12	*Gnaphalium involucratum*
	12	*Lolium rigidum*
	12	*Panicum capillare*
	12	*Paspalum distichum*
3+	12	*Spergularia rubra*
	12	*Verbena bonariensis*
	11	*Cupressus arizonica*
	11	*Digitaria violascens*
	11	*Euphorbia peplus*
	11	<u>*Wahlenbergia* spp.</u>
1+	10	<u>*Acacia mearnsii*</u>
	10	*Carduus tenuiflora/pycnocephala*
	10	*Cyperus brevifolius*
	10	<u>*Eucalyptus melliodora*</u>
	10	*Plantago coronopus*

ecologically meaningless jumble. Any future paleontologist who tried to interpret the plant species found in this modern midden in environmental terms would have the utmost difficulty in fitting them to any conceivable habitat. Although this is an extreme case, it may serve to illustrate the difficulty in reconstructing environments from biological remains much influenced by mobile human communities.

Ecesis.

It is typical of the r-selected organisms which predominate in the biota spread by man, that their strategy is one of concentration of effort on reproduction. Flowering and fruiting of many such species is probably precocious compared to that of established native plants. In the carwash flora of Canberra, described above, numerous seedlings of *Chenopodium pumilio* and *Trifolium glomeratum* flowered and fruited when the plants were barely a centimetre high. Seed of both of these species is extremely abundant in the cleanings from a wide range of crops grown on the Tablelands near Canberra. Much careful work at seed-cleaning establishments, and identification of seed loads in soils, would be necessary in order to establish that such plants dispersed by man rely on a strategy of near-saturation of the environment with their seeds, so that some seed is already present in almost any environment suddenly rendered suitable for germination and establishment of the species involved, by human or natural causes. But it seems likely that this may be the case for some native and exotic carborne species near Canberra.

Preliminary work on the flora of roadsides, and other car-frequented places, shows that many species with abundant seedlings germinated from carwash sludge rarely, if ever, establish themselves in places frequented by cars. It is no new discovery that dispersal is an extremely wasteful process, but the reconstruction of past environments from the identification of organic remains is often based on the relative numbers of different propagules found in sedimentary deposits. In the case of r-selected plants dispersed by human activities, such deposits may contain a high proportion of propagules of species which were never established at the site of deposition. Their presence in the deposit may therefore be largely irrelevant to the reconstruction of the physical or climatic nature of the environment in which the deposits were laid down. This is not a new problem in the interpretation of fossil and subfossil remains: redeposited pollen grains and seeds and other macroscopic remains washed into lowland situations from nearby highlands are commonplace in paleoecological interpretations (e.g. Cain 1944, chapter 7). But it may be particularly important in the paleoecological interpretation of deposits much influenced by human activities in mobile societies. These comments on the effects of man influencing the nature of propagule deposits need to be read together with the more general problems on the dispersal and sedimentation of seeds discussed by Watts (this volume).

Reproduction.

Baker (1965), in defining the characteristics of his "ideal weed", listed several features that contribute to the reproductive success of many plant species dispersed by man. Amongst herbs so dispersed, which most successfully exploit habitats disturbed by human activity, there appear to be two major reproductive strategies (which will be all too familiar to most gardeners): fast growing annuals or ephemerals with abundant seed production, and deeper-rooted perennials with efficient means of vegetative spread.

The abundantly-seeding annual herb is clearly the most obvious manifestation in the plant world of the r-strategist of the animal population biologists. Large seed loads of annual weeds in regularly tilled soils (Roberts 1976) are an indication of how successful (and how wasteful) such a strategy may be. But the options open to plants for vegetative spread and genome multiplication, have perhaps made plants less amenable than animals to such mathematical concepts r- and K-selection strategies. In most animals, their generation time, and the limits of an individual are easily measured and quantified: this is often not the case in plants, whose reproductive strategies may therefore be less easy to define and compare with those of other organisms.

Herbs with a capacity for vegetative reproduction are known to include a higher proportion of polyploids than either the herbaceous annuals or woody plants (Stebbins 1950, 1966). The evolutionary significance of alloploidy probably lies in its ability rapidly to fix desirable hybrid genomes in combination, even at the expense of some future genetic flexibility: it is thus a system well suited to the rapid replication of similar environments by human activities. *"Polyploidy is a very striking phenomenon in higher plants, and studies of polyploids can provide much information about the past history of plant groups ... because it is one of the very few evolutionary trends of which the direction can be determined without fossil evidence"* (Stebbins 1966). Since pollen grain size is usually related to ploidy level, it might be revealing for palynologists working with sediments containing pollen of plants affected by human activities to look more closely at pollen size of taxa now displaying polyploid series. Human activities have probably increased the relative frequency of both domesticated and non-domesticated polyploid species.

The ability of perennial weedy plants to reproduce and spread vegetatively has surely been of the first importance in promoting their spread by man because such species easily become adapted to dispersal on his tools. The success of the sterile pentaploid form of the South African *Oxalis pes-caprae* in "mediterranean" climates and *Cyperus rotundus* in the tropics and warm temperate lands is obviously related to their bulbil production associated with agriculture and other forms of soil disturbance by man.

THE ECOLOGICAL STATUS OF INTRODUCED ORGANISMS IN NATIVE COMMUNITIES

Volunteer plants spread by man are mostly species of early seral stages (e.g. Salisbury 1961). Exotic plants are characteristic of disturbed environments, and rarely invade undisturbed native vegetation (e.g. Kornas and Kornas 1967). Although obvious in its grosser forms of physical clearance of vegetation, cultivation, burning etc., "disturbance" is difficult to define in biological terms, or to measure. Some degree of disturbance is itself a feature in the regeneration of supposedly stable climax vegetation in seral succession. Conversely, native vegetation differs markedly in its resistance to invasion by exotic plants. Some communities which suffer continual gross disturbance through natural causes such as those on coastal dunes and river gravels may come to be dominated by exotic species. Lupins, boneseed, harestail grass, boxthorn, iceplant and many other species on the temperate sandy coasts in Australia and New Zealand are some examples. But some open plant communities which suffer natural disturbance have few or no species suspected as human introductions: desert dune systems in Central Australia and newly exposed coastal muds adjoining established mangrove communities in south-east Australia and in northern New Zealand are examples. Why such communities remain almost inviolate may be explained as being due to lack of human contact with environmentally suitable source areas, or to the specialized nature of plant adaptation required to live in difficult environments, or to a combination of these historical and ecological factors.

In Australia, it seems that resistance of much native vegetation to invasion by exotic plants is related to soil fertility levels as well as to disturbance factors. Heaths dominated by native Proteaceae, Myrtaceae, Restionaceae and Leguminosae in both southeastern and south-western Australia grow on soils with very low phosphate and nitrate levels, and are extremely resistant to exotic invasion, even when the vegetation is burnt or mechanically cleared. Specht (1963) has shown that the mere spreading of fertilisers in such vegetation without overt disturbance can lead to invasion by introduced annual grasses and legumes. Once clovers and medics are established, the resulting rise in soil nitrate levels may permit invasion by other exotic herbs. Moore (1967) has shown how a sequence of clearing of eucalypts, grazing by imported stock and superphosphate application on the Southern Tablelands of New South Wales, has led to the replacement of tall perennial native summer-growing grasses by different native summer-growing species from more arid areas, followed by an influx of annual winter-growing exotics native to Mediterranean lands. The management of pastures in these areas has thus led to the establishment of a winter-growing element under a rainfall regime of year-round rainfall, although a summer-growing element of mixed native and exotic annuals and perennials persists, alternating in activity with the winter-growing exotics. Many of the exotic species have become naturalized in places unmanaged as pastures.

The nitrogen fixing activity of introduced volunteer legumes has thus been associated with the success of a whole new geographical and seasonal element in the flora of much of Southern Australia.

Although most volunteer plants spread by man are colonizing species, typical of early or arrested seres, naturalized species often display a wider ecological amplitude when outside their native range than within it. Gorse, broom and bramble in New Zealand not only appear to be more vigorous (and therefore a greater nuisance to man) than in their native western Europe, but also seem to occupy a greater altitudinal range and diversity of habitats. Such casual observations are easy to make, but hard to quantify in the absence of a fairly intimate knowledge of the species concerned in different parts of the world. The explanation both for greater vigor and greater ecological amplitude outside their native ranges is generally attributed to lessened competition or predation (or both) in their new habitats. Such explanations are usually accepted for the spectacular spread of feral animals such as the rabbit, and control is attempted by the reimposition of pathogens in biological control.

Many naturalized woody plants appear to be specialized exploiters of marginal environments where the native flora into which they move apparently lacks suitably adapted species. Remote islands with naturally depauperate native floras therefore provide some of the best examples of successful invasion by woody exotics filling such untenanted niches. Mangroves and the phreatophyte *Prosopsis* (mesquite) on the coasts of the Hawaiian Islands, the large rosette and clump-forming *Furcrea* from the Carribean and *Phormium* from New Zealand on Mauritius and St Helena respectively are some examples. But a major difficulty in comparing the ecological performances of such species in their native and introduced habitats is the now widespread human disturbance of their new habitats. There are few known cases in which exotic species have been observed to move into native vegetation which has been quite undisturbed by any human activity (Wace and Dickson 1965).

In the presence of such disturbance, invasion or even total replacement of native vegetaton by exotics may be rapid and spectacular. *Tamarix decapetala* from South-West Asia replaces native riparian shrubs in south-western U.S.A. when river-flow regimes are altered (Harris 1966); boxthorn *(Lycium ferocissimum)* from southern Africa forms large thickets on exposed coasts and offshore islands in southern Australia and the North Island of New Zealand; *Hakea sericea* from south-east Australia replaces fired native proteaceous heath on Table Mountain sandstones in South Africa (Shaughnessy, in preparation). These are all examples in which the exotic species have come locally to dominate the native vegetation. But with the enormous variety of invading exotic species introduced into variously altered and disrupted native communities, it is clearly impossible to make any general statements about their ecological status in those communities.

REFERENCES

Allan, H.H. 1936 Indigene vs. alien in the New Zealand plant world. Ecology, 17, 187-193.

Allan, H.H. 1937 The origin and distribution of the naturalized plants of New Zealand. Proc. Linn. Soc. London, 150, 25-45.

Allan, H.H. 1961 Flora of New Zealand, 1., Government Printer, Wellington.

Anderson, R.H. 1939 The naturalized flora of New South Wales (excluding Graminae). Contrib. N.S.W. Nat. Herb., 1, 16-33.

Aschmann, H. 1973 Man's impact on the several regions with Mediterranean climates. Section 7, part 1, pp 363-371 in Mediterranean Type Ecosystems: Origins and Structure (eds F. di Castri and H.A. Mooney), Chapman and Hall, London.

Baker, H.G. 1965 Characteristics and modes of origin of weeds. pp 147-172 in The Genetics of Colonizing Species (eds H.G. Baker and G.L. Stebbins), Academic Press, New York.

Baker, H.G. 1972 Migrations of weeds. pp 327-347 in Taxonomy, Phytogeography and Evolution (ed D.H. Valentine), Academic Press, London.

Burbidge, N.T. 1952 The significance of the mallee habit in *Eucalyptus*. Proc. Roy. Soc. Qld., 62, 73-78.

Burbidge, N.T. 1963 Dictionary of Australian plant genera. Angus and Robertson, Sydney.

Burbidge, N.T. and Gray, M. 1970 Flora of the Australian Capital Territory. A.N.U. Press, Canberra.

Cain, S.A. 1944 Foundations of plant geography. Harper, New York.

Carlquist, S. 1965 Island Life: a Natural History of the Islands of the World. A.M.N.H., New York.

Clapham, A.R., Tutin, T.G. and Warburg, E.G. 1962 Flora of the British Isles (2nd edition). University Press, Cambridge.

Cleland, J.B. 1966 The ecology of the Aboriginal in South and Central Australia. pp 111-158 in Aboriginal man in South and Central Australia (ed B.C. Cotten), Government Printer, Adelaide.

Dony, J.G. 1953 Wool aliens in Bedfordshire. pp 56-73 in The Changing Flora of Britain (ed J.E. Lousley), Botanical Society of the British Isles, Arbroath.

Gibb, J.A. and Flux, J.E.C. 1973 Mammals. pp 334-371 in The Natural History of New Zealand (ed G.R. Williams), Reed, Wellington.

Gilbert, J.M. 1959 Forest succession in the Florentine Valley, Tasmania. Proc. Roy. Soc. Tasmania, 39, 129-151.

Godley, E.J. 1975 Flora and vegetation. pp 177-232 in Biogeography and Ecology in New Zealand (ed G. Kuschel), Monogr. Biol., 27. Junk, The Hague.

Godwin, H. 1960 The history of weeds in Britain. pp 1-10 in The Biology of Weeds (ed. J.L. Harper). Blackwells, Oxford.

Godwin, H. 1975 The history of the British flora. A factual basis for phytogeography (2nd edition). University Press, Cambridge.

Golson, J. 1971 Australian Aboriginal food plants: some ecological and culture-historical implications. pp 196-238 in Aboriginal Man and Environment in Australia (eds D.J. Mulvaney and J. Golson), A.N.U. Press, Canberra.

Good, R. 1953 The geography of the flowering plants. Longmans, London.

Harper, J.L., Lovell, P.H. and Moore, W.G. 1970 The shapes and sizes of seeds. Ann. Rev. Ecol. and Systematics, 1, 327-356.

Harris, D.R. 1966 Recent plant invasions in the arid and semi-arid southwest of the United States. Ann. Assoc. Amer. Geog., 56, 408-422.

International Roads Federation. 1976 World Road Statistics. Washington and Geneva.

Jones, R. 1975 The Neolithic, Palaeolithic and the Hunting Gardeners: Man and Land in the Antipodes. pp 21-34 in Quaternary Studies (eds R.P. Suggate and M.M. Gresswell), Royal Society of New Zealand, Wellington.

Kornas, J. and Kornas, A.M. 1967 The status of introduced plants in the natural vegetation of Poland. I.U.C.N. 10th Technical Meeting, 3, June 1966, Morges.

Laird, M. 1961 Aircraft disinsection in the jet age. Proc. 10th Pacific Science Congress, Abstracts, 232.

Lange, R.T. 1969 The piosphere, sheep track and dung patterns. Jour. Range Management, 22, 396-400.

Lawrence, R. 1971 Habitat and economy: a historical perspective. pp 249-261 in Aboriginal Man and Environment in Australia (eds D.J. Mulvaney and J. Golson), A.N.U. Press, Canberra.

Macarthur, R.H. and Wilson, E.O. 1967 The theory of island biogeography. University Press, Princeton.

Martin, P. 1967 Prehistoric overkill. pp 75-120 in Pleistocene extinctions: the search for a cause (eds P.S. Martin and H.E. Wright), Yale University Press, New Haven.

McCarthy, F.D. 1957 Habitat, economy and equipment of Australian Aborigines. Aust. Jour. Sci., 19, 88-97.

Moore, R.M. 1967 The naturalization of alien plants in Australia. I.U.C.N. 10th Technical Meeting, 3, June 1966, Morges.

Moore, R.M. and Perry, R.A. 1973 Vegetation. pp 57-73 in Australian Grasslands (ed R.M. Moore), A.N.U. Press, Canberra.

Mumford, L. 1956 The natural history of urbanization. pp 382-398 in Man's Role in Changing the Face of the Earth (ed. W.L. Thomas), University Press, Chicago.

Ogden, J. *Calystegia sepium*, and related species in New Zealand. N.Z Jour. Bot. (in press).

Praeger, R. Ll. 1911 Clare Island Survey: Botany. Proc. Roy. Irish Academy, 31.

Ridley, H.B. 1930 The dispersal of plants throughout the world. Reeve, Ashford (Kent, England).

Roberts, H.A. 1976 Weed competition in vegetable crops. Ann. Appl. Biol., 83, 321-324.

Salisbury, E.J. 1932 The East Anglian Flora. Trans. Norfolk Norw. Nat. Soc., 13, 191.

Salisbury, E.J. 1942 The reproductive capacity of plants: studies in quantitative biology. Bell, London.

Salisbury, E.J. 1961 Weeds and Aliens. Collins, London.

Sauer, C.O. 1952 Agricultural origins and dispersals. American Geographical Society, New York.

Shaughnessy, G. History of the introduction and spread of alien plants in the vicinity of Cape Town. Ph.D. Thesis to be submitted at University of Cape Town (in preparation).

Smith, A.G. 1970 The influence of Mesolithic and Neolithic man on British vegetation: A discussion. pp 81-96 in Studies in the Vegetational History of the British Isles (eds D. Walker and R.G. West), University Press, Cambridge.

Snelson, J.T. 1977 The influence of weeds on agricultural commodity, processing and marketing. Proc. Australian Weeds Conference, Melbourne, March 1976.

Southwood, T.R.E. 1976 Bionomic strategies and population parameters. pp 26-48 in Theoretical Ecology (ed R.M. May), Blackwell, Oxford.

Specht, R.L. 1963 Dark Island Heath (Ninety Mile Plain, South Australia): 7; the effect of fertilisers on composition and growth, 1950-1960. Aust. Jour. Bot., 11, 67-94.

Stebbins, G.L. 1950 Variation and evolution in plants. Columbia University Press, New York.

Stebbins, G.L. 1966 Processes of organic evolution. Prentice Hall, New Jersey.

Thomson, G.M. 1922 The naturalization of animals and plants in New Zealand. University Press, Cambridge.

Thompson, H.V. and Worden, A.N. 1956 The Rabbit. Collins, London.

Van der Pijl, L. 1969 Principles of dispersal in higher plants. Springer, Berlin.

Wace, N.M. 1972 "New Settlers", 1 and 2. (Broadcasts in Australian Broadcasting Commission "Insight" Series Nos. 377 and 378, May 1972). A.B.C. Sydney (typescript).

Wace, N.M. 1973 Naturalized plants and native vegetation in Australia. Proc. Second Victorian Weeds Conf., Melbourne, Weed Science Society of Victoria.

Wace, N.M. 1978 The carborne flora in Canberra. Proc. Ecol. Soc. Australia, 10.

Wace, N.M. and Dickson, J.H. 1965 The terrestrial botany of the Tristan da Cunha Islands. Phil. Trans. Roy. Soc. London, B, 249, 273-360.

Williams, G.R. 1973 Birds. pp 304-333 in The Natural History of New Zealand (ed G.R. Williams), Reed, Wellington.

Willis, J.H. 1972 A handbook to plants in Victoria. Vol. II - Dicotyledons. University Press, Melbourne.

Zeuner, F.E. 1963 A history of domesticated animals. Hutchinson, London.

14. THE USE OF UNIFORMITARIANISM AND ANALOGY IN PALAEOECOLOGY, PARTICULARLY POLLEN ANALYSIS

by L. Rymer

SUMMARY

Recent questioning of the validity and usefulness of the principle of Uniformitarianism has arisen through:

(a) confusion between Substantive Uniformitarianism and the Methodological Principle of Uniformitarianism, and

(b) confusion between the incorrect application of the analogical reasoning that results from accepting the Methodological Principle, and the Principle itself.

These concepts are clarified using examples drawn principally from pollen analysis. The importance of abstract 'types of event' along with real, particular, events is emphasised. The Methodological Principle of Uniformitarianism is shown to be a basic and essential feature of the conceptual framework of palaeoecology which assumes that the world is regular, lawful and therefore intelligible. Limitations of analogical reasoning are discussed. The implications for 'experimental palaeoecology' of adopting uniformitarianism are considered and the distinction is made between the explanation of an effect and its replication. Any demonstration that evolution takes place is shown to depend on the acceptance of uniformitarian principles. This means that the fact of evolution cannot be used to falsify the Methodological Principle of Uniformitarianism.

INTRODUCTION

'What we call the beginning is often the end
And to make an end is to make a beginning.
The end is where we start from'. (T.S. Eliot)

During the symposium which gave rise to this volume the concept of Uniformitarianism was strongly criticized. Several of the participants suggested that Uniformitarianism does not apply to the sorts of data, particularly the biological data, being gathered by present-day ecologists and it was even suggested that Uniformitarianism was breaking down in relation to some physical processes. Although there seemed to be a general feeling that Uniformitarianism is in some way inadequate as a tool for palaeoecological interpretation, nothing was suggested to replace it, and this despite the fact that Uniformitarian principles pervaded all of the presented papers and all of the discussion as well.

The general unease with Uniformitarianism may simply be a result of the widespread criticial appraisal of science and scientific principles that is now taking place but, as expressed during the symposium, the criticism obviously resulted from confusion over what Uniformitarianism is, and what it can do; and confusion between Uniformitarianism itself and the inadequate application of the sort of arguments employed as a result of accepting Uniformitarianism. In this paper I shall demonstrate not only that Uniformitarianism is adequate, but also that it is an essential and necessary part of the conceptual framework of any palaeoecologist.

PALAEOECOLOGY

Palaeoecology may be defined as the investigation of past organisms, their environments, the relationships between these organisms and between the organisms and their environments. As such, palaeoecology is a branch of descriptive ecology (Cushing and Wright 1967) and is based on the analysis of any relics of the past that can be found. The facts available for analysis are bound to be severely limited. As Francis Bacon (1605) clearly indicated "Antiquities are history defaced, or some remnants of history which have casually escaped the shipwreck of time". Evidence that originates in one particular area at one particular time may, through the course of time, be modified and even transported into another spatial and/or temporal context; or it may be destroyed entirely.

A palaeoecologist, being unable to directly experience or witness the past, has to seek evidence relating to it. In so doing, the palaeoecologist, or any other historical scientist, attempts to observe the results of 'natural experiments'. Whereas an ecologist is able to create experimental conditions in order to obtain results which are unknown to him in advance, the palaeoecologist must first find the results and then try to reconstruct the causes and conditions of the change which led to them. Ecologists are able both to dictate the conditions of their experiments and to observe their results directly, and this gives them an obvious advantage. Most importantly, they are able to use simple situations in which some conditions are kept constant and they are able to repeat their experiments.

The natural experiments studied by palaeoecologists are in general very complex and the degree of their complexity is not known. The very complex, multi-variate, nature of the experiments studied by palaeoecologists means that exact replication is extremely improbable. For example, although the Pleistocene is distinguished as a period (of about 2×10^6 years) during which as many as 20 warm-cool oscillations have taken place, each of the warm periods can be recognised by its distinctive flora and the sequence of immigration of plant species into particular areas (West 1970). In other words, students of the historical sciences in general, and palaeoecologists in particular, study sequences of real, individual, events. Moreover each non-recurrent event is dependent on, and/or caused by, antecedent ones and itself has causal relations with subsequent events.

This does not mean that the historical sciences are in some way different from the ahistoric sciences as argued by Simpson (1963). Any particular event (e.g. the performance of a particular chemical experiment) is unique, even if only in terms of its space-time coordinates. In using particular events (e.g. the various replications of particular experiments) to produce scientific explanations, the individual descriptions (stories) are subsumed into law-like generalizations. Scientists are therefore concerned with types of events, not with particular events, and types, being abstractions, neither occur nor recur.

The events studied by palaeoecologists may be less numerous and, because more complex, more distinctive from each other than the events studied by ecologists, but this only means that the types of events considered have to be more general, just as the work of the ecologist is less specific than that of the chemist or physicist.

The interpretation of palaeoecological data

Having found and sampled 'relics of the past' the palaeoecologist has to decide where and when they came from and then, the object of the exercise, what they mean. Two general points are important here. The first is that it can be logically demonstrated that any finite number of observations can be accommodated within an indefinitely large number of different explanations (Magee 1973). In noting this we should at the same time remember Kant's warning that, whereas pure reason is adequate in mathematics, it cannot provide, <u>independently of experience</u>, knowledge of the real world (Aaron 1977). The second point is that the value of any statement depends on its falsifiability, i.e. refutability (Popper 1959). The growth of knowledge can only proceed through the elimination of error, i.e. through the refutation of hypotheses that are either logically inconsistent or entail empirically refuted consequences. Given that different but internally consistent logical systems can be constructed (e.g. the geometries of Euclid and Riemann (Nourse 1969) it seems apparent that, if palaeoecological interpretations are to have any usefulness or real validity, they must be capable of being disproved through empirical evidence from outside the data set on which they are based.

Any interpretation of palaeoecological data has to be consistent with all the facts at hand but should also, as William of Ockham pointed out, be as direct and as simple as possible. The simplest approach to take is one that will also immediately falsify the majority of possible interpretations; this is to adopt Uniformitarianism, to take the present as the key to the past.

THE PRINCIPLE OF UNIFORMITARIANISM

The use of the present to interpret the past has a long history and distinguished users of the principle include Xenophanes of Colophon (600 B.C.), Leonardo da Vinci and Robert Hooke, all of whom used biological evidence (Berry 1968). However, it was James Hutton's championing of process studies that first gave life to the historical sciences, and it is worth quoting from his Theory of the Earth (1788):

"In examining things present we have data from which to reason with regard to what has been; and from what has actually been, we have data for concluding with regard to that which is to happen hereafter".

Since the time of Hutton, Uniformitarianism has played a very important role in all the historical sciences but it has unfortunately, come to mean different things to different people. Hooykaas (1963) has demonstrated Uniformitarianism to mean any of the following:

(1) The belief that the basic laws of physics, chemistry and (perhaps) biology remain the same through time.

(2) (1) + the belief that specifically geological processes remain the same throughout time i.e. the types of geological process have remained the same and the only geological processes are those observable at the present day, or those which are lawfully related to those acting now.

(3) (1) + (2) + the belief that the rate and intensity (energy expenditure) of geological processes remains the same through time.

Charles Lyell and Uniformitarianism.

Although the concept of Uniformitarianism is usually associated with the name of Charles Lyell (1830-33), the term itself was first used by William Whewell in his review of 'Principles of Geology'. Lyell's formulation of the principle corresponded to the third statement of Hooykaas. Because of his emphasis on rates of processes remaining the same this statement has always been contentious and would not now be accepted. Berry (1968) has suggested that Lyell's statement of the principle was governed by two factors:

(a) he needed to stress that natural laws are invariable, in order to oppose the divine creation idea, and

(b) he needed to suggest that rates of processes had been uniform in order to refute Catastrophism and in order to emphasise the great length of geological time at a period when Archbishop Usher's calculation that the earth had been formed in 4004 B.C. had not been rejected.

Lyell applied his concept of Uniformitarianism only to the inorganic world. As discussed by Mayr (1971) Lyell believed in the steady extermination of species and their replacement by newly created ones. He did not become a convert to evolution until more than 20 years after the 'Principles of Geology' were published. Lyell was a consistent Uniformitarian only because he included supernatural agencies among the causes and forces still in operation. His concept of world history included a 'perpetual intervention hypothesis' which postulated a Creator who continued to shape the world, even at the present. This concept of Uniformitarianism is obviously quite different from the consistent application of natural laws to both inanimate nature and the living world.

Substantive Uniformitarianism and the Methodological Principle of Uniformitarianism

With his principle of Uniformitarianism Lyell was making an hypothesis about nature in exactly the same way that the Catastrophists were making hypotheses about nature. In doing this Lyell was putting forward a substantive statement which could be tested against the available data. He was adopting the traditional ontological approach of what there is in the world as man knows it.

Similarly, each of the statements that Hooykaas has related to Uniformitarianism can be taken as statements about what the world is like. Taken in this sense every one of them is subject to refutation. Even the first statement, that the basic laws of science remain the same through time, if taken as an hypothesis about the real world, can be tested. The basic constancy of physical laws in both space and time is a basic tenet of scientific methodology. Yet under special circumstances, and for very brief periods of time, even such a fundamental law as the conservation of mass/energy may break down (Calder 1977). In fact this example does not disprove the Uniformitarian statement because it shows only that the law being used was not fully understood. As science works by disproof this is hardly surprising. Refutation leads to the refinement of scientific laws. The statement that water always boils at 100°C was refuted by boiling water on the tops of mountains, but this fact did not lead to a suggestion that scientific laws were only local in their application. Instead it led to a better understanding of physical chemistry.

In taking any of the statements listed by Hooykaas as being hypotheses about the real world we are adopting a concept of Uniformitarianism distinguished by Gould (1965) as Substantive Uniformitarianism. To say that a given plant species (e.g. *Eucalyptus saligna*) occupied exactly the same ecological niche and showed the same geographical distribution in the past as at the present would be an example of Substantive Uniformitarianism.

However, the general statements of Uniformitarianism need not be taken as ontological statements relating to the real world. Indeed the concept of Uniformitarianism which is essential, not only to palaeoecology but to all of the other historical sciences as well, is of an epistemological nature. It is not a statement about the real world but relates to the investigator's approach to nature. This concept of Uniformitarianism is distinguished by Gould (1965) as the Methodological Principle of Uniformitarianism. Each of the statements of Uniformitarianism distinguished by Hooykaas can be taken as part of the formal structure of science, a principle for organising our knowledge of the world, and not a dogmatic description of the way the world must be. Unlike the substantive statements of Uniformitarianism, the Methodological Principle can neither be proved nor disproved. Methodological Uniformitarianism may be considered as a working assumption to be accepted in each case for as long as its consequences throw light on the detail and interpretation of the data. Instead of a statement of the way the world must be, it is a principle for finding out how the world was, is, and will be.

As usually employed, the Methodological Principle of Uniformitarianism states only that the basic laws and types of processes remain the same, and leaves open the possibility that, given different values for the variables in these laws, rates or processes can change. In other words, the Methodological Principle of Uniformitarianism is an epistemological principle based on the second definition of Uniformitarianism as distinguished by Hooykaas (1963); it may be thought of as an extension of the basic tenet of scientific methodology: that the universe is lawful and that the basic laws of science remain constant through both space and time. It can be compared to the methodological principle of the fundamental creationists that the bible is true and that all facts have to be fitted into a biblical framework.

The Methodological Principle of Uniformitarianism serves as a means of organising our knowledge of the past by serving as an *a priori* method of limiting the interpretations that can be used to explain a given set of data. This is demonstrated rather effectively by a quotation from William Whewell (1840), himself an opponent of Uniformitarianism and a catastrophist: "The Catastrophist constructs Theories, the Uniformitarian demolishes them" but the principle also serves to provide a data set against which any hypotheses made can be tested.

Because both the ontological and epistemological statements of Uniformitarianism are contained within the same name, it may at first seem paradoxical that the Methodological Principle of Uniformitarianism can be, and is, used to refute statements resulting from Substantive Uniformitarianism. For example, the statement that processes have always operated at the same speed or intensity can be tested by means of isotopic dating methods, the use of which depends on the methodological assumption that the properties of radioactive decay were the same in the past as they are at present. Similarly, the ontological statement that plant communities have always existed in their present form, even if they have migrated from place to place, can be shown to be invalid by using the Methodological Principle of Uniformitarianism in relation to our knowledge of pollen dispersal and distribution at the present day.

Unless otherwise stated, all further references to Uniformitarianism should be taken as referring to the Methodological Principle.

ANALOGY AND UNIFORMITARIANISM

In using the present to interpret the past palaeoecologists are using analogy as their main, perhaps their only, form of interpretative argument. Analogy may be defined as 'presumptive reasoning based on the assumption that if things have some similar attributes, they will share other attributes' (Oxford University Press 1971). This sort of argument may be represented, in its simplest form, as follows:

a, b, c, d all have properties P and Q
a, b, c, all have property R
Therefore d has the property R (Copi 1953)

For example, a particular sediment, A, may have the same particle-size distribution, colour and chemical composition as a sediment B, being formed under known conditions at the present day. We might therefore conclude that sediment A formed under the same environmental conditions under which sediment B is forming.

This form of argument relies heavily on pattern recognition and the use of increasingly sophisticated models and newly-developed numerical techniques will help increase the precision with which it is applied (cf. Webb & Clark 1977). Nevertheless, no analogy is exact (unless one were to compare a particular event with itself), so no argument from analogy is certain (cf. Watson 1965). It is worth remembering that students of morphology and evolution use analogy to refer to superficial resemblance, in contrast to homology which indicates a close developmental relationship (Davis & Heywood 1963).

Only certain functions of the systems, ancient and modern, being compared may be strictly comparable. The analogy used in elementary biology between the burning of coal and biological respiration is a trivial example in which the limits to comparison are clear. In comparing fossil data with the present these limits are often not clear, and perhaps one of the most basic questions that must be asked by any palaeoecologist is:

'how similar must two sets of circumstances be for the resulting two situations also to be similar?'

A simple example which illustrates this point is related to the morphological interpretation of certain fossils. The earliest eukaryotic organisms were thought to be certain microfossils occurring in the cherts of the Late Precambrian Bitter Springs formation of central Australia (Barghoorn & Schopf 1965). These fossil algae were found as isolated cells or as loose aggregates of a few to many individuals. Some had well-defined internal structures interpreted as nuclei. A whole series of taxa of both prokaryotic and eukaryotic algae were erected on the assumption that the presence or absence of these blobs constituted valid taxonomic characteristics. Recent work has shown that the entire range of morphological variation exhibited by a dozen taxa of coccoid algae from the Bitter Springs formation can be duplicated in detail by partially degrading a single species culture of the blue-green (prokaryotic) alga *Chrococcus turgidus* (Knoll & Barghoorn 1975). The former comparison of ancient (fossil) with modern (living) algae was invalidated by the failure to take into account the variability of internal structure caused by a partial degradation of blue-green algae. The correct comparison is between ancient (fossil) and modern (dead) algae.

Much of the dissension amongst historical scientists deals with the application of analogical arguments and, in particular, the choice of suitable analogues. Do high lake-levels in the past indicate a cool, fairly-dry, climate, or did they result from a hot, but wet, climate? This sort of disagreement does not reflect on the usefulness of the Methodological Principle of Uniformitarianism and, in fact, can only be resolved through the use of this principle. To discard Uniformitarianism because incorrect analogues are being considered would be to throw out the baby with the bathwater.

When an ecologist carries out experiments to interpret present distribution patterns of vegetation in relation to plant nutrient response, the research is considered successful when the similitude between the experimental result and the natural situation is exact within the significant limits of the quantities involved. When the prediction is not verified that in circumstances similar to those of the experiment a situation will arise similar to the experimental result (e.g. the prediction that on a soil of given nutrient status species A will be more abundant than species B), then it is not the validity of this type of reasoning that is questioned (see Watson 1972). Rather, the experimenter suspects himself of having observed the experimental system incompletely, or he suspects that the two sets of circumstances are not of similar type (perhaps he paid no attention to soil water-tensions).

Analogy is successful only in so far as the analogues chosen are themselves understood. Interpretation of fossil pollen assemblages in terms of past climate implies that we understand the relative importance of climate, soil and other ecological factors in so far as they control plant distribution. A number of assumptions are made with regard to the present. If these assumptions are incorrect the analogue we are using for interpreting past climate from plant distributions is defective, and the conclusions reached are likely to be wrong. As pointed out by Davis (this volume) we assume, for example, that:

(1) plant distributions are controlled by climate,

(2) present plant distributions are in equilibrium with present climate, and

(3) fossil plants identified to present day taxa had the same ecological response and showed the same ecological tolerances as the present members of the taxon.

In most cases these assumptions have not been tested and the ease with which plants can be dispersed by man, even within quite limited areas (Rymer 1978; Wace this volume) shows that the first two assumptions are often invalid. A plant may be able to survive in a far wider range of climates than is implied by its present distribution.

A good example of the implications of the above statement, and one which is highly instructive in another sense, relates to the climate during the late Devensian (Late-glacial) period in Britain. In simplified form the story is as follows. On the basis of traditional pollen analysis this period was thought to consist of a cold period separated by a warmer period (the Alleröd interstadial) from another cold period. This interpretation was based on the facts that:

(1) deposits of the middle period contain quantities of tree *(Betula)* pollen absent from the two surrounding periods (West 1968), and

(2) at the present day, areas which are naturally without trees in general experience a more extreme climate than areas with trees.

Recently, however, this classical interpretation has been challenged because other data (fossil Coleoptera) suggest that the Late-glacial began with a sudden warming and was followed by a gradual cooling (Coope & Brophy 1972). According to this view the Late-glacial vegetation was not in equilibrium with the climate because migration rates of trees are relatively slow. By the time the trees reached certain areas the climate had already begun to deteriorate and they were then forced to move south. Coleopterans, being mobile, were able to respond much more quickly to the change in climate.

The interest of this example lies in the fact that the consideration of a further character (or attribute), namely the fossil Coleoptera found along with the pollen, immediately suggested that one feature of the analogue chosen to interpret the data was wrong, this feature being the assumption that the vegetation was in equilibrium with the prevailing climate. Determination of the degree of similitude between the data set and the analogue used to interpret it is more complicated than is often recognised; however, the greater the number of characters relating to the data we consider, the smaller will be the number of modern analogues available which meet all the characters of the data set and the more accurate will be our interpretation.

This last point can be illustrated by considering pollen analysis. If we were able to recognise only trees from their pollen grains, we would know only whether trees were present or absent at a time represented by our fossil sample. If arboreal pollen could be determined to genus or species we might gain some idea of the sorts of trees present, and this might tempt us to make environmental interpretations, perhaps in terms of possible climate as well as vegetation. If, in addition, the category non-arboreal pollen could be recognised and expressed as a percentage of the total pollen present, we might think that we could estimate the proportion of the pollen catchment covered by trees. However, using an additional attribute of the data set, e.g. the number of pollen grains deposited on each square centimetre of accumulating surface per year, we might find that, after all, the trees were not really present in the immediate area of the pollen assemblage at the time the assemblage was deposited, but that the tree pollen had all blown in from a distance (cf. Pennington 1970). The addition of this single character (pollen influx) has resulted in a very drastic change in the sort of modern analogue needed. Environmental interpretations based on the previous analogues would have been incorrect. Consideration of further characters, such as an increased number of pollen taxa based on a greater precision of pollen identification, macrofossil evidence, sediment stratigraphy and chemistry, cladocera, diatoms, archaeological evidence, and so on, will lead to a much more realistic interpretation. At the same time, as more characters are considered, the number of available analogues will be reduced. If every possible character were considered then, obviously, there would be only one possible analogue, and that is the event being studied itself.

As more and more characters relating to a particular data set are considered the reduced choice of analogues becomes more difficult to interpret because, if there are any present-day analogues, they may well be outside the experience of the scientists needing them. A particular assemblage of fossil pollen, for example, may not have any modern counterpart. This is the sort of situation that is sometimes taken to "disprove Uniformitarianism". It does nothing of the kind. As previously emphasised, palaeoecologists deal with types of events as well as with particular events, and in this they are no different from other scientists. Using models of pollen dispersal based on the present day situation, together with knowledge of the relative pollen productivities of the taxa found in the fossil assemblage and information on their present ecology, we can still interpret the fossil pollen spectrum in terms of the type of vegetation which gave rise to it. If it were possible to define exactly the present fundamental niche of each taxon present, then the overlap between the niche hypervolumes of all the fossil taxa would go some way to defining the kind of environment that must have been present when the fossil assemblage formed. Even supposing that evolution had never occurred, such a technique would never define the past environment exactly, for reasons which will be familiar to all pollen analysts but which are outside the scope of this paper.

ANALOGY, UNIFORMITARIANISM AND EXPERIMENT

The situation often arises in which a given set of palaeoecological data may have a number of possible explanations which cannot be distinguished by use of the available evidence. The sensible approach is to adopt multiple working hypotheses (Chamberlain 1965) until further information enables some of these to be rejected. A good example is the 'elm decline' of northwest Europe. Pollen sites over a very large area all show a significant decline in the proportion of their elm pollen about 5000 years ago. Numerous explanations have been offered, ranging through the agricultural activities of Neolithic man, disease, soil-development and climatic change (Smith 1970).

Palaeoecology is a branch of descriptive ecology because palaeoecologists are simply describing nature's own experiments. In trying to interpret something like the elm decline the palaeoecologist has to search for additional data to help decide between the different hypotheses. Sediment chemistry, for example, might suggest whether or not soils have changed (Mackereth 1966). The question can be asked as to why the palaeoecologist does not resort to experiment? The present outbreak of Dutch elm disease in Britain might serve as a natural experiment, and the decline of chestnut pollen in the sediments of North America which resulted from the ravages of chestnut blight has already been documented (Anderson 1974). Similarly, experiments can be carried out to simulate the effects of Neolithic man on the vegetation, to see whether the observed effects agree with a particular interpretation of the pollen stratigraphy.

Palaeoecologists do carry out this sort of experimentation but the results have implications which are different for the ecologist and palaeoecologist. Palaeoecologists do not, in general, set out to predict, but to explain; although a case can be made out that explanation and prediction are equivalent (Watson 1969), the experiments carried out by palaeoecologists do not themselves provide explanations. In effect, the palaeoecologist is trying to create new analogues which might help explain (interpret) his data. An experiment may show that a certain explanation is possible at the present time and under present conditions. Such an explanation might apply also to the past. However, the same result might be produced in different ways, even at the present and under present conditions. And we do not know the conditions of the past. Indeed, the whole point of the exercise may be to elucidate past environmental conditions! This form of experimentation is not of the same nature as that found in the physical sciences in which, as Hooykaas (1963) points out, the task is to find out how things happen, not to reproduce their effects. This is not to say that palaeoecologists should not carry out experiments, only that they should be careful that they know what their experiments can mean in relation to their palaeoecological data.

UNIFORMITARIANISM AND EVOLUTION

Substantive Uniformitarianism is ahistoric and does not allow for change. In particular, it does not allow for any unidirectional change or evolution. The fact of evolution is one of the most frequently cited objections to the uniformitarian method. However, the fact that evolution has occurred can be known only because we adopt the Methodological Principle of Uniformitarianism. Evolutionary change cannot be used to cast doubt on the general applicability of the principle because, if the methodological principle is rejected, we have no means of knowing that evolution has taken place.

Consider the fact that we find in ancient sediments plants and animals different from those now extant. These sediments are otherwise similar to sediments forming under present conditions which contain modern organisms. As there are no processes now operating able to convert, through time, dead organisms into morphological forms fundamentally different from those they already possess, the organisms found as fossils must have been different from extant organisms even before they were incorporated into the sediment. In other words, living organisms have changed (evolved) through time. Anyone not adopting this Uniformitarian argument could claim (for example) that the succession of life as exhibited in rocks of different ages resulted from diagenetic processes.

Interestingly, William Whewell (1840), one of the main critics of Uniformitarianism, objected, not because evolution had occurred, but because "...nothing in the Lyellian doctrine could explain the creation of new species by causes now in action". Publication of the first edition of Darwin's On the Origin of species... (1859) effectively removed this objection. The fact that ecological and genetical studies of present day populations can show that evolution is taking place now, does not invalidate the need for the methodological principle to state that evolution has occurred in the past. After all, the use of the present in order to interpret the past is part of what Uniformitarianism is about.

Evolution is a process that operates at the present time and has operated in the past, in exactly the same way that any geological process which operates now has also operated in the past. Although evolution is often thought of as a peculiarly biological phenomenon, this is not the case. Physical systems such as a star (Calder 1973) the atmosphere (Berkner & Marshall 1965) have also undergone, and can undergo, change. But although the systems may change, the processes, including those which effect the change, remain unchanged. Using present-day processes, theoretical models can be constructed to describe the probable consequences of such historical events as the origin of life, or the development of photosynthesis

(Bernal 1967). Such model building is a form of thought experiment in which the experimenter tries to imagine what would happen in prescribed natural circumstances, and then looks in the field for features which may support or negate his predictions (Watson 1965). Thought experiments are quite respectable; physicists often imagine what would happen under ideal conditions of perfect vacuum and so on (Nourse 1969).

The fact that evolution occurs means that the information content of fossils must decrease and become more generalized the further back in time the fossils are found. The types recognized become increasingly broad so that eventually the only taxonomic level of use for exact comparison with present organisms is the division or phylum. Although identifications within the Quaternary are often made to the level of present-day species, the context in which these species are found, and the different behaviour of the same species in the different interglacials, may suggest that some kind of ecotypic differentiation has taken place (West 1970). This is itself a uniformitarian argument and is based on the behaviour of those species at the presen

CONCLUSION

The point of the preceding discussion is that Uniformitarianism does not consist of a statement about nature. To select arbitrarily a very minute fraction of geological time, (the period in which we live) as the standard for all other geological time, would be presumptuous. The Methodological Principle of Uniformitarianism does not state dogmatically that the past had to be like the present. The principle is simply a part of the structure of science, a conceptual framework which science imposes on the phenomena that it is studying. There are other ways of organising our knowledge of nature based on different methodological assumptions. Fundamental creationists for example, have an internally consistent logical system for organising geological and biological information. Their system is based on the assumed truth of the Bible and because this is the base on which they organise their knowledge, creationist systems can never be disproved. They are organized around a principle which states that they (the systems) are true, and which allows only one hypothesis to explain all data.

Historical scientists, even when using uniformitarian methodology, always begin with the effects that they are studying, and not with the present. The effects studied must themselves be used to teach us the nature and the intensity of the causes which have operated. Processes may have operated in the past which do not operate now. As previously explained, the application of uniformitarian principles shows that evolution has taken place and that in times earlier than the Devonian the land surface would not have been clothed with vegetation. As a result, erosion processes and processes of mass transport may have been different from those now operating because humid areas now have vegetation. Nevertheless, adopting the Methodological Principle of Uniformitarianism enables us to state that, although the processes operating under such conditions may not operate now, such processes would be related to presently acting processes in some lawful way. In effect this enables us to create modern analogues. Such analogues might be created literally, by stripping areas of vegetation (as in mining operations), or by means of thought experiments or by building models.

When considering the historical sciences one should bear in mind that although the present may be the key to the past, knowledge of what has happened is important for understanding the present and even, as Hutton (178 stated 'for concluding with regard to that which is to happen hereafter'. To use the adage common among palaeoclimatologists: if it has happened, it can happen. Study of the effects of erosion under humid conditions at a time when there was no land vegetation might provide information about what could happen if we now set about removing vegetation from humid areas. This idea of Uniformitarianism as a means of linking past, present and future, was the concept used by Hutton and is made very effectively by a quotation from 'Burnt Norton', one of the Four Quartets of T.S. Eliot:

'Time present and time past
Are both perhaps present in time future,
And time future contained in time past'.

REFERENCES

Aaron, R.I. 1977 Epistemology. pp 925-947 in Encyclopedia Britannica Macropedia, volume 10, Encylopedia Britannica Inc., Chicago.

Anderson, T.W. 1974 The chestnut pollen decline as a time-horizon in lake sediments in eastern North America. Can. J. Earth Sci., 11, 678-685.

Bacon, F. 1605 The advancement of learning. (Edited and with an introduction by G.W. Kitchin in 1861) J.M. Dent and Sons, London.

Barghoorn, E.S. and Schopf, J.W. 1965 Microorganisms from the Late Precambrian of central Australia. Science, 150, 337-339.

Berkner, L.V. and Marshall, L.C. 1965 On the origin and rise of oxygen concentration on the earth's atmosphere. J. atmos. Sci., 22, 225-261.

Bernal, J.D. 1967 The origin of life. Weidenfeld and Nicolson, London.

Berry, W.B.N. 1968 Growth of a prehistoric time-scale based on organic evolution. W.H. Freeman, San Francisco.

Calder, N. 1973 Violent universe. B.B.C. Publications, London.

Calder, N. 1977 The key to the universe. B.B.C. Publications, London.

Chamberlain, T.C. 1965 The method of multiple working hypotheses. Science, 148, 754-759.

Coope, G.R. and Brophy, J.A. 1972 Late Glacial environmental changes indicated by a coleopteran succession from North Wales. Boreas, 1, 97-142.

Copi, I.M. 1953 Introduction to logic. Macmillan, New York.

Cushing, E.J. and Wright, H.E. 1967 Introduction. pp 1-8 in Quaternary Palaeoecology (eds E.J. Cushing and H.E. Wright), Yale University Press, New Haven.

Darwin, C. 1859 On the origin of species by means of natural selection, or, the preservation of favoured races in the struggle for life. J. Murray, London.

Davis, M.B. 1978 Climatic interpretation of fossil pollen data (this volume).

Davis, P.H. and Heywood, V.H. 1963 Principles of Angiosperm taxonomy. Oliver and Boyd, Edinburgh.

Gould, S.J. 1965 Is uniformitarianism necessary? Am. Journ. Sci., 5, 223-228.

Hooykaas, R. 1963 The principle of uniformity in geology, biology and theology. E.J. Brill; Leiden.

Hutton, J. 1788 Theory of the earth, or investigation of the laws observable in the composition, dissolution, and restoration of land upon the globe. Royal Society of Edinburgh, Transactions, 1, 209-304.

Knoll, A.H. and Barghoorn, E.S. 1975 Precambrian eukaryotic organisms: a reassessment of the evidence. Science, 190, 52-54.

Lyell, C. 1830-33 Principles of geology, being an attempt to explain the former changes of the earth's surface by reference to causes now in operation. 3 vols. J. Murray, London.

Mackereth, F.J.H. 1966 Some chemical observations on Post-glacial lake sediments. Phil. Trans. R. Soc., B250, 165-213.

Magee, B. 1973 Popper. Fontana, London.

Mayr, E. 1971 The nature of the Darwinian revolution. Science, 176, 981-989.

Nourse, A.E. 1969 Universe, earth, and atom. Harper and Row, New York.

Oxford University Press, 1971 The compact edition of the Oxford English Dictionary. 2 vol. Oxford University Press, U.S.A.

Pennington, W. 1970 Vegetation history in the north-west of England: a regional synthesis. pp 41-79 in Studies in the vegetational history of the British Isles (eds D. Walker & R.G. West), Cambridge University Press, Cambridge.

Popper, K.R. 1959 The logic of scientific discovery. Hutchinson, London.

Rymer, L. 1978 Ethnobotany and the native distribution of Gorse (*Ulex europaeus* L.) in Britain. Environmental Conservation (in press).

Simpson, G.G. 1963 Historical science. pp 24-48 in The Fabric of Geology (ed C.C. Albritton). Addison-Wesley Inc., Reading, Massachusetts.

Smith, A.G. 1970 The influence of Mesolithic and Neolithic man on British vegetation: a discussion. pp 81-96 in Studies in the vegetational history of the British Isles (eds D. Walker & R.G. West), Cambridge University Press, Cambridge.

Wace, N.M. 1978 Human modification of the natural ranges of plant and animals (this volume).

Watson, R.A. 1965 Similitude in direct and thought experiments in cave geology. Bulletin of the National Speleological Society, 27, 65-76.

Watson, R.A. 1966 Discussion: is geology different? a critical discussion of "The fabric of geology". Philosophy of Science, 33, 172-185.

Watson, R.A. 1969 Explanation and prediction in geology. Journal of Geology, 77, 488-494.

Watson, R.A. 1972 Limitations on substituting chemical reactions in model experiments. Z. Geomorph. N.F., 16; 104-109.

Webb III, T. and Clark, D.R. 1977 Calibrating micropalaeontological data in climatic terms: a critical review. Annals of the New York Academy of Sciences, 288, 93-118.

West, R.G. 1968 Pleistocene geology and biology with especial reference to the British Isles. Longmans, London.

West, R.G. 1970 Pleistocene history of the British flora. pp 1-11 in Studies in the vegetational history of the British Isles (eds D. Walker & R.G. West), Cambridge University Press, Cambridge.

Whewell, W. 1840 <u>The philosophy of the inductive sciences founded upon their history</u>. 2 vols. John W. Parker, London.

15. ENVOI

by D. Walker

SUMMARY

Errors in the reconstruction of Quaternary terrestrial environments are discussed arising from (a) the incomplete specification of an organism by its fossil remains, (b) the difficulty of securing comparisons with the limited range of biotic and biotic-environmental complexes available now, and (c) the use of inappropriate statements of organism-environment correlations. Suggestions are made for cautious progress and frequent reassessment of earlier conclusions in the light of continuing developments in related experimental fields.

ORGANISMS AND THEIR FOSSIL REMAINS

All biologically-based reconstructions of Quaternary environments derive ultimately from morphological comparisons between fossils and similar living organisms. A few Quaternary fossils can be compared exactly with a living species or even lower taxonomic unit. Where such exact comparisons cannot be made the reason may be

(a) that there is insufficient (known) variation amongst the potentially fossil parts of the various living species to allow their taxonomic separation on these grounds, or

(b) that, although the variation of the living species is sufficient and known, the fossil falls outside it.

In the latter case, argument then concentrates on the possible adaptive significance of the morphological divergence from the living form. It is particularly applicable to vertebrates, notably mammals, in which the commonly fossilized parts (e.g. bones and teeth) are likely to have been directly related with aspects of the animal's environment. It may seem more speculative than direct species to species analogy because it starts from differences rather than similarities, yet it may in fact beg fewer questions. Thus the fossil remains allocated to an extinct genus *Smilodon* may allow identification with living animals no closer than the family (Felidae) yet its teeth leave no doubt that it was a carnivore because of the basic and taxonomically widespread adaptive relationship between teeth and food (Martin and Guilday 1967).

The level to which a fossil can be identified often depends on the part which is fossilized and our technical capacity for recording the details of morphology. Some commonly occurring fossils, such as some pollen grains, are not identifiable below the genus, or sometimes below the family, because they are insufficiently varied or because the variation is continuous. Nor have the details of pollen morphology, apart from some attributes of water plants for instance, much direct adaptive significance in environmental reconstruction.

It is perhaps in those cases where a fossil exactly matches its living counterpart that the most insidious difficulties lie. In fact, practically every attribute of a potential or actual fossil of a species has some variance (e.g. the number of tubercles on a fossil seed); exactness of match will be greatest where there are several attributes on which to base comparisons, each of which has small variance. Matching morphologies of fossilized parts, however, do not necessarily imply identity of other parts or functions in the organisms they represent. There is always the possibility that the fossil, though exactly resembling a part of a living organism, may have come from a related but extinct form of different physiological requirements. More importantly, however, it is evident (Franklin, this volume) that genetic changes with profound ecological implications can arise sufficiently frequently to have changed the within-species tolerances in as short a period as the Quaternary. This means that a fossil might well identify with a living species but that the environmental correlates of that species might have been different in the past.

Errors from this source might well be rare. But any species is composed of a spectrum of genotypes which confer on it a range of environmental tolerances and, in any one time and place, a modal tolerance displayed by the majority of its members. At best, and without allowing for the kinds of genetic tolerance changes alluded to earlier, a fossil certainly identified to a species can only indicate an environment somewhere between the extreme limits demonstrated by the living population. That the fossil may be numerous is no indication that the species it represents was growing near its modern modal tolerance for this may rapidly shift in response to changing selection pressures.

Biologically elementary though these considerations are, they have not inhibited spuriously precise environmental derivations from supposedly 'identified' fossils in the past. There is, however, some comfort to be gained from those occasions when a number of fossil taxa all indicate the same environmental range. Nevertheless, it is the range that has significance, not any value between its extremes.

ANALOGIES BETWEEN PAST AND PRESENT

As Rymer (this volume) points out, the only way in which we can hope to deduce the environmental correlates of a fossil assemblage is by recourse to what we can observe and experiment with at the present day. This means that we must be guided, not only by the patterns and processes which exist today but also by our knowledge of how these may change. Indeed, fossil associations may themselves contribute otherwise unavailable information about the possibilities of change. Two species found in incontrovertible association as fossils but not now living together (in the sense of being able to contribute to the same set of remains) furnish information about their species which would not otherwise be known. The environmental implications of such a find may be baffling if the two species now live in mutually exclusive environments, or unusually precise if their fundamental niches have some known overlap.

Processes are known which would have allowed today's species to have lived in mixtures which do not currently exist and in environments from which they are now absent. So it is unnecessary to restrict Quaternary interpretations by the associations and tolerances which are immediately apparent today. On the other hand it would be unwise to invoke processes which are not known to operate today in deriving past environments from fossil data. Yet, because knowledge of these processes is continually increasing, it will sometimes be useful to re-examine old fossil data in the light of such new findings.

Hypothetical mixtures of plants different from anything known at the time of investigation have punctuated the reconstruction of Quaternary vegetation for many years. Thus the European "Eichenmischwald", at one time thought to have clothed western Europe for much of the Holocene, was a hypothetical construct. It was never clear whether the hypothesis involved the component taxa in more-or-less uniform mixture or whether they were individually and patchily distributed in different ecological situations; the matter was first clearly aired in relation to Ireland by Morrison (1959). Iversen (1951) drew attention to steppe plants in the otherwise tundra flora of the Danish Late-glacial, an association not easily reconcilable with living vegetation types then known. Bell (1969) has analysed the occurrences of ecologically even more varied plants from southern Britain in the last glacial period.

It has long been evident, therefore, that plants have grown in associations in the past in which they do not grow today. The biological implications of this fact are considerable but environmental reconstruction is made less easy by it. Such reconstructions are most confidently made (within the constraints referred to in the previous section) when qualitative and quantitative correspondence between fossil assemblage and living group is more-or-less exact, leaving no information unaccounted for. When this is not the case, it is important to assess the differential significance of the components of the fossil assemblage. It may be, for instance, that one trophic or structural group corresponds with a living one whilst the rest do not. If this group is pivotal to the rest (e.g. plants in a mixed plant:herbivore assemblage or canopy trees in a mixture from all vegetation strata) and their rates of responses to environmental change are similar, its environmental correlates are probably the ones to invoke.

Environmental arguments which turn on the location of major vegetational and faunistic discontinuities (e.g. rainforest: savanna boundary) are more likely to have acceptable and useful modern analogues than are those whose recognition is based on correspondence between a few taxa. This is especially the case if the balance between them depends not only on subtle environmental factors but also on competition and availability. The exclusion of environmentally well-adjusted species from a locality by another species which is slightly better adapted to the total ecological situation is a common-place of ecology. Availability, however, is a factor of greater significance than we often credit. The relatively rapid climatic changes which have punctuated the Quaternary (Chappell, this volume) must have favoured those species with wide tolerance limits and those which, though with much narrower limits, had the capacity to react to change quickly. So rapid has environmental change been that these characteristics, and distance from prior location, must significantly affect the composition of a flora at any time in the Quaternary, including the present day (cf. Davis, this volume). Perhaps some established assemblages of organisms operating as a balanced ecosystem are so interdependent that they can only survive if their integrity is maintained; if they move they must move together. Most, however, are unlikely to be so conservative so that, under the stress of environmental change, each species moves according to the capacity of its reproductive mechanism or its ability to reach and exploit locations beyond the limits of its prior distribution. An ecosystem might well develop at such a location identical with one formerly existing nearby but only after such a time as will allow the most slowly migrating and establishing member to arrive there. In the meantime the location will be occupied by a changing assemblage of organisms which are able to exploit it. Particularly in those parts of the world (e.g. formerly glaciated regions) where environmental changes directly affecting plants and animals have been most extreme, many ecosystems are likely to have been 'incomplete' during the greater part of Quaternary time and may still be so. Moreover, during a period (such as the first third of an interglacial) in which organisms are mostly expanding their ranges, distance from prior location (refuge) would exercise strong selection on the plants and animals which came to occupy a location. At most times during the Quaternary, therefore, utilization of the environment by plants and animals is likely to have been opportunistic. This would have selected for dominance some species or biotypes which would not have such importance in the more fully balanced utilization of the environment by a more complete ecosystem.

Indicator species are of two kinds. Community indicators are those which, singly or as a group, are supposed to specify the presence of a larger number of taxa, an ecosystem, community or 'formation'. When such are encountered as fossils it is tempting to use the apparent environmental correlates of the whole group 'indicated' for interpretative purposes. The folly of this procedure is clear from what has been written above about inconsistency in the correlation of organisms through time.

Environmental indicators are taxa for which the environmental tolerances are well authenticated by observation and experiment with living organisms. Despite the frequency with which the apparent extreme tolerances, particularly of plant species, are used to estimate environments of the past, there are dangers in doing so. For example, distribution limits observed today are related to a complex of environmental and biotic factors, a change in any one of which may grossly affect apparent correlations with others. Only laboratory and transplant experiments can establish the true tolerance limit of a genotype with respect to a particular environmental variable. The number of genotypes for which such information is unequivocally available is very small. Because of the rapidity of environmental change during the Quaternary, those indicator species which can travel fast and far will be the most useful during expansionist periods. By contrast, long-lived species will be poor indicators of local conditions during periods when they and other species might be expected to be losing ground. But perhaps the biggest snag is the range of tolerance exhibited by a species' total genotypic spectrum, all of which range may not be included within the living populations from which the correlations are derived.

MEASURES OF ORGANISM-ENVIRONMENT RELATIONSHIPS

The tolerance limits of organisms are usually stated in terms of simple environmental parameters, such as the mean temperature of the hottest month of the year or the mean annual rainfall. The unreality of such measures for the organisms involved is immediately apparent. Thus animals have many ways of temporarily adapting to high temperatures whilst what matters to terrestrial plants is not so much rainfall as precipitation:evaporation ratio confounded with soil conditions. Once even this degree of complexity is admitted, however, the reconstruction of the past environment of a fossil becomes much more problematical and multifactorial. An animal skeleton may imply one temperature if there were suitable soil for burrowing and another if there were not. A plant fossil is evidence of one rainfall intensity under one temperature regime but of another rainfall if temperatures rendered a different proportion of the rain available to the plant.

Experience, largely with agricultural plants, increasingly shows that the kinds of measures routinely recorded by meteorological instruments have little direct meaning for plants and presumably also for animals. Rather do growth and reproduction more commonly depend on 'syndromes' of environmental circumstances; of the coincidence of, say, ready availability of water during the period of critical day length before which a certain total amount of heat had been received by a plant. At best, therefore, a fossil identifies a complex of environmental variables in which one factor ma compensate for another and in which the length of exposure to the variables probably compensates for deficiencies in the intensity of some of them.

A special case of this problem is that of seasonality. It might well be expected that climatic change would often have taken the form of changes in seasonal characteristics rather than in annual means. The results of this on flowering, and on animal reproduction through changes in food supply, could be as drastic, if not more so, than a substantial change in, say, mean annual temperature. From the fossil record as we currently read it, it would be impossible to distinguish between the two causes. We badly need to identify, in commonly available fossil materials, features which are the products of particular seasons and which have characteristics measuring aspects of the physical environments of those seasons, however complex they may be.

Even an apparently significant environmental measure based on the coincidence of the edge of an organism's geographical area with an instrumentally established meteorological parameter, may only be an ephemeral statement of the value of that parameter in the syndrome of environmental conditions limiting the organism's distribution at the present day. This raises the possibility that it might be more useful to identify the syndrome to which plants, animals or their joint communities respond and to state the nature of former environments in these terms. In practise it would use the 'minimum set' of variables defined by Nix (this volume) and would include a statement of the interrelations between them; for example that the effect of increasing temperature above a certain threshold would be offset if an increase in available water accompanied it. It might well prove that the factors most commonly included in such minimum sets could be more useful for correlation with other measures of the environment (e.g. ice volumes) than the parameters currently used.

It is likely that present concern about the economic effects of possible climatic variation and the desire to identify crop plants most appropriate for growing in many regions, particularly biologically marginal regions, of the world, will lead to more meteorological recording. It would be gratifying if, as a matter of course, derivatives of these data with agreed biological significance could be reported.

PROSPECT

Biological Quaternary research is currently undergoing a reassessment of its assumptions and of the contributions that it can make to related disciplines. The emergence of geochemical methods of determining some environmental measures (e.g. temperatures) and their application particularly to ocean sediments, has provided an independent set of data against which to compare the often less precise and always less continuous indications of fossil plants and animals. Yet fossils remain the most important source of information about changes in Quaternary <u>terrestrial</u> environments available to us, so that it is important to maximise their contribution and, where possible, to improve its precision. Ecophysiological, genetical and population ecological research on living organisms has by now demonstrated the complexity of the organism-environment relationships from which we wish to derive simple statements about changes in major environmental variables. Given this complexity, how should we modify our statements about the environments indicated by our fossils?

First we should try to establish, by recourse to the best living analogues available, the factors in the syndrome which controls the occurrence of a species or community and the basic nature of their interactions. Should we then wish to select one factor for quantification (e.g. temperature from a pollen curve) we should explicitly state the limits within which the other factors must fall without vitiating the correlation. We should use the total range of the factor tolerated by the taxon as a whole and so provide a span of quantities within which the one true for the fossil may lie. We should use the separate indications of the various fossils which are found together to refine the estimate by mutual exclusion. All this should be done explicitly and quite separately from any more intuitive assessment which may be based on the results and the experience of the investigator. One of the great difficulties in comparing the reconstructions of different authors lies in discovering where analysis ends and intuition begins; both have their uses but it is important to know which is which.

Secondly, we must be aware that our analogues are likely to be imperfect partly because of the potential diversity of associations between organisms, and between them and their environments, but also because of the mutability of these relationships. Knowing also that our fossil assemblages represent but fragments of the biota of their times, we should perhaps hesitate confidently to identify them, even by implication, with total ecosystems of the present day. In the long run it might be more valuable to point out similarities <u>and</u> differences and to use <u>both</u> as sources of environmental information. Above all we should use current knowledge about the relative genecologies, ecophysiologies and population dynamics of the living taxa represented by fossils to inform the use we make of the fossils in environmental reconstruction. Thus a plant with few ecotypes might provide better information than one with many; in some circumstances an organism with a capacity for rapid migration and establishment would give a better measure of climate than one with greater inertia. Moreover, we should be more sensitive to the capacity of different groups of organisms to measure different things, or the same thing with differing precision, depending on their individual biological endowments and on their places in their ecosystems (Tüxen 1974).

Much of this, as with much of the message of the other papers in this volume, is little more than a restatement, in terms of the current level of knowledge about living organisms, of the principles which have always guided the best reconstructions of Quaternary terrestrial environments. Their restatement and discussion, the main purposes of the Symposium from which this book derives, is appropriate at a time when the desire to understand, and thereby predict, environmental changes is quickening in several disciplines and amongst a broader public. Doubtless, reconstruction will proceed, as in the past by successive approximations. But it is important that each step in the process should be informed by the limitations inherent in the biological nature of the material from which it derives.

It is also timely to point out that the benefits to be obtained from association between Quaternary research workers and their ecologist, physiologist and geneticist colleagues do not all flow one way. Where morphological change can be seen amongst Quaternary fossils (e.g. bear, man) it exemplifies the amounts, directions and rates of evolution which have occurred in nature and with which any rules derived from the manipulation of living breeding systems or the statistical study of variation must be consistent. As Macfarlane (this volume) demonstrates, the distribution of fossil relatives of some of today's animals helps explain both similarities and differences in aspects of the physiologies, and hence ecological behaviours, of the living forms. And as we resist the temptation to over-simplify the comparisons we make between fossil and living assemblages, we increasingly provide evidence for the dynamic nature of plant and animal populations both in the past and now.

ACKNOWLEDGEMENTS

Some of the assertions in this paper were amongst the considerations which led to the organisation of the Symposium. Others clearly derive from the papers presented at it and from discussion which took place there. I make no claims for their originality and thank all those, known and unknown, who contributed to them. Dr J.H. Calaby and Dr N.M. Wace helped improve the original manuscript.

REFERENCES

Bell, F.G. 1969 The occurrence of southern, steppe and halophyte elements in Weichselian (Last-glacial) floras of southern Britain. New Phytol., 68, 913-922.

Iversen, J. 1951 Steppeelementer i den senglaciale Flora og Fauna. Medd. dansk geol. Foren., 12.

Martin, P.S and Guilday, J.E. 1967 A bestiary for Pleistocene biologists. pp 1-62 in Pleistocene extinctions: the search for a cause (eds P.S. Martin and H.E. Wright Jr.), Yale University Press, New Haven.

Morrison, M.E.S. 1959 Evidence and interpretation of "Landnam" in the north-east of Ireland. Botaniska Notiser, 112, 185-204.

Tüxen, R. 1974 Synchorologie einzelner vegetationseinheiten in Europa. pp 267-292 in Handbook of vegetation science Part VIII Vegetation dynamics (ed R. Knapp), Junk, The Hague.